Strategies for Sustainability

Series Editors

Rodrigo Lozano, Faculty of Engineering and Sustainable Development,
University of Gävle, Gävle, Gävleborgs Län, Sweden
Angela Carpenter, Faculty of Engineering and Environment,
University of Gävle, Gävle, Gävleborgs Län, Sweden

Subseries: Spatial Planning and Sustainable Development

Subseries Editor

Zhenjiang Shen, School of Environmental Design,
Kanazawa University 2C718, Kanazawa City, Japan

The series focuses on "implementation strategies and responses" to sustainability problems – at the organizational, local, national, and global levels.

Our objective is to encourage policy proposals and prescriptive thinking on topics such as: sustainability management, sustainability strategies, lifestyle changes, regional approaches, organisational changes for sustainability, educational approaches, pollution prevention, clean technologies, multilateral treaty-making, sustainability guidelines and standards, sustainability assessment and reporting, the role of scientific analysis in decision-making, implementation of public-private partnerships for resource management, regulatory enforcement, and approaches to meeting inter-generational obligations regarding the management of common resources.

We favour trans-disciplinary perspectives and analyses grounded in careful, comparative studies of practice, demonstrations, or policy reforms. This largely excludes further documentation of problems, and prescriptive pieces that are not grounded in practice, or sustainability studies. Philosophically, we prefer an open-minded pragmatism – "show us what works and why" – rather than a bias toward a theory of the liberal state (i.e. "command-and-control") or a theory of markets. We invite contributions that are innovative, creative, and go beyond the 'business as usual' approaches.

We invite Authors to submit manuscripts that:

- Document and analyse what has and has not worked in practice;
- Develop implementation strategies and examine the effectiveness of specific sustainability strategies;
- Propose what should be tried next to promote greater sustainability in natural resource management, energy production, housing design and development, industrial reorganization, infrastructure planning, land use, business strategy, and organisational changes
- Prescribe how to do better at incorporating concerns about sustainability into organisations, private action, and public policy;
- Focus on trans-disciplinary analyses grounded in careful, comparative studies of practice or policy reform; and
- Provide an approach "…to meeting the needs of the present without compromising the ability of future generations to meet their own needs," and do this in a way that balances the goal of economic development with due consideration for environmental protection, social progress, and individual rights.

More information about this subseries at http://www.springer.com/subseries/15804

Lanchun Bian • Yan Tang • Zhenjiang Shen
Editors

Chinese Urban Planning and Construction

From Historical Wisdom to Modern Miracles

Springer

Editors
Lanchun Bian
School of Architecture
Tsinghua University
Beijing, China

Yan Tang
School of Architecture
Tsinghua University
Beijing, China

Zhenjiang Shen
School of Environmental Design
Kanazawa University
Kanazawa, Japan

ISSN 2212-5450 ISSN 2452-1582 (electronic)
Strategies for Sustainability
ISSN 2522-8463 ISSN 2522-8471 (electronic)
Spatial Planning and Sustainable Development
ISBN 978-3-030-65561-7 ISBN 978-3-030-65562-4 (eBook)
https://doi.org/10.1007/978-3-030-65562-4

This Springer imprint is published by the registered company Springer Nature Switzerland AG
The registered company address is: Gewerbestrasse 11, 6330 Cham, Switzerland

Foreword

In current China, the biggest reform in the planning field is undoubtedly the formation of a new territorial spatial planning system by integrating previously decentralized land resource planning, urban and rural planning, etc. The impact of this reform is so great that the boundary of the previous planning systems is basically broken, most planning methods are under revolution, and the new system is in the process of complete reconstruction. Thus, the uncertainty in urban planning and construction in China comes along with it.

Amid this critical transition period, the book does not focus on the reforming itself, but rather reviews and summarizes some of the successful experience of China's urban planning and construction, such as the ancient capital region planning, the new development in megacities and small villages, and the trends in sustainable and healthy city construction, which will be a great reference to the current planning reform. And undoubtedly, China should inherit and maintain these successful experiences in its new territorial spatial planning system. As the previous planning systems are primarily focused on urban physical planning and constructions, which do not cope with the need of post urbanization era anymore, the reform is greatly expanding the planning scope, for land-use control, environmental protection, urban governance, and other related issues that seldom considered in urban planning before being merged together into the new system.

From a global perspective, the book also provides worldwide readers an important window to understand Chinese urban planning across ancient and modern times. It seems that the modern urban planning and construction theories and practices born in western developed countries, especially European and South American countries, have incrementally imported into China and greatly influenced its urban planning practice since the twentieth century. However, China's unique political, economic, and cultural context proves that the driver of its miracle absolutely contains the native wisdom, which is the oriental urban planning and construction philosophy and approaches created by Chinese planners.

By illustrating the Chinese planning experience and convey the thoughts behind the planning and practices to the international readers, the book shows a significant value for promoting global readers to better understand the mechanism about the reform and miracle of China's urban planning and development. As the authors came from different departments and focused on different fields, they have offered multiple views about China's urban planning development, such as the remarkable ancient achievements in urban construction and the current development characterized by environmental and technological friendliness. Obviously, Chinese urban planning and construction reform is still on the way, but its unique and previous successful experiences can also enlighten other countries and planners overseas.

(Translated from Chinese)

President, China Association of City Planning Kai Tang
Beijing, China

Foreword

As an ancient oriental civilization with long history, China is a mysterious and distinctive country in the world. The urban theories and practices developed in the ancient China still have an impact on the present urban constructions. The ancient Chinese planning philosophy has been not only embedded in the axes, grid structure, ordered neighborhood, and the spatial hierarchy of a city, but also in the "mountain-water-city" relationship at the regional scale. Being a rural society for a long time, China had always paid more attention to the development of agriculture and rural areas, and therefore its urbanization had remained at a low level until recent decades. By 1990, the urbanization rate in China had only reached 26.44%. Since then, Chinese economic development has begun to take off, and the urbanization rate has reached 60.6% by the end of 2019. For such a populous country, this is a remarkable achievement represented as the booming megacities and super high-rise buildings everywhere.

The successful growth of Chinese economy and urban construction has caught the world's eyes, making people in other countries curious about what is the wisdom driving China to reach such a miraculous development. However, it is really pity that there are not many English literatures that have summarized Chinese planning wisdom and/or have interpreted it to the world. This book will be such a piece to bring worldwide readers an overview of the wisdom embedded in the Chinese urban planning and construction ranging from ancient to modern history and contemporary time. And meanwhile, the book is readable due to the fact that it uses many case studies to illustrate the transformation and planning practices in China. It covers the ancient wisdom, modern miracle, and future vision in urban planning and development in China, with each chapter on a specific topic in various cities or regions at different scales.

Chinese urban planning is mainly centralized and promoted by the government bodies in the past half-century. The process of the top-down system has also shown some weakness in recent years, causing the urban planning and development to focus more on the market drive and public participation. In the new era, community governance and social response are gradually speaking loud on Chinese urban planning system. Facing the urgent challenges in improving human settlements and solving the severe climate change, etc., Chinese government has also paid more attention to deal with these emerging issues. Simultaneously, new information technologies have been widely applied in planning, while sustainable and environmental concepts have been highly valued in urban growth. All the discussions on the topics in the book tend to resonate in some other parts of the world.

Since the pursuit for a better life and liveable cities is always the same in various countries, I believe that the Chinese experiences presented in this book can also shed some light on the planning systems of other countries and regions.

President, China Academy of Urban Planning and Design Kai Wang
Beijing, China

Contents

About the Editors

Yan Tang is an associate professor at the School of Architecture, Tsinghua University, Beijing City, China. She is an editor in charge of *China City Planning Review*, deputy secretary general of Urban Regeneration Committee of China Urban Planning Society, committee member of Urban Design Committee of Architectural Society of China, and committee member of Eco-city Committee and Resilient City Committee of Chinese Society for Urban Studies. She has also worked as the secretary of 7th Architecture Discipline Appraisal Panel, Academic Degree Committee of the State Council of China. Her research interests include urban design and planning, urban regeneration, and urban and rural governance. She is a registered planner in China, and her work always highlights the combination of research and practice.

Lanchun Bian is a professor at the School of Architecture, Tsinghua University, Beijing City, China. He is an executive director of China Urban Planning Society; a vice-chairman of Urban Design Committee, vice-chairman of Urban Regeneration Committee, and committee member of Historical and Cultural City Preservation of China Urban Planning Society. He is also a committee member of the Expert Advisory Group of Beijing Historical and Cultural City Preservation. His interests mainly focus on urban design theory and practices, historical and cultural heritage preservation, and residential planning. He has published dozens of academic papers.

Zhenjiang Shen is a professor at the School of Environmental Design, Kanazawa University, Japan. He is a member of the Engineering Academy of Japan from 2018, whose research interests include smart city construction, policy-making support system for planning and design using GIS, VR, and Information Communication Technology (ICT). He is a commission member of Commission on Geospatial Analysis and Modeling of International Cartographic Association (ICA), Research Committee on Information Systems Technology, Architecture Institute of Japan (AIJ), City Planning Institute of Japan (CPIJ), and also work as a joint member of

Fudan University and PhD Instructor in Tsinghua University, China. Dr. Shen is Editor-in-chief of IRSPSD International (Indexed in SCOPUS, ESCI), editorial member of the AIJ Journal of Technology and Design (EI, SCOPUS), and editor of subseries of Spatial Planning and Sustainable Development, in Strategies for Sustainability, Springer.

Chapter 1
Overview from Historical Wisdom to Modern Miracles: Experience from Chinese Urban Planning and Practices

Zhenjiang Shen

Abstract In ancient China, the *Kaogong Ji* defined traditional Chinese urban form based on wooden building technology and adherence to traditional Chinese culture. Due to the might of western armies and their occupation of oriental countries in the 1800s, there was a need for serious social transformation. Furthermore, as Western culture seeped its way into the lives of the Chinese populace, so did Western theory on urban planning. Thus, western planning systems were introduced into Chinese society. However, as has always been the case throughout the long history of China, historical wisdom has been preserved throughout all sorts of modern technological miracles and social change. Emerging Chinese scholars will be discussing how Chinese urban planning and management has changed Chinese urban form throughout this book. For the contemporary Chinese city, planners learn from western planning system and apply planning theory to planning practices. The terms related to planning and design theory are hotly debated among Chinese planners, the most widely accepted being modern Western thinking, bound by traditional oriental values, through the lens of their personal opinions on Chinese urban form.

Keywords Traditional Chinese urban form · Oriental tradition value · Western planning system · Contemporary Chinese city · Planning practice · Planning concept

1.1 Introduction

In this book project, we are collecting the experiences from Chinese planning and practices from the view of sustainable cities and communities. The planning and practice introduced in this book prove that the Chinese planners can benefit from the

Z. Shen (✉)
School of Environmental Design, Kanazawa University, Kanazawa, Japan
e-mail: shenzhe@se.kanazawa-u.ac.jp

L. Bian et al. (eds.), *Chinese Urban Planning and Construction*, Strategies for Sustainability, https://doi.org/10.1007/978-3-030-65562-4_1

experiences of western planning practices while developing their own planning system wholly distinct from the western planning system for sustainable cities and communities, and providing a road map for developing countries looking to achieve sustainable urban form. As our contributors to this book, Chinese planners demonstrate how they support economic, social, and environmental links between urban, peri-urban, and rural areas by strengthening city and regional planning. They learn from western experiences, but in regards to urban form, they try to keep historical wisdom for traditional environmental resilience, road networks, and community identity for sustainability.

In recent years, urban planning has focused more on solving the practical problems of development by Chinese planners, in which planners have imitated the city models of Western countries. Urban zoning is often used as a policy tool to control development, and the Chinese socialist system often oversees the urban planning process in China, and is largely left to the discretion of each individual municipality. Most Chinese scholars agree that the original Chinese urban form started from the fantastic form of the imperial city described in the *Kaogong Ji*, known in English as *Records of Examination of Craftsman* (Theobald, 2020). Ancient urban form of the imperial city reflected Chinese social structure and land use was partitioned in traditional grid communities, by which ancient Chinese planners showed remarkable sustainability toward contemporary urban planning theory. In modern China, the concept of urban planning is often taught in universities as an invention of Western civilization, followed by the image of the western cities ideals such as garden city, eco city, green city, and smart city. Those concepts related to urban structure reflect the social structure and land use activities of Western cities and Western cultures.

Even though Chinese planned urban form is similar to their Western counterparts, Chinese planners are seeking their own vision to urban planning and management of the Chinese planning system. This book aims to provide a unique thinking perspective and a coherent collection of expert opinions that present the facts and operation methods of urban planning and practices in various projects and cities across China, based on China's diverse political, economic, cultural, and social background. We will explore the effects of the "Reform and Opening-up" policy that began in 1978 and was carried out with great financial assistance and influence from foreign powers. China launched a massive urban development program across the entire nation. The construction of huge amount of industry factories, commercial business districts, and residential urban development projects were implemented under increasingly large economic demand and exponential population growth. As a result, Chinese planners at the time favored high-density urban form and high-rise buildings to alleviate the tension between rapid economic growth while protecting agricultural land use and existing ecological networks in China. There are many critical views on the high rate of implementation of Western-style urban development projects during this time, however, Chinese planners at the time are able to strike a balance between metropolitan growth and protecting countryside villages for agriculture and ecological tourism.

1.1.1 Historical Wisdom and Modern Miracle in China

Each of the 15 chapters will focus on a specific Chinese urban development topic, with an overarching focus on urban form transformation. The content spans from ancient to modern times, and nearly all authors are experts in Chinese urban planning. Through integrating valuable studies across thousands of years of history, we explore the wide variety of urban planning and practices experience in China and the diverse views and interpretations by both domestic and international scholars and policy makers. For each topic discussed in this book, three keywords will be used to frame our understanding of the content: historical wisdom, modern miracles, and future planning for sustainable cities and communities.

Regarding historical wisdom, The Ming and Qing dynasty imperial city of Beijing is often referenced in Western urban planning courses. Modern architects and planners look to the traditional palaces and settlements when approaching current Chinese projects. Such uniqueness of historical wisdom comes from the concepts outlined in the *Kaogong Ji* as the gold standard for urban planning and thinking, and was strictly adhered to for nearly 2000 years until the western concept of urban planning was introduced. The particulars of ancient Chinese urban planning are the reflection of the social structure and its activities on the layout of cities. Urban blocks and their classified land use functions as the guidelines for gridded community areas and palaces. This book intents to present a unique angle for academic discussion on the understanding of the oriental concept of urban form in China and how it has assimilated and is pioneering the future of urban design. Authors in this book aim to find the focal point for potential Chinese contribution in terms of new knowledge and experience from both theoretical and practical perspectives.

Regarding modern miracles, by admiring the elegance of city planning and construction in historical periods, we can acquire a new perspective on how Chinese cities were organized and planned and how Chinese planners presented the leading ideas of the time that guided development. The impressive scope and scale of recent urban planning projects in China like the merging of Shenzhen, Shanghai Pudong, and Zhujiang New Town of Guangzhou. However, while Chinese planners are enthusiastic about optimizing urbanization and utilizing Western urban planning theory, too often they lose sight of any sense of local Chinese culture and environment in the search for efficiency. At the same time, international scholars and planners are keen to investigate the success of not only China's successful economic development strategy, but also innovative urban planning and construction theories. There is an attempt to understand the reasoning and mechanisms behind this success, but English language analyses and reports are few and far between.

In 2019, a new planning system in China has begun, namely a spatial planning system which will not be discussed in this book. It is important to summarize what has been achieved by the urban and rural planning system leading up to the establishment of the new spatial planning system. Thus, this book focus on the urban and rural planning system in China as it has been previously, despite the Chinese government's recent decision to usher in a new era of urban and rural planning. The new spatial planning system in China is a tool for making the planning management and planning organization more logical, covering not only urban areas, but agricultural areas, forest areas, and natural preservation areas. It is likely the new system will be more easily understood by international readers if the contents of this book have already been read.

1.1.2 Literature Review

1.1.2.1 Chinese Planning System in China

The concept of Chinese urban form was birthed from the *Kaogong Ji*, and over time Western concepts of garden city, eco city, green city and smart city became the default style. This book outlines why and how Chinese urban planning and practice has achieved success over time, and how the Western world can better understand the Chinese experience as a whole.

While virtually every journal on urban development has published articles on Chinese urban development, there are few books that have discussed in detail the transformation of Chinese urban form from the perspective of urban planning and practices. Foremost of these works is "*Planning for Growth: Urban and Regional Planning in China*" written by Fulong Wu (2015), in which he systematically introduces China's planning system, policy, and practices. Li Yu's book, "*Chinese City and Regional Planning Systems*" (2014) focuses mainly on the institutional arrangements that govern planning and development, but is light on the leading ideas driving planning and the results on the ground. Remi Curien's "*Chinese Urban Planning*" (2014), briefly discusses many of the topics covered in this book, but lacks the detailed results of projects and changes. The more ambitious book by Chen Yuanzhi, Alan Hudson, and He Lisheng, "*Chinese Urban Transformation,*" focuses on planning and development in six cities and offers considerable detail for these examples but lacks a wider perspective of Chinese urban planning practice. Springer also appears to have several books that deal with aspects of urban planning in China, including Zhang and Wang's (2019) "*Urban Planning and Development in China and Other East Asian Countries*," and Wong, Han, and Zhang's (2015) "*Population Mobility, Urban Planning and Management in China*." Campanella (2012) in the book of "*The Concrete Dragon: China's Urban Revolution and What it Means for the World*" surveys the driving forces behind the great Chinese building boom, traces the historical precedents and global flows of ideas and information that fused to create a bold new Chinese cityscape, and considers the social and environmental implications on China's urban future. Due to the socialist political structure and particular social systems in China, urban planning system has played a crucial role in the urbanization progress. So far, few English books take into consideration the transformation of historical Chinese urban form when evaluating modern Chinese urban planning.

1.1.2.2 Transformation of Contemporary Chinese Cities

When comparing "*From Historical Wisdom to Modern Miracles: Learning from Chinese Urban Planning and Practices*" to books on similar topics, we focus on the wisdom and successful experiences of Chinese urban planning and practices, in order to more deeply debate the Chinese urban form related to each individual issue or topic in various cities or regions at different scales.

In the book edited by Wu Weiping, and Piper Gaubatz (2013) "*The Chinese City*," an overview is provided about China's historical urban planning system, and urbanism beliefs, offering a critical understanding of China's urbanization, exploring how the complexity of the Chinese city both conforms to and defies conventional urban theories when compared to the experiences of cities elsewhere around the world. Ren (2013) in the book "*Urban China*" identifies the past trajectories, present conditions, and future prospects of Chinese urbanization in the modern era, and presents the social phenomena and underlying issues of rapid urbanization in China from the perspectives of society, economy, and culture, focusing on urban-rural duality and social inequality. Some chapters mention the changes of urban landscape in China over time, the range of years discussed is small relative to the storied history of the Middle Kingdom as a whole.

In the following book, we attempt to provide a comprehensive perspective of the transformation of urban form in urban planning and practices of China while combining the urban system, planning concepts with planning theory behind them into a single coherent strategy. We introduce the pillars of Chinese urban planning and summarize its successes at different factors of scale. In addition, most other literature is written about issues with the urbanization process in China, while few scholars address successful use cases of urban planning and practices in China.

1.2 Overall Approach Goals

As discussed above, this book is organized into three keywords, namely historical wisdom, modern miracles and future transformation, with a focus on sustainable urban form. When looking at historical wisdom and modern miracles, the authors in this book review the sustainable urban form at the district, city, and regional scale, and identify ancient concepts of Chinese city planning relays the traditional wisdom of resilience from ecological, engineering, social and economic dimensions. Regarding the urban form discussed in this book, at the regional level, the regional design of ancient cities in China is proposed to be functional layout, spatial structure and environment protection leading growth toward functional circles, infrastructure space, and stable and orderly development based on the concepts of historical wisdom. At the city and district level, Xi'an in the Han Dynasty can be seen as an early example of Supergrid and Superblock structures, with naturally formed Supergrids with a number of Superblocks within the city.

This book has four parts and 15 chapters; however, the content of each chapter is relatively independent. First, 3 classic use cases are selected to identify the unique wisdom of ancient Chinese city planning related to regional design, water conservation systems, and urban districts, which continue to play a significant role in current urban-rural practices. From the "Regional Design" of ancient Chinese capitals to the spatial pattern of traditional settlements that can still be seen today at the world cultural heritage town of Pingyao, the ancient Chinese construction experience provides useful practical tools and paths to realize long-term sustainable development. It is the special Oriental philosophy, civilization and way of thinking that has cultivated such wisdom, which is noticeably different from modern western theory. The Supergrid and Superblock structure need not have to be totally replaced by the western street blocks structure, and the removal of all walls around residential areas may not be a final solution as previously believed in the planning practice of Chinese cities.

Throughout the nineteenth century, as Chinese planners learned from the planning experience from the western countries, three chapters focus on the historical periods of how the Beijing-Tianjin-Hebei Capital Region, Tsingdao and Chongqing city changed and grew. These chapters together provide a well-rounded analysis on the main modern issues of urban development in China at varying scales, including government administration, planning systems, urban investment, social impact, construction models and so on. Existing lessons and problems will also be discussed to foster dialectical thinking. These chapters explain in detail the particular process and institutions that led to the achievements, and how government, market and society play a role during the process. There are also case studies like community development and urban development projects used as examples in this book.

Apart from discussing the influence of Western theory on Chinese planning history, planners in China are likely to also import new technologies from the western countries. In part three, the sustainable urban form archived by planning practices in China is discussed for various scales, like eco-city practices, healthy city at city scales, green low-carbon high-density area at urban district scales, and green building at the building level.

The final part of the book considers the Chinese experience of modern urban planning and development in the modern era of information technology. Various Chinese urban planning achievements related to the new rising demands of information technology that have drawn significant international attention are explored, such as the rapid growth of urban new districts and special urban zones, the evolution of Taobao online-shopping business villages, and the development of Chinese hi-tech parks.

1.3 Historical Wisdom and Urban Planning

In this book project, we wish to share our understanding on the historical wisdom and modern miracles of Chinese cities with the world. The current Chinese urban form has been transformed under the development pressure alongside economic and technical innovation. Meanwhile, the modernism movement of urban planning simultaneously changes the traditional urban form and its community, thus Oriental urban form has also transformed with the global shift in urban form style and strategy. In this sense, urban regeneration projects and historical conservation projects have transformed traditional society and community structure in China. However, planners and policy makers attempt to maintain an adherence to the ancient green system while accepting the concept of green city.

The conservation of heritage and the protection policy is an important aspect of traditional urban form, best seen in Chang'an during the Tang Dynasty or Beijing at the time of Yuan and Qing Dynasty. In the chapter, "*Regional Design of the Ancient Chinese Capitals: A Case Study of Chang'an in Tang Dynasty*" (Guo, 2021), ancient Chinese cities are shown to be highly integrated with and closely related to the extramural area. "Regional Design" was a design philosophy implemented throughout the entire capital region to meet the functional needs of the capital and evoke grandiose spatial imagery. However, little research has paid attention to its planning and design at the regional scale. "Regional Design" is defined by three aspects: Functional Layout forming several functional circles, Spatial Structure taking the regional geographical condition as the backbone of urban spatial structure, and Protection Policy ensuring stable and orderly development in a capital region. The authors here argue that preservation of ancient traditions on contemporary urban and regional planning are discussed, as planners and policy makers continue to keep ancient urban system beliefs alive when considering "Regional Design" philosophies.

It is important to rethink the applicability of traditional values and incorporate traditional planning concepts in modern city design in meaningful ways. Chen (2021) presents a chapter, "*Looking into the Supergrid and Superblock Structure in Chinese Cities: Taking Xi'an and Nanjing as Examples.*" Supergrids are a large-scale network of wide roads that defines a series of cells called Superblocks, which is similar to the concept of traditional *lifang,* the boundary of traditional communities in ancient Chinese cities. Unique to Chinese city design, each Superblock contains a number of walled living quarters next to each other assisted by few narrower streets to minimize cross traffic. The State Council of China released a set of planning guidelines in 2016 that suggests a transformation of the current Super grid-and-block structure to a finer network, but it was strongly opposed by people who emphasize the importance of the gated community walls as an important safety measure. This chapter provides a comprehensive exploration of the development of the community and road network in Chinese planning history and the modern planning practice using Xi'an and Nanjing as case studies. The use of Superblocks in combination with traditional walls and gates around living quarters is a unique traditional value that reflects Chinese spatial planning logic overruling the encroachment of modern western planning ideals.

Zhang, Xu, and Chen in their chapter, "*Adaptative Resilience: Traditional Wisdom of the Mulan Weir Water Conservancy System in Song-Yuan Period,*" present that the Mulan Weir in southeast China, a World Heritage Irrigation Structure that has been in service for nearly a thousand years, adapts to natural disasters such as drought, floods, and tidal shifts while ensuring consistent and reliable agricultural irrigation. Recently, local economic strategy has further provided a long-term plan to ensure the continued existence of the Mulan Water Conservation System. The chapter analyzes the traditional wisdom of adaptive resilience instated during the Song-Yuan Period, based on the ancient framework of resilience analysis involving ecological, societal, and economic systems. The ecosystem of the Mulan River basin was in a relatively dynamic balance using a three-level water conservancy system, consisting of a weir—sluice gate—culvert, subtly improved local irrigation conditions, and flexible disaster scenario responses.

Traditional values constitute an important aspect when understanding Chinese wisdom, and wisdom in urban planning is no different. To extract traditional values and understand its wisdom, these three essays have been given different scopes, physical scales and aspects of attention related to the revitalization and adaptive resilience of the Chinese planning practices.

1.4 Transformation of Traditional Values into Urban Planning

Chinese planners have learned from the past; however, they have struggled to protect traditional urban form under the massive pressure from social and economic development in planning practice. For example, when Tiananmen Square in Beijing was being proposed, it was argued that new city center should be located outside of the historical downtown areas. Even though there have been great achievements in modern day urban projects, the overall tone of the urbanization process in any country must be counterbalanced by protecting tradition, and it is from this tension that most conflicts arise. In most cases, planners can have a theoretical understanding of their traditional culture, and attempt to keep the core values that mark their city as different from others, but in the era of modernism many projects can begin to blend together into an unrecognizable shape.

Chinese planners have also been very flexible in accepting new concepts from Western countries for sustainable development, however practices were under serious developmental pressure. Tang and Meng (2021), in their work, "*From Concentration to Decentralization, The Spatial Development of Beijing and the Beijing-Tianjin-Hebei Capital Region,*" introduce the decision to construct Tongzhou, the Sub-City Center of Beijing, and the Xiong'an New Area, the new magnetic development center in Hebei province, these are the two most significant actions taken for exploring a more sustainable approach for the capital region. Since Beijing and Tianjin have absorbed most of the regional investments, resources, and talent pool, the emerging "mega-city diseases" of Beijing caused by over-dense urban areas and overpopulation, severe air pollution, water resource shortage, serious traffic jams, and continued environmental deterioration. The capital region of Beijing-Tianjin-Hebei (BTH) has developed in increasingly unbalanced way over the past four decades. National strategies of decentralizing the non-capital function of Beijing have been proposed and coordinative development of Beijing, Tianjin, and Hebei has been launched in the BTH region.

Another on the colonization of the industrial city of Qingdao is taken as case study in this book. In the chapter, "*From Tsingtau to Qingdao*" (Kunzmann and Zhan, 2021), the authors argue that in contrast to many other cities in China, the old city center of Qingdao has been fully preserved. It has been well protected in the past, but Qingdao is looking forward to utilize professional and economically viable urban regeneration policies. From 1897 to 1914, the city had been a German mandated territory, named Tsingtau. The German administration had curbed excessive land speculation 100 years ago based on then-innovative land regulation. In this chapter, achievements of modern city development, such as the Sino-German Eco-development Park are presented and end with a brief account of challenges the city will face in the years ahead to conserve its historical heritage while still meeting the requirements of a modern Chinese metropolis.

The following chapter discussed community development in China, community engagement is a new prospective way to push for modernization as in the case of Chongqing. Huang (2021) proposes in "*Planning Approach for Urban Community Development and Regeneration: Chongqing Practice*", that community care for the lower class has increasingly become the location of innovative social governance, which is already completely different than the traditional social structure in ancient Chinese cities. This chapter uses Chongqing urban community development planning practices since 2010 as the case study, and selects 2–3 practice cases to summarize and reflect how to best apply the American theory and approach of community engagement for community planning practices in Chongqing for the purpose of community regeneration.

Urban planning is a technical and professional process concerned with land use development, transportation, communication infrastructures and the built environment (McGill University, 2008) in the urbanization process. In China, urban planning operates under nationalized land ownership, which lacks the negotiation process between stakeholders typically seen in other countries. Recently, the urban planning system in China has been transformed because the lack of negotiation process between all stakeholders, such as developers, agricultural industries, forest, and ecological functionaries. However, more time is needed to see the full ramifications of new urban planning and practices will change in the future.

1.5 Green and Eco City Planning for Sustainability

The concept of green city and eco city coming from Western countries has deeply influenced Chinese planner communities. Chinese planners have been accepting new concepts from Western countries and enthusiastically putting them into practice. They have made great strides to combine the experiences of eco city and green city planning practices from Western countries with their own practices in China. At the same time, the original concept of Oriental traditional urban form combined with modern urban structure of Chinese cities continues to progress.

In the chapter, "*Green Low-carbon High-density Urban Center Planning: Wuhan Wangjiadun Area*" (Xu and Xu, 2021), authors claim that a city is the intersection of human social civilization and technological progress. In China, due to the rapid urbanization process, the conflict between cities, population, and the environment is becoming more and more serious. In high-density urban centers, how can green low-carbon planning and practices be carried out under such intensive use of land? Wuhan Wangjiadun Central Business District provides a relevant case for discussion and analysis. In view of the green and low-carbon planning and development concept, Wangjiadun carried out the overall planning, in which Wuhan Center became a green super high-rise commercial complex of 438 meters. The central block of the surrounding area was promoted by green transportation, and a continuous corridor of underground traffic makes full use of the underground space.

Chai (2021) offers an overview of the eco city construction in China in his chapter, "*Towards a Sustainable City: A Scoping Review of Eco city Construction in China.*" With the rapid development of urbanization, China is facing many challenges regarding environmental issues, such as air pollution, water deterioration, urban heat island effect, etc. To cope with urban ecological crises, China is increasing the focus on eco city construction. In this chapter, the origin of ecological urban planning in China has been traced to its genesis, then several typical eco cities were taken under review to evaluate the efficacy of these improvements. Even though there are already hundreds of cities on their way to becoming eco cities or low-carbon cities, many more all over China are attempting to transform themselves into smart cities with ecological considerations, thereby making the urban systems operate more efficiently and ecologically.

To elaborate on the theoretical development and urban practice of healthy city planning in China and its implication for other countries, Wang (2021), presents "*Healthy City Planning: Insights from China.*" Here it is proposed that urban planning, as an important tool for allocating resources and shaping urban form, has the authority to promote public health. It is followed with a demonstration of healthy city planning in China at the district and community level. Innovation of institutional design within the current planning system of China is explored to illustrate the impacts of decision-making in the planning process on the physical and mental health of local residents and the advantages of integrating considerations for public health in that planning process.

In the chapter, "*From Green Building to Green City:The Practice in Jiangsu Province China*" (2021), Wang discusses that the Yangtze River Delta urban agglomeration formed by Jiangsu and Shanghai and Zhejiang Province has become one of the top six world-class urban agglomerations. In the field of urban and rural construction, up to now, Jiangsu Province has set up 66 green ecological urban areas with a planned area of 10,000 square kilometers and more than 110 million square meters of green buildings in the demonstration area. At the same time, the construction of green transportation, green energy, comprehensive utilization of solid waste, and comprehensive utilization of water resources has been carried out with remarkable results. These improvements are best seen in the demonstration zones of Suzhou Industrial Park, Nanjing Hexi New City and Wuxi Taihu New City. Jiangsu's total green building area accounts for a quarter of the country's total, and the total quantity and size of green urban areas rank first in the country.

This series of chapters focuses on China's own practices with the Western experiences introduced into China for promoting green buildings and green eco city, including green ecological special planning in new urban development areas. The planning practices and learning experiences from the Western countries not only carry great significance for modern Chinese urban development, but also have equal significance for the development of any modern city around the world, as well as renewal of green cities and green city development.

1.6 Modern Miracles Brought Forth by Technological Innovation and Economic Growth

In this part, modern planning and practice miracles based on the rapid urbanization process in the recent 30–40 years in China are introduced, also shown are the newest trends and planning methods of urban development in modern China. Foremost of these miracles is Beijing and its innovations that regenerated the capital city; followed by Suzhou's miraculous transformation that forged and cast the successful Powerhouse model for the rest of the Chinese cities. The most amazing case study is the challenges brought to the new urban space, namely e-commerce space in an urban area, particularly the smart cities of China. The socio-economic transformation of those cities and villages which have embraced e-commerce and have collectively formed the so called "Taobao Communes."

"*New Countryside in the Internet Age: The Development and Planning of E-commerce Taobao Villages in China*" (2021), Luo and Qiao present that Taobao villages have become a special economic geography phenomenon in Chinese rural regions under the influence of economic globalization and the explosion of internet-based companies. The rapid proliferation of Taobao villages from 2013 to 2018, located in the urban edge and the edge of the metropolitan area, are almost solely led by grassroots entrepreneurs. For coping with the chaos due to combined rapid development and weak regulation, the authors advocate that the planning of Taobao villages should focus on the bottom-up construction of rural living environments.

Innovation power is crucial for urban economy development. In "*Innovation Districts in Beijing: Evolution, Distribution, Development Mechanisms and Planning Policies*" (2021), Yuan, Liu, Ren, Chen, Wang and Shi present that Beijing has developed diverse types of innovation districts, which support all kinds of innovation activities giving a new impetus for urban growth. The distribution of innovation districts in Beijing recently is composed of six main types varied from science and technology parks, university parks, science cities, or entrepreneurship communities. In general, the innovation districts in Beijing are experiencing a transformation from traditional science and technology parks to innovation districts with enterprise as its core, from government-led enterprise to community-led. Beijing has built up its innovation space hierarchy to better support and encourage a wide variety of innovation activities.

Yang (2021) in his chapter, "*Suzhou's 'Growth Machine': Tracking the Driving Force behind a Fast Growing Chinese Urban Region,*" introduces Suzhou's fast economic growth in the past 30 years. During this timeframe, society benefitted from an economic boom mixed with tough challenges and choices associated with diminishing land resources and sudden environmental deterioration. This chapter reveals how a pro-growth coalition involving state and market powers formed a "Growth Machine," which accelerated socio-economic changes. Because Suzhou is one of the most successful fast-growing cities in China, other cities experiencing similar "Growth Machine" effects can gain insights regarding the implications of these strategies for community development and the evolution of modern urban form.

Besides planning projects in traditional urban areas, modern planning and practice miracles based on the rapid urbanization process in the recent 30–40 years in China are specifically looked into, also showing some of the newest trends in urban development areas across China, providing rich planning experiences and references of modern miracles for planning innovation. Chinese planning practices and exploration aimed at sustainable development to meet future demands illustrates the transformation of Chinese cities to cope with challenges like sudden economic growth, informatization, climate change, and localized adaption of international planning concepts.

1.7 Discussion

Ancient China achieved much in the world of urban planning and practices that deserves spreading and promoting, and modern China's rapid development in economic and urban construction recently has also attracted worldwide attention. This book aims to introduce Chinese urban planning and practices from historical wisdom to modern miracles and future transformation, and provide a comprehensive overview of Chinese urban planning and practices to international readers in fully modernized and developing countries alike. There is more and more interest in Chinese urban planning and practices, and much to learn about China's development of economic and urban construction.

In this book, the authors use various case studies to interpret the urban planning and practices in China based on their own experiences and expertise. In this way, they are able to objectively summarize the existing situations and experiences, and propose solutions surrounding the relevant problems. The authors come from different fields such as academic circles, planning and design institutions, consulting companies, and more, helping ensure the diversity and comprehensiveness of the viewpoints of the book. Compared with ancient Chinese planning styles, the four-part organization for past, present and future is a major leap forward. Here the practices of present day during the process of urbanization are focused on as well as the organization and planning processes at several governmental levels.

Where this book differs from others is the variety of authors from a variety of institutions and cities. They each introduce Chinese urban planning and practices with case studies in their own area of expertise. Furthermore, this book presents the absolute latest transformation of urban planning happening in China, such as e-commerce Taobao villages and health city planning. Such innovations are not mentioned in previous books, and this book provides the comprehensive coverage and in-depth case examples of what is happening right now in the most modern Chinese cities. Both the breadth and depth of leading ideas are covered here.

Regarding the details of our findings in this book project, firstly, economic innovation supporting urban development in metropolitan areas and new countrysides has been discussed in this book, and economic and technical driving forces in urban region improve that urbanization, while traditional Chinese social structure is completely transformed. Secondly, reviewing the practices of the regional spatial structure for protection and cultivation of regional natural environments in China, planners paying attention to natural water environment conservation, keeping traditional resilience systems in major function zones in place, all while considering possible regeneration of human settlements and the power of urban communities. Thirdly, while learning from the planning practices and concepts of Western cities, creating detailed economic and community plans, Chinese planners have been able to conserve the urban form of ancient Chinese cities while putting those planning methods into practice for transformation of regional structures, urban forms, road networks, such as attempts to decentralize some metropolitan areas, and block scaling up of transportation in urban design. Finally, Chinese planners have been very open to accepting new concepts learned from Western countries, such as healthy, green, and eco cities for sustainability, proving that Chinese planning system is capable of accepting Western planning experiences while simultaneously developing their own planning system improvements for sustainable cities and communities.

This book targets students and professionals wishing to understand current Chinese urbanization as well as businesses and investors wishing to get involved in projects in Chinese cities. Finally, it may also serve as a guidebook for the considerable number of foreign students studying in China looking for understanding of the country's impressive urban planning practices.

References

Campanella, Thomas J. (2012). The concrete dragon: China's urban revolution and what it means for the world. Chronicle Books.

Chai, Zhi(2021) Towards a Sustainable City: A Scoping Review of Eco-cities development and practices in China, *From Historical Wisdom to Modern Miracles: Experience from Chinese Urban Planning and Construction, in Spatial Planning and Sustainable Development edited by Bian, Tan and Shen*, Springer

Chen, Yuanzhi; Alan, Hudson & He, Lisheng. (2018). Chinese urban transformation: a tale of six cities, London: RIBA Publishing.
Chen, Xiaofei (2021) Looking into the Supergrid and Superblock Structure in Chinese Cities: Taking Xi'an and Nanjing as Examples, *From Historical Wisdom to Modern Miracles: Experience from Chinese Urban Planning and Construction, in Spatial Planning and Sustainable Development edited by Bian, Tan and Shen*, Springer
Rémi Curien, (2014). Chinese Urban Planning, *China Perspectives,* Vol. 2014/3, 23-31. https://doi.org/10.4000/chinaperspectives.6528
Li Yu. (2014). Chinese City and Regional Planning Systems. Cambridge University Press, Farnham and Burlington: Ashgate. https://doi.org/10.1017/S0305741014001350
Luo, Zhengdong; Qiao, Yibo (2021) New Countryside in the Internet Age: The Development and Planning of E-commerce Taobao Villages in China, *From Historical Wisdom to Modern Miracles: Experience from Chinese Urban Planning and Construction, in Spatial Planning and Sustainable Development edited by Bian, Tan and Shen*, Springer.
Guo, Lu (2021) 'Regional Design' of the Ancient Chinese Capitals: A Case Study of Chang'an in Tang Dynasty, *From Historical Wisdom to Modern Miracles: Experience from Chinese Urban Planning and Construction, in Spatial Planning and Sustainable Development edited by Bian, Tan and Shen*, Springer
Huang, Lin (2021) Three Stages of Urban Community Development and Regeneration Planning in Chongqing, *From Historical Wisdom to Modern Miracles: Experience from Chinese Urban Planning and Construction, in Spatial Planning and Sustainable Development edited by Bian, Tan and Shen*, Springer
Kunzmann, Klaus R. and Zhan, Erpeng (2021) From Tsingtau to Qingdao, From Historical Wisdom to Modern Miracles: Experience from Chinese Urban Planning and Construction, in Spatial Planning and Sustainable Development edited by Bian, Tan and Shen, Springer
McGill University (2008). "What is Urban Planning". *Archived from https://www.mcgill.ca/urban-planning/planning*, on April 8, 2020
Ren, Xuefei. (2013) Urban China. John Wiley & Sons.
United Nation (2020), "Sustainable Development Goals", *Archived from* https://www.un.org/sustainabledevelopment/cities/, on April 8, 2020
Ulrich Theobald (2020), Kaogongji, *Archived from* http://www.chinaknowledge.de/Literature/Classics/kaogongji.html on April 8, 2020
Wang, Lan (2021) Healthy City Planning: Insights from China, *From Historical Wisdom to Modern Miracles: Experience from Chinese Urban Planning and Construction, in Spatial Planning and Sustainable Development edited by Bian, Tan and Shen*, Springer
Wang, Dengyun (2021) From Green Building to Green City——The Practice in Jiangsu Province, China, *From Historical Wisdom to Modern Miracles: Experience from Chinese Urban Planning and Construction, in Spatial Planning and Sustainable Development edited by Bian, Tan and Shen*, Springer
Wong, Tai-Chee; Han, SunSheng; Zhang, Hongmei (2015). Population Mobility, Urban Planning and Management in China, Springer.
WTANG Yan and MENG Xiangyi (2021) From Concentration to Decentralization: The Spatial Development of Beijing and the Beijing-Tianjin-Hebei Capital Region, *From Historical Wisdom to Modern Miracles: Experience from Chinese Urban Planning and Construction, in Spatial Planning and Sustainable Development edited by Bian, Tan and Shen*, Springer.
Wu, Weiping, and Piper Gaubatz. (2013). The Chinese city. Routledge.
Wu, Fulong. (2015). Planning for Growth: Urban and Regional Planning in China. RTPI Library Series.
Yang, Yifan (2021) Suzhou's "Growth Machine": Tracking the Driving Force behind a Fast Growing Chinese Urban Region, *From Historical Wisdom to Modern Miracles: Experience from Chinese Urban Planning and Construction, in Spatial Planning and Sustainable Development edited by Bian, Tan and Shen*, Springer

Yuan, Xiaohui; Liu, Xiyu; Ren, Junyu; Chen, Jun; Wang, Liang; Shi, Xiaodong (2021) Innovation Districts in Beijing: Evolution, Distribution, and Development Mechanisms, *From Historical Wisdom to Modern Miracles: Experience from Chinese Urban Planning and Construction, in Spatial Planning and Sustainable Development edited by Bian, Tan and Shen*, Springer

Xu, Sheng (2021) Green Low-carbon High-density Urban Center Planning Wuhan Wangjiadun area, *From Historical Wisdom to Modern Miracles: Experience from Chinese Urban Planning and Construction, in Spatial Planning and Sustainable Development edited by Bian, Tan and Shen*, Springer

Zhang, Tianjie; Xu, Qiuyin and Chen, Shuangchen (2021) Adaptive Resilience: Traditional Wisdom of the Mulan Weir Water Conservancy System in Song-Yuan Period, *From Historical Wisdom to Modern Miracles: Experience from Chinese Urban Planning and Construction, in Spatial Planning and Sustainable Development edited by Bian, Tan and Shen*, Springer

Zhang, Guanzeng; Wang, Lan (2019), Urban Planning and Development in China and Other East Asian Countries, Springer.

Part I
Historical Wisdom and Urban Planning

Chapter 2
"Regional Design" of the Ancient Chinese Capitals: A Case Study of Chang'an in Tang Dynasty

Lu Guo

Abstract Ancient Chinese Cities are highly integrated with and closely related to the extramural area. The planning and design of the capital was not restricted to the area within the city wall; rather it was extended to the surrounding areas. "Regional Design" was implemented in the whole capital region to meet the functional needs of the capital and create a grand spatial image. There has been plenty of research on the urban planning and design of Chang'an in Tang Dynasty (618–907), which was the capital of the most prosperous dynasty of ancient China. However, little research has paid attention to its planning and design at the regional scale. This article aims to solve this problem using the "triple-evidence" method. The "Regional Design" of Chang'an would be expounded from two aspects: (1) Functional Layout. The layout of the capital function was carried out at the scale of the Guanzhong Basin, forming several functional circles. (2) Spatial Structure. Regional-scale axes connecting the city and the natural mountain was established to form a spatial backbone, and the high points with good view of the natural terrain were occupied to build important buildings forming several controlling "belts" of the regional space. The spatial law and the implementing subjects of regional design in ancient Chinese capital area would be discussed as a conclusion.

Keywords Regional design · Chang'an · Guanzhong Basin · Regional spatial order

2.1 Introduction

Ancient Chinese Cities are highly integrated with and closely related to the extramural area (Mote 1977). "China's cities, despite achieving great size, continued merely a 'denser in quality' version of the greater agrarian reality" (Kotkin 2005).

L. Guo (✉)
School of Architecture, Tsinghua University, Beijing, China
e-mail: guolu2014@tsinghua.edu.cn

L. Bian et al. (eds.), *Chinese Urban Planning and Construction*,
Strategies for Sustainability, https://doi.org/10.1007/978-3-030-65562-4_2

Therefore, research on spatial planning and design of ancient Chinese cities should extend beyond the city wall, and consider the city and its surrounding extramural areas as an integrated whole. The planning and design of the capital was not restricted to the area within the city wall; rather it was extended to the surrounding areas. "Regional Design" was implemented in the whole capital region to meet the functional needs of the capital and create a grand spatial image.

Tang Dynasty (618–907) was a period with strong national power and prosperous culture in the Chinese history. Chang'an, which was the national political and cultural center then, has received a lot of academic attention in the aspect of its urban planning and design. Many scholars have carried out extensive studies on its spatial layout (Yan 1948; Ma 1978; Boyd 1962), planning concepts (Zhang 1987; Heng 1999; Xiong 2000) and planning methods (Fu 2001).

However, it should be noted that, just like the other capitals in imperial China, in Tang Dynasty, various large-scale constructions were conducted in the surrounding area of the capital, including mausoleums, palaces, temples, and villas. As a result, a capital area with distinctive regional spatial order was formed—the capital as center and Guanzhong Basin (Central Shaanxi Plain) as hinterland. However, most existing research on spatial planning and design of Chang'an is limited to the urban scale, and research on regional scale is mainly done by historical geographers, which concentrated on terrain (Shi 1995), vegetation (Shi et al. 2006), traffic (Wang 1985) and hydraulic engineering (Li 2004). In other words, there has been almost no research on the construction of regional design of Chang'an Area.

This article aims to solve this problem using the "triple-evidence" method. The "Regional Design" of Chang'an, i.e., the construction of regional spatial order of Chang'an Area, would be expounded from two aspects: (1) Functional Layout. (2) Spatial Structure.

2.2 Method and Materials

2.2.1 "Triple-Evidence" Method

The "triple-evidence" method is used in this paper, i.e., based on the evidences from archeology, literature, and field survey. The double-evidence method (i.e., unearthed relics and historical records) is a widely used method in the study of Chinese history. This paper develops this method. The natural geographical environment is the basis and spatial platform of urban planning, and the difference between ancient and modern natural terrain is small, which could be used as an important reference for the study of urban planning history.

Among the literature materials, the image materials with clear spatial attributes and the literary works with both objectivity and subjectivity have received special attention. Meanwhile, the investigation and analysis of the natural landform and geographical conditions are also the important focus of this study.

2.2.2 Relevant Historical Documents and Archeological Findings

There are many types of historical documents related to the regional design of Chang’an Area in Tang Dynasty, including: official history and political books, category books, geographical records, local gazetteers, and literary works, etc. (Table 2.1). There are also some notes of the current people (Hiraoka 1989 & Hiraoka and Imai 1991; Li and Xu 1996; Liu 1997; Chen 2006). These documents

Table 2.1 Relevant historical documents

Type	Name	Main content
Official history Political books	*Sui Shu* [Book of Sui] *Xin Tangshu* [New Book of Tang] *Jiu Tangshu* [Old Book of Tang] *Tang Liudian* [Six Statutes of Tang] *Tongdian* [Comprehensive Institutions] *Tang Huiyao* [Institutional History of Tang]	Regional geographical conditions, urban site selection and spatial layout, regional social culture, etc.
Category books	*Taiping Guangji* [Extensive Records of the Taiping Xingguo Period] *Taiping Yulan* [Imperial Reader of the Taiping Era] Cefu Yuangui Yuhai *Wenyuan Yinghua* [Finest Blossoms in the Garden of Literal]	
Geographical records	*Kuodizhi* Vol.1 *Yuanhe Junxian Tuzhi* Vol.1–4 [Yuanhe Maps and Records of Prefectures and Counties] *Lidai Zhaijing Ji* Vol.1–6 [The Imperial Capitals Through History]	Historical evolution natural environment, cities, historic sites, and etc.
Local gazetteers	*Liangjing Xinji* [New Records of the two Capitals] *Chang’an Zhi* [Gazetteer of Chang’an] *Yong Lu* [Records of Yong] *Leibian Chang’an Zhi* [Supplement to the Gazetteer of Chang’an] *Guanzhong Shengji Tuzhi* [Illustrated Record of Famous Sites in Guanzhong] *Tang Liangjing Chengfang Kao* [Research on the Wards of the Two Capitals of the Tang Dynasty] *Nanshan Gukou Kao* [A Survey of the Valleys of the Southern Mountains] *Xi’an Fuzhi* [Gazetteers of Xi’an Prefecture] *Shannxi Tongzhi* [Comprehensive Records of Shaanxi] The gazetteers of various counties	Natural environment, social life, palaces, temples, imperial mausoleums, wards, and streets

(continued)

Table 2.1 (continued)

Type	Name	Main content
Literary works	*Quan Tangshi* [Complete Tang Poems] *Quan Tangwen* [Complete Works of Tang] *You Chengnan Ji* [Tour to the South of the City] *Mingshan Shenggai Ji* [Records of Famous Mountains] *Shimo Juanhua* [Engravings in Ink and Stone]	Urban layout, spatial imagery, social life, historical sites, etc.

Table 2.2 Relevant archeological findings

Archeological findings		Sources
Urban spatial structure	Outer City Wall, imperial city, palace city, streets, markets, wards, etc.	Institute of Archaeology, Chinese Academy of Social Sciences 1961, 1963, 1974, 1987
Sites and layout of the important buildings in and around Chang'an	Daming palace, Xingqing palace, Qinglong Temple, Yuanqiu (the altar for heaven), etc.	Institute of Archaeology, Chinese Academy of Social Sciences 2007; Ma 1989; An and Li 2000
Sites and layout of the temporary palaces and temples in Chang'an area	Jiucheng palace, Huaqing palace, Yuhua palace, Cuiwei palace, etc.	An and Ding 1995; Luo 1998; Lu 1978; Li et al. 1991
Sites and layout of the imperial mausoleums	Zhaoling mausoleum, Qianling mausoleum, Qiaoling mausoleum, etc.	Liu and Li 1987; Wang and Lou 1980
Important traffic sites	Ba bridge, eastern Wei bridge, regional road across Qinling Mountain	Compilation Committee of Local Chronicles of Shaanxi Province 1995

constitute a rich historical basis for this paper, but there are also corresponding challenges in extracting and integrating a large number of scattered information (Table 2.1).

The archeological investigation and research on Chang'an area in Tang Dynasty began in the early twentieth century and has achieved substantial results so far, providing a solid empirical basis for the study of regional spatial order construction (Table 2.2). The basic spatial pattern of the capital city and the layout of some important buildings are generally clear, but the various archeological relics that have been discovered in succession over a long period of time and scattered in large-scale regional space are still isolated, and a clear system has not been outlined (Table 2.2).

2.2.3 *Historical and Geographical Conditions of Chang'an Area*

Chang'an Area was the capital region that has the longest history in ancient China. Before the Tang Dynasty, most capitals of China in different dynasties were established in the middle part of Guanzhong Basin on the banks of Wei River, which are in and near current Xi'an City (Fig. 2.1). In total, Chang'an Area was the capital for 13 dynasties, which altogether lasted for more than 1200 years. Tang Dynasty, which was the last among the 13 dynasties, had inherited rich heritages from previous dynasties. In the meantime, Tang Dynasty was "the highpoint of imperial China, both politically and culturally" (Lewis 2009). During this period, the construction in Chang'an Area had reached its historical peak in various aspects, which had a profound influence on later generations.

Geographically, Chang'an City was located at the most open part in the middle of Guanzhong Basin. The basin was a relatively closed olive-shaped geographic unit. It is around 300-km long from east to west; and is about 20 km at the narrowest and more than 100 km at the widest from south to north. In the mind of the ancient Chinese, the Guanzhong Basin was closely connected to Chang'an City. For example, in Volume 7 of *Tang Liu Dian* [Compendium of Administrative Law of the Six Divisions of the Tang Bureaucracy] (739), it is recorded that: "the Yellow River and Hua Mountain were on the left of the capital, Long Mountain (current south

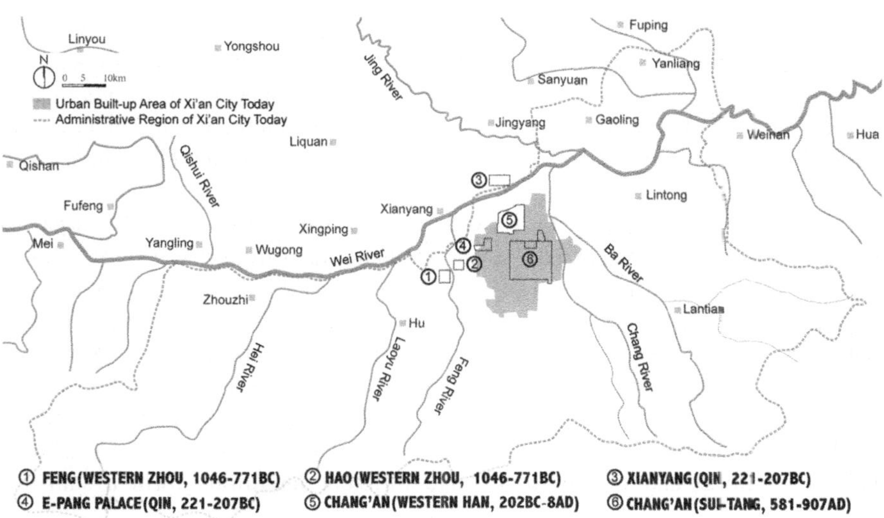

Fig. 2.1 Locations of capitals in various dynasties in Chang'an Area. Note: There has been no consensus on the spatial scope of Xianyang City—the Capital of the Qin Dynasty (221–207 BC). The site of Xianyang City marked in the map was the historic site of Xianyang Palace

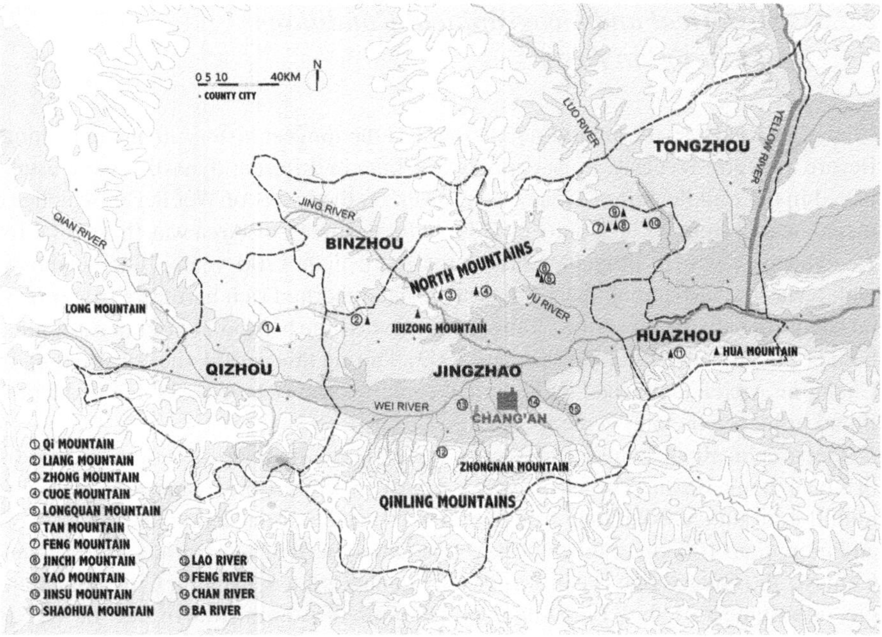

Fig. 2.2 Natural topography and administrative regionalization of Chang'an Area in Tang Dynasty. Source: Guo and Rong 2020

segment of Liupan Mountain) was on the right, Zhongnan Mountain[1] was in the front, and Jiuzong Mountain was in the back." In fact, this is the boundary of Guanzhong Basin. In Tang Dynasty, this was also the boundary of the administrative region consisted of the prefecture of Jingzhao (including Chang'an and more than 20 surrounding counties) and its surrounding "Four Assistant Prefectures," i.e., Qizhou, Fenzhou, Tongzhou, and Huazhou (Fig. 2.2).

The terrain of Guanzhong Basin has distinctive characteristics. Wei River passes through from west to east, and the terrain gradually rises from riverbed to the south as well as to the north, forming a stair shape on both sides. From north to south, the terrain can be divided into five landform areas: (1) the north mountain and hill tableland area (above 1000 m), (2) loess tableland area to the north of Wei River (500–1000 m), (3) terrace plain area along Wei River (below 500 m), (4) loess tableland area at the northern foot of Qinling Mountains (500–1000 m) and (5) Qinling Mountains (above 1000 m) (Li 1984). The terrain to the north of Wei River is high in north and low in south, and different landforms form a stair-like

[1] In history, Zhongnan Mountain refers to two different ranges. In the broad sense, it refers to the middle segment of Qinling Mountains, which is within current Shaanxi Province. In the narrow sense, it refers to the middle segment of the Zhongnan Mountain in the broad sense, mainly consisted of the segment from Wugong in the west to Lantian in the east. In this paper, Zhongnan Mountain refers to the broad sense if not specified.

distribution with significant altitude differences and clear boundaries. To the south of Wei River, a slope inclines from southeast—the northern foot of Qinling Mountains to northwest, on which different landforms are intertwined, i.e., without obvious boundaries in between. There are many rivers on the northern foot of Qinling Mountains, the scouring of which has resulted in the uneven surface of the slope, and the loess tablelands are intertwined with depressions. Longshou Tableland, Leyou Tableland, Shaoling Tableland and Shenhe Tableland, which will be mentioned later, are the most famous loess tablelands.

2.3 Concentric Regional Function Layout Centered on the Capital City

In Sui-Tang Dynasties, China was rebuilt and unified, which ended the segmentation and conflicts that lasted for more than 300 years. During this period, comprehensive reform was carried out, which strengthened national power economically and politically. In Tang Dynasty, the total population of the prefecture of Jingzhao and the Four Assistant Prefectures was more than 3.1 million, and the population density was around 64.7 people per km^2.[2]

The population composition of the imperial capital was particularly complex and changeable. Chang'an area was a home where all social strata coexist, such as the royal family, ordinary citizens, officials and literati, monks and nuns etc. Taking the capital city as the center, they carried out different social activities in the surrounding areas. Correspondingly, the layout of the capital function was carried out at the scale of the Guanzhong Basin, forming several functional circles.

2.3.1 Royal Families: Tour and Banquet in Natural Landscape, Mausoleums on High Mountains

The mausoleum was the emperor's post-mortem residence and played a very important role in the function layout of the capital. There were 18 imperial mausoleums of Tang Dynasty in Guanzhong Basin, including Xianling of Gaozu, Zhaoling of Taizong, Qianling of Gaozong, Dingling of Zhongzong, Qiaoling of Ruizong, Tailing of Xuanzong, Jianling of Suzong, Yuanling of Daizong, Chongling of

[2] The population of Jingzhao and the four Assistant Prefectures is about 3.1 million according to Vol.43 of Jiu Tangshu [Old Book of Tang], and the land area is about 47.9 thousand km^2 estimated based on Tan's Historical Atlas of China. This population data only included the registered population, and the actual population of the Guanzhong Basin was higher than this figure. For example, a large population was not under the management of prefectures and counties. but of various departments of Central Government, including the members of the royal family, eunuchs, maids in the imperial palaces and household troops. Besides, there was also a high migrant population.

Dezong, Fengling of Shunzong, Jingling of Xianzong, Guangling of Muzong, Zhuangling of Jingzong, Zhangling of Wenzong, Duanling of Wuzong, Zhenling of Xuanzong, Jianling of Yizong, Jingling of Xizong. These mausoleums are all located on the southern slopes and foothills of the North Mountains to the north of Chang'an City. The westernmost is Qianling Mausoleum (about 70 kilometers from Chang'an), and the easternmost is Tailing Mausoleum (about 105 kilometers from Chang'an), forming a fan with Chang'an as the center. The distance between the Mausoleums and Chang'an City is more than 100 li,[3] but they maintained a close relationship with certain counties. In Tang Dynasty, "Cichi County" was set up for the purpose of serving the mausoleum. After Fengxian County was declared Cichi County in 729 AD, the counties such as Liquan, Fengtian, Fuping, Sanyuan, and Yunyang were successively designated as the Cichi Counties. Thus, the eighteen mausoleums were divided into 6 groups centered on certain counties, which were basically distributed within about 30 li from these counties (Jianling and Zhenling were slightly over 40 li) (Fig. 2.3).

Tour and feasting were important components in the life of the royal families. The temporary palace was the most definitive imperial recreation place. During the Sui and Tang Dynasties, a large number of palaces were built around the capital. There were about 18 palaces, 10 of which were built in the Tang Dynasty, namely: Qingshan Palace and Longyue Palace, Yong'an Palace, Yuhua Palace, Cuiwei Palace, Huaqing Palace, Jincheng Palace, Wanquan Palace, Youlong Palace, Wangchun Palace. The other 8 palaces were inherited from Sui Dynasty, they are: Changchun Palace, Xingde Palace, Fengquan Palace, Jiucheng Palace, Taiping Palace, Qiongyue Palace, Jincheng Palace, and Shentai Palace. These palaces were distributed in the vicinity of Chang'an City to more than 200 li away from the city. The nearest ones were within 60 li of Chang'an City, while the distant ones were within 30 li from certain counties. In addition to the palaces, the venues for the feasting activities include private houses and villas, temples, public scenic areas, natural landscapes, etc., mostly within 60 li from the capital (Fig. 2.4).

2.3.2 Ordinary Citizens: Recreation in the Suburbs

In the Tang Dynasty, the prosperity of economy, the development of culture, and the increase in population had made the masses of the citizens gradually prosper and their lives were more colorful. The recreation activities of the ordinary citizens mainly included four types: landscape appreciation, climbing, waterfront recreation, and religious activities, which required beautiful landscape, high terrain or towers, waterfront areas and Buddhist or Taoist temples. Based on these requirements, several public scenic spots had been formed around Chang'an City: (1) Qujiang Pool, Ci'en Temple and Xingyuan Garden; (2) Leyou Tableland, Qinglong

[3] 1 li = 531 m

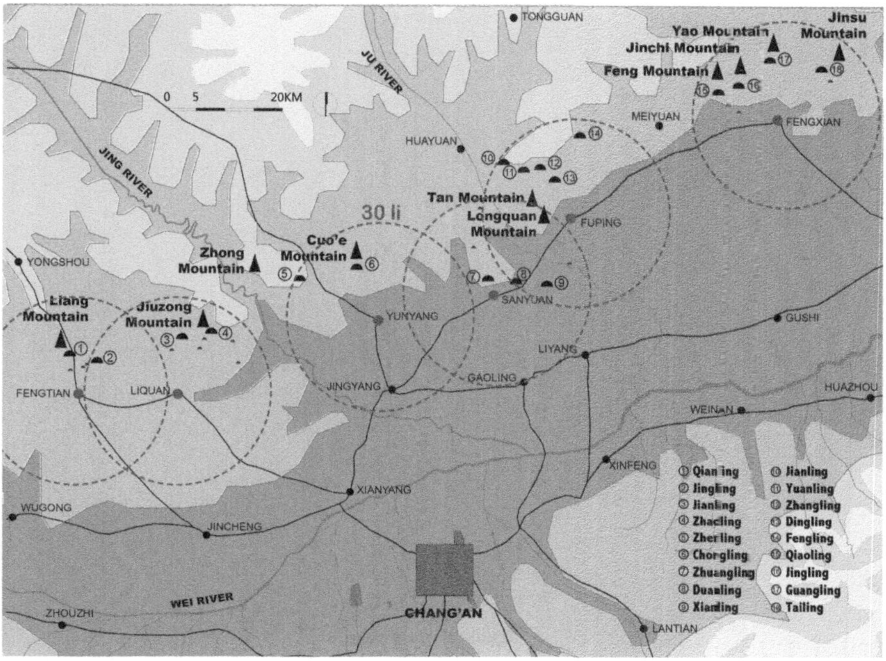

Fig. 2.3 Distribution and Zoning of Imperial Mausoleums in Guanzhong Basin of Tang Dynasty. Note: The administrative division in the figure is in Kaiyuan Era, but the construction of the mausoleums is a continuous process, which was not complete at this time. For ease of understanding, they are all marked in the same figure

Temple and Chan River; (3) Kunming Pool and Dingkun Pool. In addition, there were also temples to the south of the city, Ba River, Wei River and etc. A common feature of these scenic spots was that they were close to Chang'an City, within 30 li, with convenient transportation and close connection with the city (Fig. 2.4).

2.3.3 *Officials and Literati: Banquet in the Suburbs, Study in the Mountains and Forests*

Although the officials and literati lived in Chang'an City, they still yearned for the natural landscape. The private gardens and villas were perfect places to meet the needs of participating in political life in the capital and enjoying the natural beauty at the same time. They often had one or more villas near the city.

According to Li (2005)s' *Tangdai Yuanlin Bieye Kaolu* [Gardens and Villas of Tang Dynasty], Shi (1996)s'\ *Xi'an Lishi Dituce* [Xi'an Historical Atlas] and Zhang Li (2006)s' *You Chengnan Ji* [Tour to the South of the City], there were 101 famous villas around Chang'an, including 24 in the east, 71 in the south, 4 in the west, and

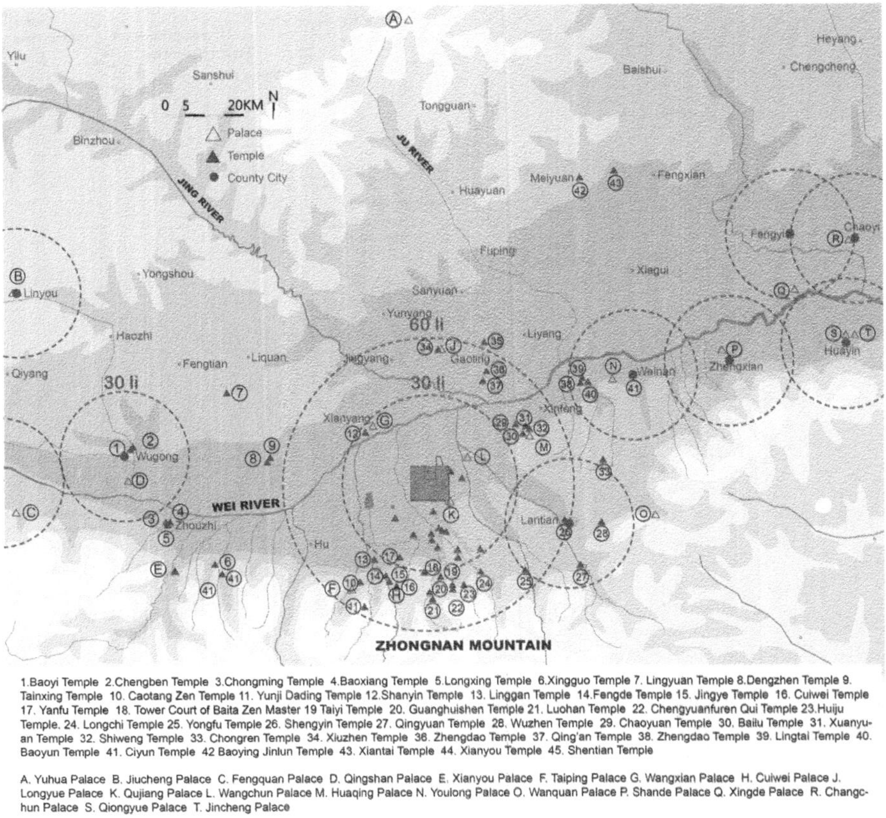

Fig. 2.4 Distribution of palaces and temples in Chang'an Area of Tang Dynasty

2 in the north. The southern suburb of Chang'an was the most concentrated area of villas, among which Fanchuan and Zhongnan Mountain were the areas with the highest density. And the eastern suburb was also an important villa distribution area. The villas in Fanchuan and eastern suburb were mainly distributed within 30 li from the city, which were mostly the resting places for the dignitaries and officials. The villas in Zhongnan Mountain were mainly located in the range of 50–60 li from Chang'an City, most of which were places for the scholars to study and live in seclusion.

2.3.4 *Monks and Nuns: Practice in the Mountains*

Buddhism and Taoism in the Tang Dynasty were closely integrated with politics and social life. The temples outside Chang'an City were mainly located in the quiet scenery as the long-lasting tradition, furthermore most of them had convenient traffic connections with the capital city. According to the records of *Xu Gaoseng Zhuan* [Continued Biographies of Eminent Monks], *Song Gaoseng Zhuan* [Biographies of Eminent Monks of Song], *Chang'an Zhi* [Gazetteer of Chang'an] and *Leibian Chang'an Zhi* [Supplement to the Gazetteer of Chang'an], there are 55 temples around the Chang'an City, among which 11 were in the east, 32 in the south, 12 in the north. The southern suburb of Chang'an was another religious center besides Chang'an City. Eight of the 32 temples to the south of the city were located within 30 li away from the outer wall of the city, and 4 of them were located near certain counties, and 20 of them were located in Zhongnan Mountain (Fig. 2.4).

In summary, the layout of the capital function was carried out at the scale of the Guanzhong Basin, which could be divided into three spatial levels, forming several functional concentric circles: (1) less than 30 li: the most active area with large public scenic spots, temples, and private gardens and villas, etc.; (2) 30–60 li: area for more specialized activities with temporary palaces of the royal families, villas of the literati and temples; (3) More than 100 li: the distant area with temporary palaces and imperial mausoleums.

2.4 Regional Spatial Structure Based on Geographical Pattern

Chang'an area was a large area with a great number of artificial constructions. In order to form an overall spatial order, on the one hand, regional-scale axes connecting the city and the natural mountain was established to form a spatial backbone by taking the regional geographical condition as the frame of reference; on the other hand, the high points of the natural terrain with good view were occupied to build important buildings forming several controlling "belts" of the regional space.

2.4.1 Regional-Scale Axis Connecting City and Mountain

The axes of Chang'an were the spatial backbone of the city. They were not limited to the scale of the city, but extended to the regional scale, connecting the city with the surrounding natural environment as a close whole.

1. The axis of Taiji Palace-Shibie Valley. Tang Dynasty had inherited the capital city of Sui Dynasty, and changed its name from Daxing to Chang'an. At the beginning of the construction of Daxing, there was a very clear central axis, which began from the Daxing Palace (Taiji Palace in Tang Dynasty) in the north, extended along the Tianmen Street, went through Zhuque Gate, the south gate of the Imperial City and arrived at Mingde Gate, the south gate of the outer city wall. This axis extended all the way beyond the city wall. The documents of the Tang Dynasty, such as *Tang Liudian* and *Xijing Ji*, recorded that Chang'an City

was facing south to Ziwu Valley of Zhongnan Mountain. However, in *Yonglu* of Cheng Dachang, a scholar of Northern Song Dynasty (960–1127), the city was facing Shibie Valley. According to the contemporary topographic map, the longitude of Mingde Gate is E108°56′, the longitude of the mouth of Shibie Valley is also about E108°56′, while the longitude of the mouth of Meridian Valley is about E108°53′. Obviously, there was a regional axis from the Taiji Palace to the mouth of the Shibie Valley (Fig. 2.5).

2. The axis of Daming Palace-Zhongnan Mountain. In Gaozong era of Tang Dynasty, the political center moved to the Daming Palace in the northeast corner of the city. The Daming Palace formed a new axis along with the surrounding urban areas. There was an axis from south to north in Daming Palace, which ran through the core palaces including Zichen Hall, Xuanzheng Hall and Hanyuan Hall. The southern end was Danfeng Gate, the southern gate of Daming Palace. Outside the Danfeng Gate, a broad street was constructed crossing the original Yishan and Yongchang Wards, as the continuation of the axis. This axis extended southward, thus an axis on a regional scale was established, starting from the Daming Palace in the north and facing the Zhongnan Mountain in the south. The south end of this axis was the mountain peak named Niubeiliang today, at an altitude of 2802 m, which is the highest peak in the eastern section of the Qinling Mountains. From the documents of Tang and Song Dynasties, we can see clearly the visual connection between Daming Palace and Zhongnan Mountain. There are also many poems in *Quan Tangshi* which recorded the scenery of Zhongnan Mountain viewing from Daming Palace (Fig. 2.5).

2.4.2 *Important Buildings Taking the High Ground*

It is a common pursuit for the ancient Chinese to climb high to achieve a panoramic view. There were many such descriptions in ancient Chinese literatures, while traditional Chinese paintings also often adopted the "bird's-eye view." In terms of planning and design in ancient China, the tradition was to take the high ground to build important building. The so-called "high ground" refers to the location on relatively high natural terrain with large open space, which could provide a wide view. As mentioned above, North Mountains to the north of Wei River and Qinling Mountains to the south, were undoubtedly the high grounds in the region. The terrain between the two mountains was not smooth, where tablelands were the high grounds. In Chang'an Area in Tang Dynasty, most important constructions were built on these high grounds.

1. Building city on plain terrace. Chang'an City was built on the loess tablelands and ridges extended from the northern foot of Zhongnan Mountain to the south of Wei River. In accordance with related records in *Yuanhe Junxian Zhi* [Illustrated descriptions of the administrative regions in the Yuanhe period]

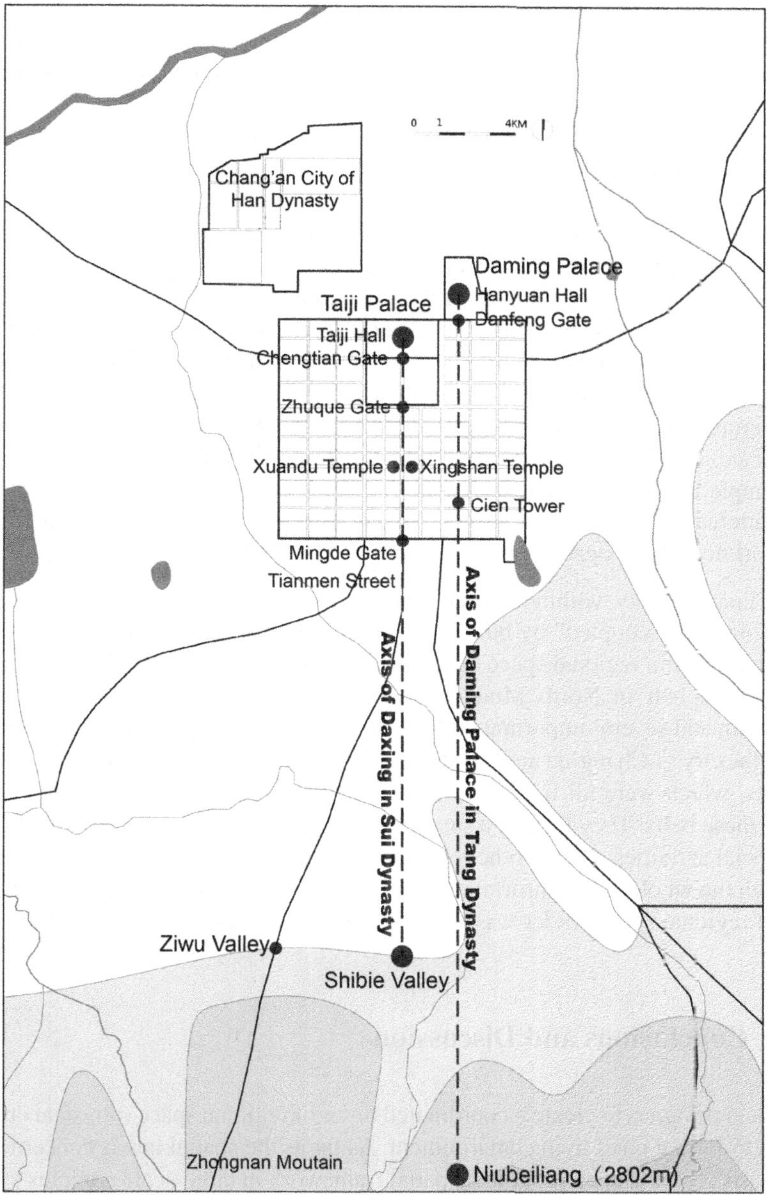

Fig. 2.5 Regional axes of Chang'an in Tang Dynasty

(814), when planning Chang'an City, Planner Yuwen Kai (555–612) started with selecting "six high slopes" within the region for a general layout, these six slopes were regarded as key areas for urban planning, on which important buildings were built.

2. Building the imperial mausoleums on high mountains. There were 18 imperial mausoleums of Tang Dynasty in Guanzhong Basin, 14 of which were all at the south face of North Mountains. The mausoleums were built on mountains at an altitude of 750–1200 m, while Chang'an City only had an altitude of around 450 m. The group of imperial mausoleums extended through the north part of Guanzhong Basin, which was nearly 150 km from west to east. With mountains behind them, these mausoleums faced toward the plain, occupying commanding positions. Overlooking the capital city, they formed a high barrier to the north of Chang'an City along with North Mountains.
3. Building palaces and temples in the low-mountain zone. There were many palaces and temples in Chang'an Area, which were mainly concentrated on Zhongnan Mountain, south of Chang'an City. In the low-mountain zone of Zhongnan Mountain, there were about 10 palaces, such as Fengquan Palace, Taiping Palace, Cuiwei Palace, Wanquan Palace, Huaqing Palace and Qiongyue Palace, and 20 temples, such as Xianyou Temple, Yongfu Temple and Wuzhen Temple, all with a grand scale and gorgeous and impressive appearance. They scattered in the beautiful scenery of Zhongnan Mountain, just like a string of pearl necklace extending to the south of Chang'an City.

In Tang Dynasty, within the Chang'an Area, almost all the high places that could be used were "occupied" by buildings. These constructions formed several "belts" expanded within regional space conforming to the landform, including the imperial mausoleum belt in North Mountains, the temple and palace belt in Zhongnan Mountain and several important building belts formed along the tableland boundaries in the city of Chang'an and its suburbs. On the one hand, the projects can be seen as dots, which were all located in areas with beautiful sceneries and open vision along these belts. They formed a small-scale physical environment suitable for various social activities. On the other hand, these "dots" formed various belts that went through the whole area conforming to the landform, so that the comprehensive control of regional spatial order was achieved (Fig. 2.6) (Guo and Rong 2020).

2.5 Conclusions and Discussion

Regional design is to create a coordinated order of regional space (physical environment) to form a good living environment. As far as the spatial law is concerned, the emphasis is on the construction of spatial framework in large-scale complex system. Concentric regional function layout centered on the capital city and the regional spatial structure based on geographical pattern both focused on this. As far as implementing subject is concerned, the central government controlled the large structure and key points of regional space, which was the leader of regional design. The joint participation of all social strata played a role of enrichment and maintenance. Therefore, it could be concluded that the regional design is not easy to be controlled by the specific needs and tastes of individuals, but closely related to the material and

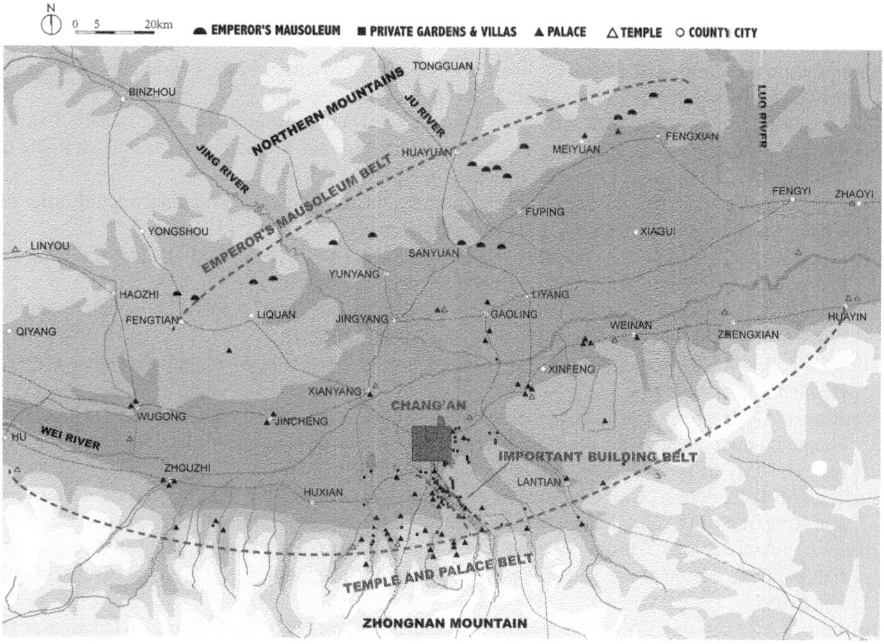

Fig. 2.6 "Belts" within the regional space of Chang'an Area in Tang Dynasty. Source: Guo and Rong 2020

spiritual needs of the mainstream of a regional society, and promoted simultaneously with the process of regional development.

Compared with the well-ordered and magnificent regional space formed in history, the chaos of the contemporary regional space order is an indisputable fact and a severe challenge facing the times. The experience of the construction of regional spatial order in history still has important enlightening significance today.

Acknowledgments This work was supported by Beijing Municipal Social Science Foundation [grant number 19YTC037] and National Natural Science Foundation of China [grant number 51978362].

References

An J., & Ding X., 1995. Sui Renshougong Tang Jiuchenggong 37 Hao Dianzhi De Fajue [Excavation of No.37 Palace Site of Jiucheng Palace of Tang and Renshou Palace of Sui Dynasty]. Archaeology, 1995 (12)

An J., & Li C., 2000. Shanxi Xi'an Tang Chang'an Cheng Yuanqiu Yizhi De Fajue [Excavation of Yuanqiu Site of Chang'an City in Tang Dynasty in Xi'an, Shaanxi]. Archaeology, 2000 (07).

Boyd, A., 1962. Chinese Architecture and Town Planning 1500 B.C-A. D 1911. Chicago: University of Chicago Press.

Chen X., 2006. Guanzhong Yizhi Jizhu [Annotated Notes on Collection of The Lost Chronicles of Guanzhong]. Beijing: Science Press.

Compilation Committee of Local Chronicles of Shaanxi Province, 1995. Chronicles of Shaanxi Province ·Chronicles of Cultural Relics. Xi'an: Sanqin Press, 1995.

Fu, X., 2001. Zhongguo Gudai Chengshi Guihua Jianzhuqun Buju Ji Jianzhu Jianshe Fangfa Yanjiu [Study on Urban Planning, Architectural Layout and Architectural Design Methods in Ancient China]. Beijing: China Architecture & Building Press.

Guo, L. & Rong X., 2020. Taking the high ground: construction of the regional spatial order of Chang'an Area in Tang Dynasty, Planning Perspectives, 35 (1), pp: 115-141.

Heng, C. K., 1999. Cities of Aristocrats and Bureaucrats: The Development of Medieval Chinese Cityscapes. Honolulu: University of Hawai'i Press.

Hiraoka T., & Imai K., 1991. Index of Chang'an and Luoyang of Tang Dynasty. Shanghai: Shanghai Ancient Books Publishing House.

Hiraoka T., 1989. Chang'an and Luoyang in Tang Dynasty: Materials. Shanghai: Shanghai Ancient Books Publishing House.

Institute of Archaeology, Chinese Academy of Social Sciences, 1987. Tang Chang'an Huangcheng Hanguangmen Yizhi Fajue Jianbao [Excavation Report of Hanguang Gate Site of Chang'an Imperial City in Tang Dynasty]. Archaeology, 1987 (05)

Institute of Archaeology, Chinese Academy of Social Sciences, 2007. Archaeological Discovery and Research of Daming Palace Site in Tang Dynasty. Beijing: cultural relics press.

Institute of Archaeology, Chinese Academy of Social Sciences. 1961. Tang Chang'an Cheng Xishi Fajue Ji [Excavation of the Site of Western Market of Chang'an in Tang Dynasty]. Archaeology, 1961 (05).

Institute of Archaeology, Chinese Academy of Social Sciences. 1963. Tang Chang'an Cheng Kaogu Jilue [Archaeological records of Chang'an City in Tang Dynasty]. Archaeology, 1963 (11).

Institute of Archaeology, Chinese Academy of Social Sciences. 1974. Tang Chang'an Mingde Men Yizhi Fajue Jianbao [Brief Excavation Report of the Mingde Gate Site of Chang'an in Tang Dynasty]. Archaeology, 1974 (01).

Kotkin, J. 2005. The City: A Global History. New York: The Modern Library, p. 53.

Lewis, M. E., 2009. China's Cosmopolitan Empire: The Tang Dynasty. Cambridge: Harvard University Press, 1.

Li H. 2005. Tangdai Yuanlin Bieye Kaolu [Gardens and Villas of Tang Dynasty]. Shanghai: Shanghai Ancient Books Publishing House.

Li J. et al., 1991. Tang Cuiweigong Kaogu Diaocha Jianbao [A brief survey of Archaeology in Cuiwei palace of Tang Dynasty]. Archaeology and cultural relics, 1991 (03)

Li J., & Xu S., 1996. Zengding Tang Liangjing Chengfang Kao [Revised Research on the Wards the Two Capital Cities of Tang Dynasty]. Xi'an: Sanqin Publishing House.

Li, J., 1984. Shanxi Dili [Geography of Shaanxi Province]. Xi'an: Shaanxi People's Publishing House, pp: 20–37.

Li, L., 2004. Guanzhong shuili kaifa yu huanjing [Water Conservancy Development and Environment in Guanzhong Area]. Beijing: Science Press.

Liu Q., & Li Y., 1987. Shaanxi Tangling Diaocha Baogao [Investigation report of Tang mausoleum in Shaanxi Province] See Archaeology collection 5. Beijing: China Social Sciences Press. pp: 216-263.

Liu W., 1997. Hantang Fangzhi Jiyi [The Collection of Chronicles of Han and Tang Dynasties]. Beijing: Beijing Library Press.

Lu J., 1978. Shaanxi Tongchuan Tang Yuhuagong Yizhi Diaocha [Investigation of Yuhua Palace Site in Tongchuan], Shaanxi. Archaeology, 1978 (06).

Luo X., 1998. Tang Huaqing Gong [Huaqing Palace of Tang Dynasty]. Beijing: Cultural Relics Press.

Ma, D., 1989. Tang Chang'an Qinglongsi Yizhi [Qinglong Temple Site of Chang'an in Tang Dynasty]. Acta Archaeologica Sinica, 1989 (02).

Ma, Z., 1978. Sui Tang Chang'ancheng [Chang'an City in Sui-Tang Dynasty]. Chengshi Guihua [City Planning Review], 1978 (1), pp.37–44.

Mote, F. W., 1977. The Transformation of Nanking, 1350–1400. In G. W. Skinner, ed. The City in Late Imperial China, Taipei: SMC, pp. 101–155.

Shi, N., 1995. Xi'an Diqu Dixing De Lishi Yanbian [The Historical Evolution Of The Terrain In Xi'an]. Zhongguo Lishi Dili Luncong Journal of Chinese Historical Geography 1995(4).

Shi, N., 1996. Xi'an Lishi Dituce [Xi'an Historical Atlas]. Xi'an: Xi 'an Map Publishing House.

Shi, N., Zhu, S. & Cao, E., 2006. You Chengnan Ji Jiaozhu [Collation and Annotation to South City Travel Notes]. Xi'an: Sanqin Press.

Wang S., & Lou Y., 1980. Tang Qiaoling Kancha Ji [Exploration Record of Qiaoling of Tang Dynasty]. Archaeology and Cultural Relics, 1980 (04).

Wang, K., 1985. Shanxi Gudai Jiaotong Gaishu [A Brief Account of Ancient Traffic in Shaanxi]. Renwen Zazhi The Journal of Humanities 1985(3), pp. 94–97.

Xiong, V., 2000. Sui-Tang Chang'an: A Study in the Urban History of Late Medieval China. Ann Arbor: University of Michigan Press.

Yan, W., 1948. Tang Xijing Kao [Study on the West Capital of Tang Dynasty]. Xi'an: New China Press.

Zhang, Y., 1987. Tangdu Chang'an [Capital of Tang Dynasty: Chang'an]. Xi'an: Northwest University Press, 1987.

Chapter 3
Looking into the Supergrid and Superblock Structure in Chinese Cities: Taking Xi'an and Nanjing as Examples

Xiaofei Chen

Abstract Supergrid is a large-scale network of wide roads that defines a series of cells as Superblocks. Uniquely in China, each Superblock contains a number of walled living quarters next to each other assisted with a few narrow streets to minimize through traffic. Together, they make a Global-local transportation system as one of the most predominant city skeletons of modern Chinese cities. While Barcelona's planning movement has been converting street blocks to Superblocks since the 1990s, the State Council of China released a set of planning guidelines in 2016 and suggests the opposite: transforming the current Super grid-and-block structure to a finer network by disassembling existing gated communities within Superblocks. This was strongly opposed by the people, who emphasize the importance of the wall to their safety. Such conflict between the policy and people's reaction reveals a lack of understanding of the structure in Chinese social and planning context.

This chapter provides a comprehensive exploration of the development of the Superblock structure in Chinese planning history and elaborates the modern planning practice by examining one of each Superblocks from Xi'an and Nanjing as case studies. Through investigation, this chapter points out that the use of Superblocks in combination with walls and gates around living quarters is a unique Chinese spatial planning logic despite the influence of modern western planning ideas. It is important to rethink the applicability of the proposed guidelines and incorporate the traditional planning concept in modern city design.

Keywords Supergrid · Superblock · Urban structure · China

X. Chen (✉)
College of Architecture and Urban Planning, Qingdao University of Technology, Qingdao, China
e-mail: chenxiaofei@qut.edu.cn

L. Bian et al. (eds.), *Chinese Urban Planning and Construction*,
Strategies for Sustainability, https://doi.org/10.1007/978-3-030-65562-4_3

3.1 Introduction

Supergrids and Superblocks in China constitute a grid-and-cell urban structure that is uniquely different from Superblock structures in other parts of the world: a coarse-grained network of wide roads forming a Supergrid (see Fig. 3.1). It further defines a number of Superblocks, each containing a number of residential neighborhoods ranging in varying sizes, surrounded and protected by a mixture of walls, gates, buildings, fences, plantings, and guard houses, providing different levels of accessibility to the public and local residents (Chen 2018). Interestingly, in 2016, the state council of China released a set of planning guidelines that suggest to both demolish the walls around residential quarters and also to transform Superblocks to smaller street blocks (Ministry of Housing and Urban-Rural Development of the People's Republic of China 2016). Those suggestions are strongly against by most people even though academics claim that this change would greatly improve the current traffic condition.

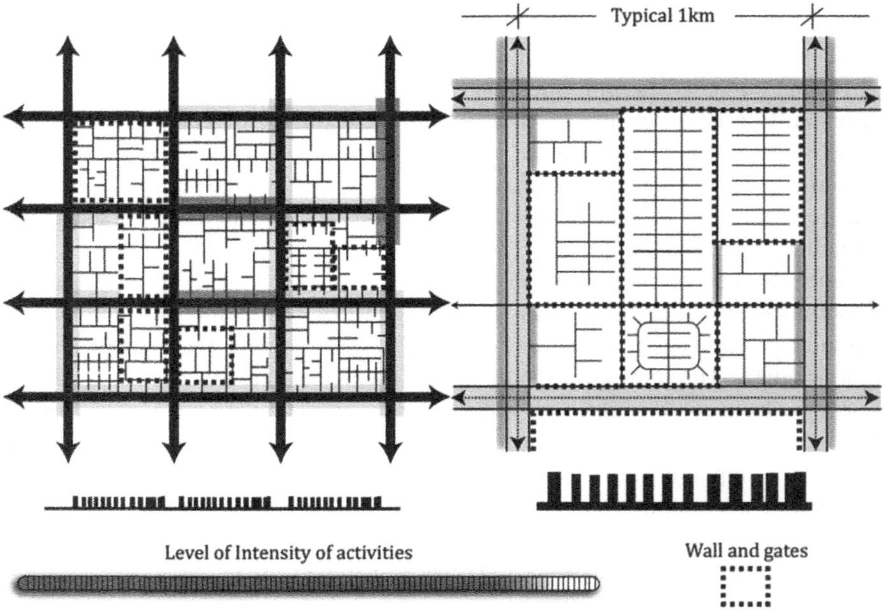

Fig. 3.1 A typical Chinese Supergrid and Superblock structure in modern cities (by author)

Under the influence of the road network planning regulation (No.GB 50220–95),[1] the major arterial roads of Chinese cities were required to be set with an interval between 750 m and 1200 m depending on the varying sizes, which naturally forms a Supergrid with a number of Superblocks in cities (Xu 2007, 202–205). Supergrid and Superblock are therefore interdependent and mutually supportive parts of a single urban system. After the release of the guideline, a new regulation has been published (GB/T51328–2018)[2] to replace the old one. In this document, the length of block edges is reduced to less than 300 m for residential use, and less than 100 m for commercial use to allow better movement. However, the regulation is not sufficient to guide current city redevelopment as it only provides a general framework and reveals a lack of understanding of the structure in Chinese context.

The central idea of this chapter is to investigate how the Supergrid and Superblock structure has been evolved through history to its modern state in China following a chronological order. It further elaborates this modern planning practice by using two Superblocks as examples. The findings indicate that a similar Supergrid and Superblock structure has been consistently and pervasively used in the construction of Chinese cities since ancient times and the structure shows a strong correlation to a "wall-and-gate" system that contributes to the formation of a grid with cul-de-sacs street systems within Superblocks.

3.2 Method and Materials

3.2.1 Literature Review Method

This study adopted both qualitative and quantitative methods for investigation. The first aim of this study is to explore how this Supergrid and Superblock urban structure transforms in various periods of time. The timeline was set to trace its origin in the ancient time until contemporary China. The second aim is to investigate how this structure operates in contemporary Chinese cities, and it required first-hand spatial data from fieldwork. The case study method was also applied for comparison purposes to provide more specific evidence to illustrate how this structure has been evolved and became the modern structure.

Because the first aim of this study is to illustrate the change of certain urban patterns of the structure, the qualitative method was adopted for investigation. Three steps were carried out:

---

[1] Policy No.GB 50220-95: This regulation is called "*Code for City Road and Traffic Planning and Design*" (《城市道路交通规划设计规范》). It was used to regulate and standardize traffic and road planning, and terminated in March first, 2019.

[2] GB/T51328-2018: This regulation is called "*Urban Transportation Planning Standard*" (《城市综合交通体系规划标准》). It has been published in 2018, and replaced the GB 50220-95 as the newest regulation for road planning Chinese cities on March first, 2019.

Step One: this study collected a number of historical books, old maps, and newspapers, articles and documents that discuss the related topics from local archives, online libraries, and book shops.

Step Two: this study further analyzed the collected materials and converted them into maps and diagrams, which can provide a strong visual supplement to words.

Step Three: this study summarized those materials and revealed some patterns about structural transformation.

Moreover, this study adopted the historical mapping method which is also a qualitative method that can visualize historical text and spatial data for graphic-oriented academics and researchers, such as Architect, urban designers, and planners (Win 2014). Applying this method can make a strong historical study about the structural change of the Chinese city planning.

The second aim is to exhibit and visualize the modern structure by using the data collected from site visits. Hence, both qualitative and quantitative methods were adopted for a better understanding of the physical characteristics. Traditional urban design mapping methods are also used to create maps and diagrams to interpret and quantifying various features of the structure in the two case study sites. In order to understand the contemporary structure, a number of maps and visual data are needed as listed below:

- Cadastral Map.
- Building Layout.
- Transportation Map.
- Road and Street Network.
- Block and Plot layout.
- Neighborhood Map.
- Landscape Map.

Furthermore, there are several specific map types and series are focused in searching for related visual data from maps and those includes:

- Bird's eye view.
- Aerial Photos and Ortho-photographs.
- Plot Maps and Surveyors Notes.
- Plans and sections.
- Drawings of ancient topography and city views.
- Infrastructure maps.

3.3 Review of Existing Studies

Before the release of the guidelines, studies of the Superblock structure in China is rather limited, and most of them discuss the negative impacts of Superblocks on transportation by Chinese (i.e., Liao 2004, p.46–49; Xu and Yang 2008a, p.24–28). At least five studies (Zhao 2002, Zhao 2006, Monson 2008, Zhao 2008, Xu and Yang 2009), directly review the problem of Supergrid and Superblock structure because of the lack of connectivity and properly sized land subdivision. At least 23 studies[3] review the Superblock structure indirectly. They study the gated community and claim that gated communities have created barriers to existing road networks and pedestrian movement. Street block and Superblock types of gated communities are both discussed. Among the 23 studies, at least four directly point out that the gated community at a Superblock scale can create disconnection problems in cities. Xu and Yang (2008a, 2008b) assert that this type of Superblock is detrimental to China's urban environment because they have very limited connections to the surrounding areas and wider city network. Wang (2010) and Nieminen (2012) claim that one of the major problems of the Chinese Superblock structure is the missing connection between the inside and outside of residential superblocks because of fences and walls acting as major barriers. From urban planning perspectives, they regard the gated community as a source of social, economic and environmental problems by creating barriers and isolating movement.

After 2016, an increasing number of studies start to appear and focus on the modification of current Supergrids and Superblocks. Those studies can be classified into four types: (1) Studies that discuss the advantages and disadvantages of Superblock and street block. They conclude that the latter is better than the former, and Chinese cities should convert the current Supergrid and Superblock system into street block system (Chen 2016a, 2016b; Chen 2017; Gong et al. 2017; Xiao 2016; Peng 2017); (2) Studies that discuss the change of Supergrids and Superblocks into street blocks to improve the disconnection and traffic efficiency from a transportation perspective (Shen et al. 2016, Zhang and Xie 2017, Jiang 2017; Wu 2017; Tan et al. 2017; Tan, Zhu, Wang 2016; Zhang et al. 2016; Li 2017; Chu 2017); (3) Studies that discuss practical examples of the places where Superblocks are converted into street blocks and the corresponding problems that they encountered in the process of modification (Xu and Yang 2010; Calthorpe Associates 2011; Wang 2016; Chen 2016a, 2016b; Lu 2017; Zhang 2017; Wu 2018); (4) Only a few studies (i.e., Rowe and Guan 2016; Kan et al. 2017) disagree with the total abolishment of the Supergrid and Superblock structure and argue that this change can bring a series of social problems, and it is better to reconsider the suggestion and adopt case-by-case modification method.

3.4 The Importance of the "Wall-and-Gate" Structure

"A wall is a key element in the formation of space in the Chinese conception" (Zhu 2004, 46). Any Chinese Architecture that is composed of a number of buildings would not exist without walls to create a series of architectural experiences (Cheng

[3] Those studies are: Zou and Bian (2000), Wang (2003), Liao (2004), Wu (2005), Huang (2006), Li and Yu (2006), Xu and Yang (2007), Li and Li (2007), Yang and Min (2008), Liu and Li (2009), Song and Zhu (2009), Xu (2009), Xu et al. (2009), Hu (2010), Wang (2010), Dou (2010), Song (2010), Liang and Su (2011), Feng et al. (2011), Wei and Qin (2011), Hu et al. (2011), Liang and Su (2011), Deng (2013) and Wallenwein (2013)

2009, 348–408). "At a structural level," Chinese cities are "constructed by a system of walls, not by a field of open spaces that are naturally related and congregated to each other" (Zhu 2004, 46). Xu and Yang (2009) also points out that walled enclosure and closed space are distinctive morphological features in traditional Chinese cities (p.100). They both reflect a "long-existing tradition of walled urban morphology" on a macro and micro scale and an "introverted living patterns during over two thousand years of highly centralized feudal monarchy" (p.100–101). Some studies specifically point out that the walls and gates are the special characteristics of Chinese urban design and planning in demarcating space and they are the inseparable parts of Chinese culture and religion through thousands of years of evolution (Lu and Stamenovic 2004; Lu 2006a, 2006b, p.369–392; Nieminen 2012, p.6–19). In other words, a strong sense of spatial enclosure forming a "wall-and-gate" structure is one of the most fundamental features of Chinese cities. Walls are not only a physical but also functional and psychological features that construct Chinese cities and society (Mote 1977, p.20–49; Steinhardt 1999, p.10–60).

3.5 Transformation in China: From a Closed to Open and to Closed Urban System

It is important to point out that the Supergrid and Superblock structure as a way of organizing space has been used in Chinese cities since the ancient times. The earliest Supergrid and Superblock structure concept can be traced back to the proposed ideal planning paradigm in *Kao Gong Ji* (see Fig. 3.2). The chessboard street network and major wide roads (as Supergrid) in combination with walled residential quarters (as Superblocks) as spatial unit are the fundamental design principles from *Kao Gong Ji*. Cities in *Qin* and *Han* Dynasties loosely follows the rules and the very first application of those principles is the *Chang'an* city in *Han* Dynasty (He 2012). Those design principles were further adapted in the design of larger and grander

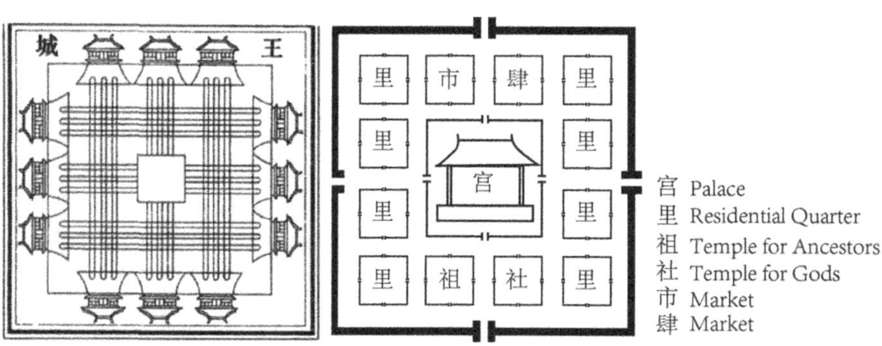

Fig. 3.2 The Classical model in Kaogongji (left, Steinhardt 1999, p. 34); A diagrammatical interpretation of the ideal Chinese planning model (right, re-drawn by author based on Liu 1936, p. 575)

cities, such as the city of *Ye* in *Caowei* Dynasty, the city of *Ping* and *Luoyang* in Northern *Wei* Dynasty, each marks a progress in the update of those design principles to a higher level (Wang 1992; Zhang and Meng 2011; Yao 1992).

In the process, the development of those cities paved the way to the formation of the well-known "*Chang'an* model" in *Sui* and *Tang* Dynasty (Chen 2004), which formulated "a classical planning model, prescribing a grand, centric, Confucian order" (Zhu 2004, p. 6). The whole city was divided into 108 walled Superblocks by a system of wide roads as a Supergrid. Each walled residential quarter (one *Lifang*) became a Superblock with a system of cross streets (see Fig. 3.3). It is this Supergrid and Superblock model of *Chang'an* in *Tang* reached a peak and mature state and has an International influence on many neighboring countries (such as Korea and Japan) at the time (Wang 2014, p.173). The ideas of using regular and symmetrical super grid-and-block layout with walled quarters (*Lifang*) to create closed urban systems gradually become a norm in city design.

In the *Song*, *Yuan*, *Ming*, and *Qing* dynasties, a Supergrid can still be found in many cities, as cities became more street-oriented with smaller streets linking the major wide roads to the residential courtyards like a fishbone structure[4] (Wu 1999; Zhu 2004). Even though the walled Superblock structure was in decline because of the development of market economy (Heng 1999 & Friedmann 2005), walls around cities and courtyard compounds within each Superblock were left largely intact.

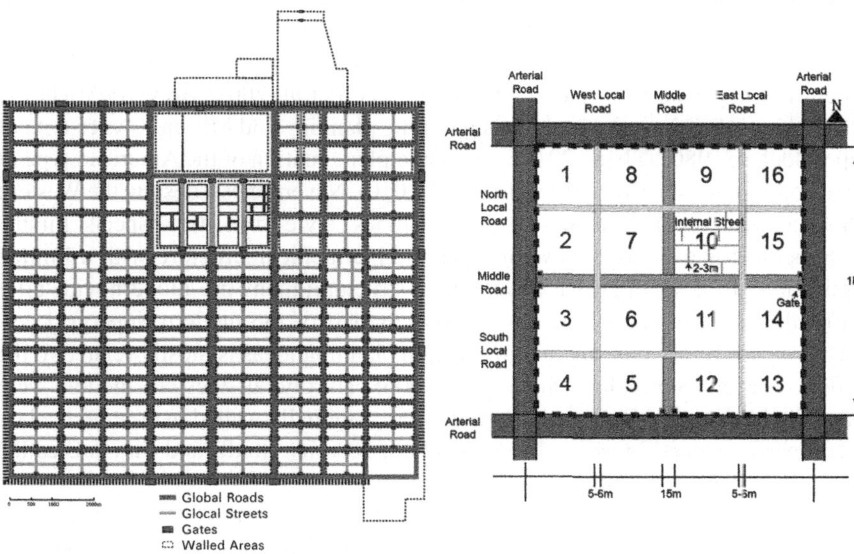

Fig. 3.3 The Chang'an Supergrid and Superblock model (left) and a Superblock example in the city (right) (Drawn by author according to He 2012, Shanxi Archaeology Institute 2008; Global Road and Arterial Road are the same)

[4]The orientation of buildings based on the south-facing building principle meant that the Supergrid network contained more north-south Global roads and smaller streets running in an east-west direction (Zhu 2004, see Figure 3.4).

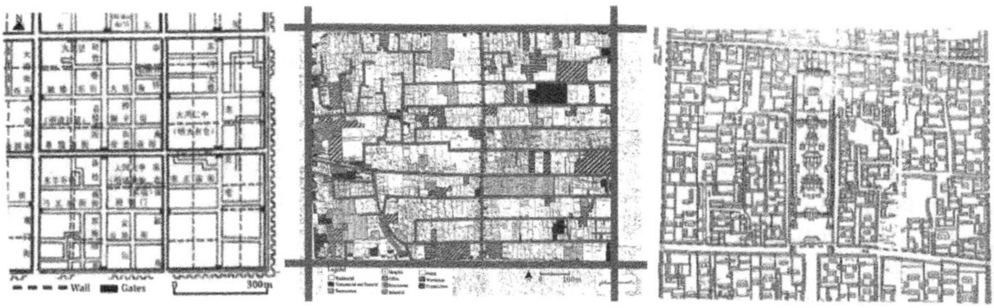

Fig. 3.4 A Superblock in *Datong* in *Song* Dynasty—approx. 7–9 AD (Fu 2001, p.11, left); A Superblock in *Beijing* in *Qing* Dynasty (adapted by author from Wu 1999, p.161, middle); A part of *Beijing* in *Qing* Dynasty to show the wall structure and courtyard buildings in the urban context in 1750 (Zhu 2004, p.85, right)

Many parts of the city had a number of walled spaces with fences and gates to demarcate the boundary of certain properties or groups of people, and the level of openness started to decline again (ibid). As a result, a Supergrid still mostly remained to connect between gates, creating Superblocks with a cross street plus laneway street layout (see Fig. 3.4). The system during this period was not an open system, but relatively opened only at some level.

In modern times, the Supergrid and Superblock structure came under four stages of Western influence in general: experimentation with the neighborhood unit concept from America from the 1920s to the 1940s (Lu 2005 & Ren 2008); introduction of the Russian company town and microdistrict models and the formation of *Danwei* system in the period from the 1950s to 1970s (Lu et al. 2001; Bray 2005; Hartog 2010); the application of comprehensive traffic planning and Milton Keynes type of Superblock in 1980s (Shane 2014); and the implementation of the American type of gated community since the 1990s (Liu & Li 2010). Nevertheless, despite the Western influence, the structure maintained a clear Chinese style from three aspects. Firstly, the Supergrid is used as the overarching and hierarchical structure with a limited application of the small grid and radial road structure; many pre-modern wide arterial roads were preserved and often reused as part of the modern Supergrid networks. Secondly, the introduction of Western planning ideas is mostly related to the modern Western Superblock model (Lu 2006). Thirdly, all Western models were adapted with a special Chinese twist: they were eventually walled in order to declare the property ownership and enhance levels of security (ibid), and they are called Xiaoqu or gated communities (see Fig. 3.5).

Thus again, the wall structure is used pervasively in combination with western Superblock models and the major structure of cities becomes that of Superblocks; and the street layouts within each Superblock of this Supergrid becomes a closed system with culs-de-sacs. This particular transformation of the city morphology has given a very special characteristic of the modern Chinese Superblock structure and it is the wall and gate structure that decides the level of openness and performance of the urban Supergrid and Superblock structure in contemporary China.

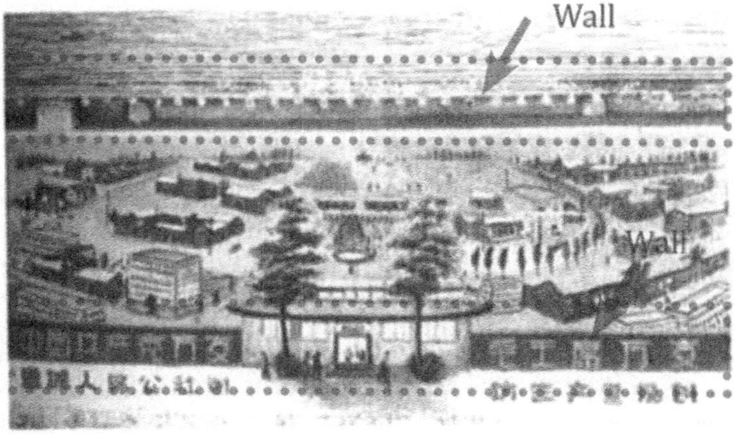

Fig. 3.5 A walled *Danwei*, plan for *Xiefang* production team, *Suicheng* People's Commune, *Henan*, 1958 (Adapted by author from Lu 2006, p.110)

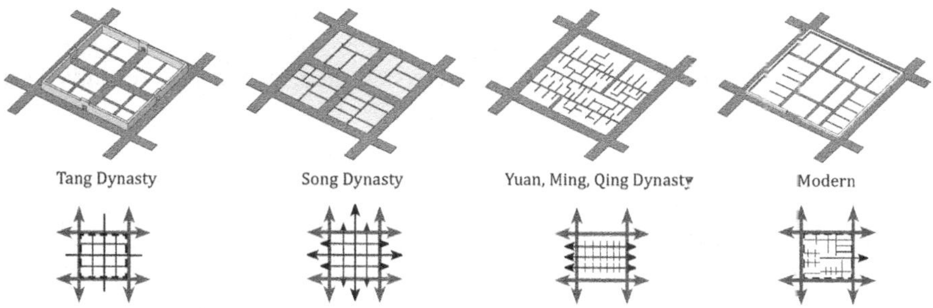

Fig. 3.6 Historical transformation of Chinese and Japanese Superblock structure (Drawn by author based on Wu 1999, He 2019, Zhu 2004, Friedmann 2005, Lu 2006)

Through various transformations, the use of walls around cities and Superblocks in structuring cities became more unified and systematic, and changes from a close to semi-open and back to a closed system as discussed above (see Fig. 3.6). Moreover, they were not only given more symbolic and political meaning to represent law and order (Steinhardt 1999, p.34), but are also used as a device to manage and control social security and movement by creating a number of closed systems (Wang 2014, p.173). More importantly, a symbolic feature of the structure at the time (Liu and Lai 2008, 41–69) In other words, the social structure persists with walled communities acting as self-contained units within Superblocks, indicating that strong social value is attached to this structure. Spaces are hence "fragmented, dissected, enclosed, localized and relativized" by using walls (Zhu 2004, 50–51) It is not only a device to protect, but also a physical structure that shapes the order of space, to create unity through diffusion, centrality through decentralization, and concordance through separation (ibid).

3.6 Case Study: The Supergrid and Superblock Structure in Contemporary Xi'an and Nanjing

3.6.1 Supergrids of Xi'an *and* Nanjing

Xi'an and *Nanjing* are representatives of typical Chinese cities and they both have long histories as being ancient capitals and they are highly influenced by their ancient predecessors. *Xi'an* and *Nanjing* in their modern conditions are about 3866 km^2 and 924 km^2, including highly urbanized inner-city areas of 540 km^2 and 180 km^2 respectively, where the Supergrids are located (see Fig. 3.7). Supergrids of the two cities have average intervals about 1 km and create Superblocks that are typically around 1 km^2. Both of them are built on relatively flat land with mountains surrounding them according to the principles of Fengshui of choosing sites, and they are affected by numerous heritage sites. *Xi'an*'s Supergrid is very regular and expands in all directions, whereas the natural geographical features have strong impacts in places to distort the regularity and formality of the Supergrid in *Nanjing*, which tend to be broken where it gets close to water elements and mountains. The Supergrid in *Nanjing* starts to deform in the north part and where it is close to water elements and mountains. The Supergrid in *Nanjing* also has a more regular but smaller grid interval (0.5–1 km^2) in the riverside area, which is a recently built district, mostly in the twenty-first century (see Fig. 3.6).

3.6.2 Jinyuan *and* Daguangli *Superblocks*

The case study sites in *Xi'an* and *Nanjing* are named as *Jinyuan* and *Daguangli*. They represent two types of spatial locations in a Supergrid. *Jinyuan* Superblock is the second Superblock on the west side of the city wall in the inner-city area, and the *Daguangli* Superblock, is located on the edge of the Supergrid (see Fig. 3.8 and Table 3.1). The two selected Superblocks are relatively the more regular one among their neighboring Superblocks, and *Daguangli* Superblock has a river channel separating the Superblock into two parts with a few connections.

Jinyuan Superblock is about 0.75 km^2 with average widths and lengths of 0.8 km and 0.94 km. *Daguangli* Superblock, in comparison, is about 0.45 km^2 bigger than *Jinyuan* with 1.4 km on the long edge and 0.8 km on the short edge. The total built area of Jinyuan is about 0.21 km^2 giving a 26.3% building coverage. Daguangli is slightly larger with 0.34 km^2 (28% building coverage). The population density of Jinyuan is estimated between 243 to 300 people/ha (a figure calculated from the information contained on an on-site poster that records the population in the southeast quarter). Daguangli Superblock has a population density of approximately 210 people/ha (estimation based on the Editorial Board of local Chronicles of the City of Nanjing 2008 & 2009). The populations of both sites are mixed with people of different ages, including local residents and migrant workers.

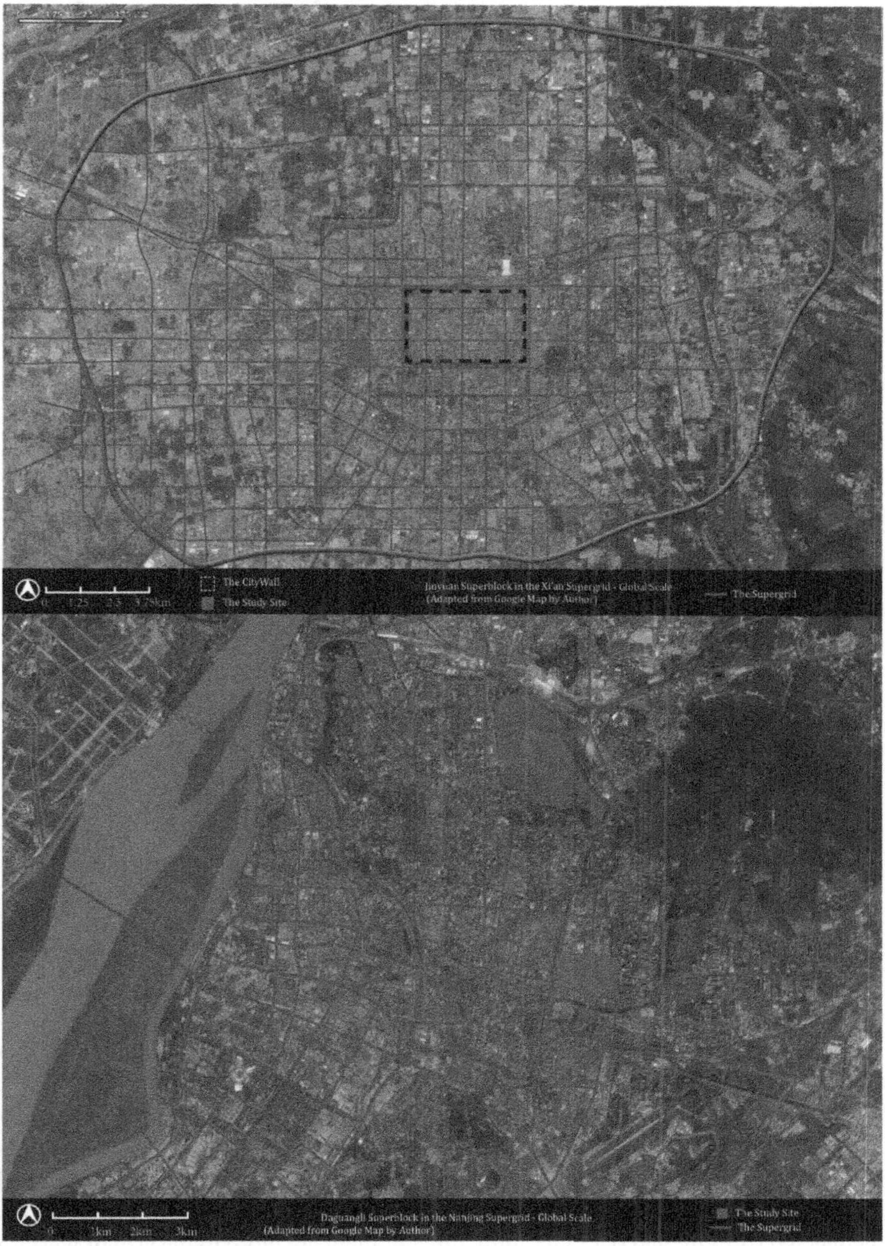

Fig. 3.7 The Supergrids of *Xi'an* (*top*) and *Nanjing* (*bottom*) (Drawn by author based on Google Map 2017)

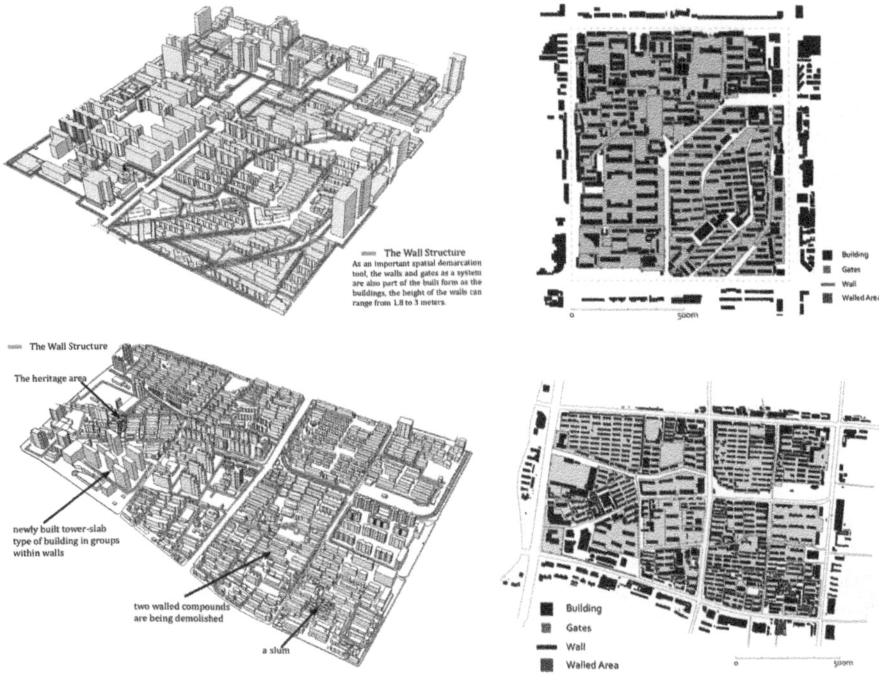

Fig. 3.8 Jinyuan (top two) and Daguangli (bottom Two) Superblock—3D Model (Left) and Built Form (Right) (by author)

Table 3.1 Basic Measurement of Jinyuan and Daguangli Superblocks (by author)

Superblock	Size (km2)	Width (km)	Length (km)	Sub-areas	No. of walled area	No. of gates	Density of gates	No. of buildings	Population density
Jinyuan	0.75	0.8	0.94	3	72	118	1.63	1121	272
Daguangli	1.2	0.8	1.4	14	59	111	1.88	1786	210

Jinyuan and *Daguangli* Superblock also represent two types of structures that can be commonly found in Chinese cities. *Jinyuan* represents one type of Superblock that is very regular and only subdivided into three sub-areas, which ranges greatly between 0.0154 and 0.4964 km^2 in size. On the other hand, *Daguangli* represents another type of Superblock, which is composed of 14 smaller sub-areas and intersected by a river channel. Although those areas have a wide range of sizes between 0.012 km^2 and 0.23 km^2, only three of them are close to the size of western street blocks. The other bigger ones are more common in Chinese cities. Moreover, those sub-areas are smaller in the area to the south of the river channel than in the north part, creating a more permeable structure in the south.

The sub-areas contained a maze-like wall structure in both Superblocks. *Jinyuan* Superblock had 72 walled irregular areas with very little streets in between, containing 1121 individual buildings. In *Daguangli*, the walls subdivided the

Superblock into 14 smaller sub-areas with 1786 individual buildings and mostly aligned consistently with the edges. There were 118 gates in 72 walled areas giving an average distribution density at 1.63 in *Jinyuan* Superblock. Similar density can also be found in *Daguangli* Superblock, where 111 gates can be found around 59 walled areas presenting a density at 1.88.

The general morphologies and structures of the two Superblock present four important observational characteristics: (1) there is the pervasive use of wall and gate structure in this Superblock. Pavilions and one-story row houses are mostly located on the edge of the Superblock or around the walled compound as part of the wall to enclose space. (2) Individual buildings in different sizes are generally high, homogeneous and oriented to face south to get the maximum sunlight. They are also grouped collectively within walls and gates and stand in big open spaces. (3) Edge buildings are generally of lower heights and with fewer walls than those of the inner area. Even though the lower edge is gradually replaced by higher towers, most of them still have a one to four level commercial podium at the bottom. (4) The locations of the gates in the wall structure determine a system of the private streets within each walled compound. In other words, the spatial order is mainly organized by the wall structure rather than the street network.

3.6.3 The Formation of the Supergrids in Xi'an *and* Nanjing

As the origin of the earliest Supergrid and Superblock structure, *Xi'an* in the *Han* Dynasty was still a prototype. It was further developed to its full maturity as the famous Eastern planning model, which included a regular Supergrid with 108 walled Superblocks in the *Sui* and *Tang* Dynasties and leaves a long-standing influence on the development of the modern structure. However, this structure was destroyed through numerous wars. The city was then reconstructed and shrank to the size of what was previously the central Royal Palace area (about 3×4 km), and this is from the *Wudai* to *Ming* Dynastic periods, when shifts in the location of gates in the city walls affected the positions of Global roads and the numbers and sizes of Superblocks. It is the structure in the *Qing* Dynasty that inherited characteristics from previous dynasties to set the basis of the modern Supergrid. The modern expansion of the city starts from the two Global roads constructed in *Qing* Dynasty (see the dotted blue lines in Fig. 3.9). Many Global roads and the drainage canal from the *Tang* period were also reused and rebuilt as part of the modern Supergrid system.

In the modern period, the city went through a long period of social chaos and wartime instability. As an alternative capital, the redevelopment of the city into three major stages: the Pre-wartime capital (1927–1932), the Alternative Capital (1932–1945), and the Post-wartime reconstruction (Ren 2009). There are ten major planning documents that have impacts on the development of the city to various degrees (see Table 3.2), and they all emphasize heritage conservation as the fundamental planning principle in city planning. The idea of opening up the gates of

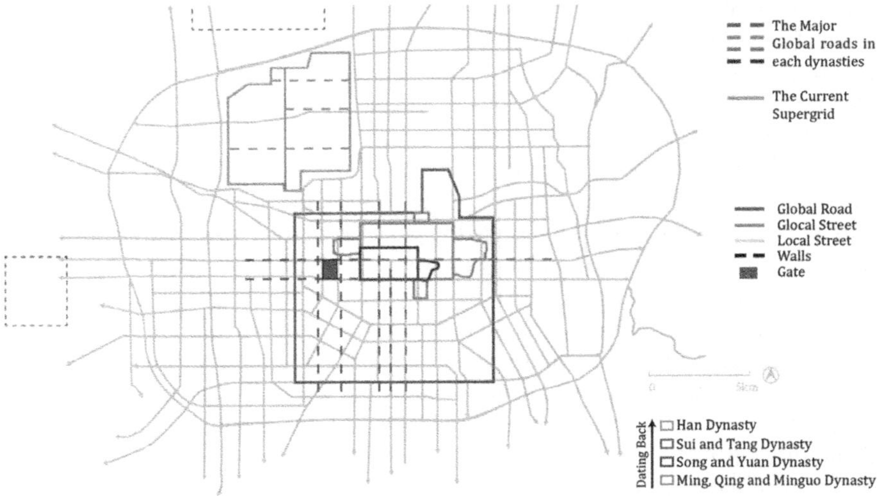

Fig. 3.9 Xi'an Supergrid History – Structural Transformation (Maps are drawn by the researcher based on the Editorial board of the District of Lianhu of Xi'an 2001, Shi 1996, He 2012)

the city wall to address the conflict between modern transportation and heritage protection has a deterministic impact on the formation of the modern Supergrid and allows the major roads within the city wall to expand.

Similarly, the modern Supergrid of *Nanjing* has been formed under the influence of three major ancient cities that were built in this area. Through the analysis of historical maps of the city in different periods, imprints of the city structures of the Six Dynasties (229–529), Southern *Tang* Dynasty (937–975), and *Ming* Dynasty (1368–1644) remain clear on the modern Supergrid (see Fig. 3.10). Those capital cities were also constructed with Supergrid and Superblock structures in which many major roads were related to the city gates. In other words, major roads between the gates of those ancient capitals were used as city-wide or "global" connections (indicated as dotted green, purple, and orange lines in Fig. 3.10(1)), and these roads divided the cities into Superblocks. Those Global roads of ancient times set an embryonic framework for the future planning of the city road network.

The redevelopment of *Nanjing* was highly influenced by the western planning ideas and theories at the time. The Capital Plan in 1929 was directed by two American planners, who were trying to transform *Nanjing* into an "Eastern Washington" by adopting what was considered to be the most advanced modernist planning at the time (Editorial board of local chronicles of the city of Nanjing 2008 & 2009). The city was designed with some radial roads and a small grid structure as was commonly used in Western cities (see Fig. 3.10(2) in both black and red lines). The grid is particularly small in the northwest and the east part of the city, where little historical development had occurred. However, the small grid network was never fully realized because of the triangular shape land parcels that are generated by a radial-grid structure and were considered as a waste of land in the Chinese

Table 3.2 Planning Documents of Xi'an during 1920s to 1940s (by author)

Planning Document	Year	Planner	Source
Municipal Construction Plan of Chang'an in Shanxi	1927	Department of Construction	Shaanxi Provincial Government Construction Department Construction Reporting and Editing Office, construction report. The Republic of China 16 years (1927) November
Outline of the Construction Project of the 20th Anniversary of Shaanxi Province	1931	Department of Construction	Xi'an Municipal Archives, Northwest Development of the Republic of China (internal information) 2003, 11
Administrative Plan of the Construction Department of Shanxi Province from the year 22 to 24 (Municipality B)	1933		
Discussion on the Division of *Xijing* City	1934	*Ping Ji*	Prepare for the selection of the historical materials of the Xijing Accompanying Capital Archives, Northwest University Press, 1994, 10
Xijing Municipal Construction Plan Guidelines		*Jingtian Sui* in *Yisushe*	
Xijing City Zoning Plan Description	1937		
Rural Construction Plan Outline	1939	Xijing Municipal Construction Committee	
Xijing Planning	1941		
Xi'an Municipal Government's Meeting on the Width of the Roads around the Four-lane Road of the City's Clock Tower	1946		Republic of China Archives
Xi'an District and Road System Plan	1947		Republic of China Archives

spatial conception (ibid, p.20 & Wang 2014). As a consequence, the modern plan was adapted from this Capital plan by Chinese planners, and they formulated the idea of using squares to expand concentrically in all directions (ibid.), which by its nature is a Supergrid and Superblock structure (See red lines and lighter red lines in Fig. 3.10(2)).

3.6.4 *Formation of* Jinyuan *and* Daguangli *Superblocks*

Jinyuan Superblock was mostly formed through the expansion of Global roads from the 1950s to the early 2000s (see Fig. 3.11(1)). Four major stages of the historical transformation of *Jinyuan* can be detected (see Fig. 3.11(2 to 5)). The *Liquan* Superblock of the *Tang* Dynasty is the earliest predecessor of modern *Jinyuan*. Over 50% area of *Jinyuan* Superblock lies on top of the *Liquan* Superblock, as discovered

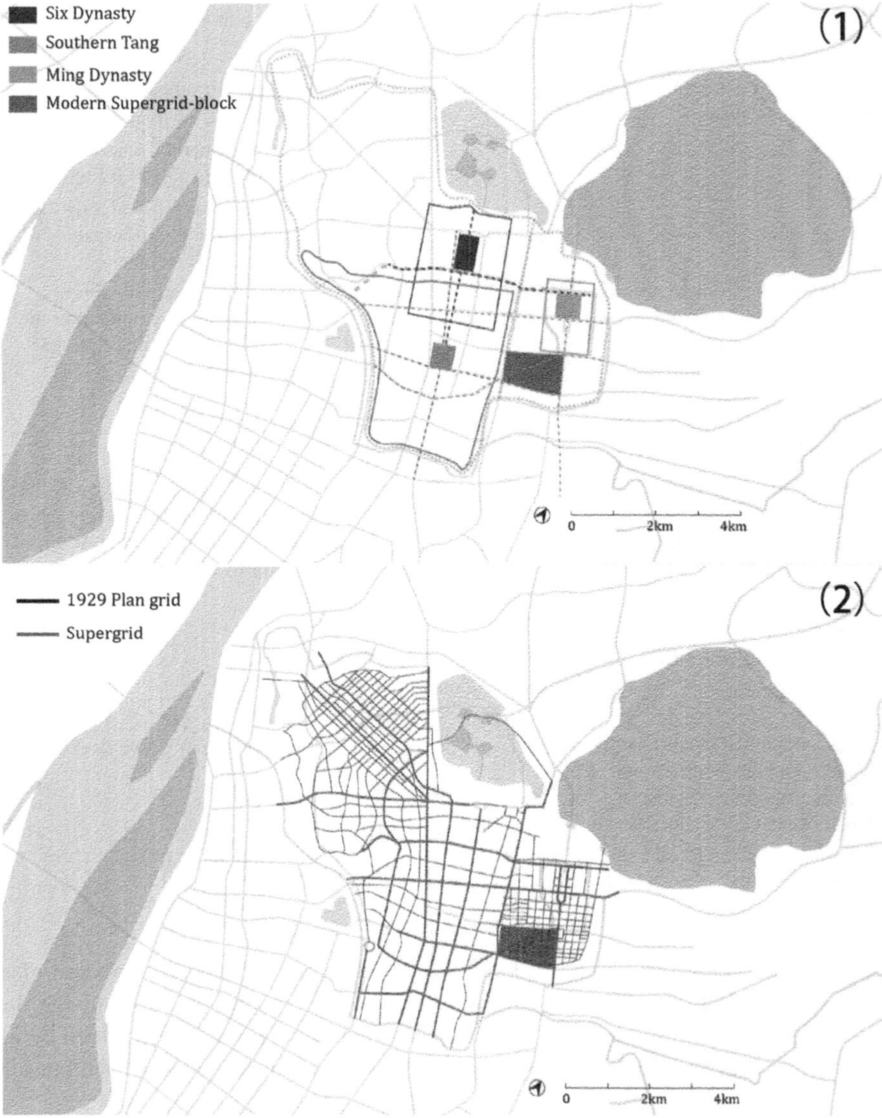

Fig. 3.10 (1) The Historical Transformation of the Nanjing Supergrid (Drawn by author based on the records from Wu 2011, Yang 2009, editorial Board of local Chronicles of the City of Nanjing, 2008 & 2009, Chen & Gazzola 2013, Zhi 2014); (2) The Comparison between the Capital Plan 1929 and Modern Supergrid Network (Drawn by the researcher based on the records from the Editorial Board of local Chronicles of the City of Nanjing, 2008 & 2009, Zhu 2014)

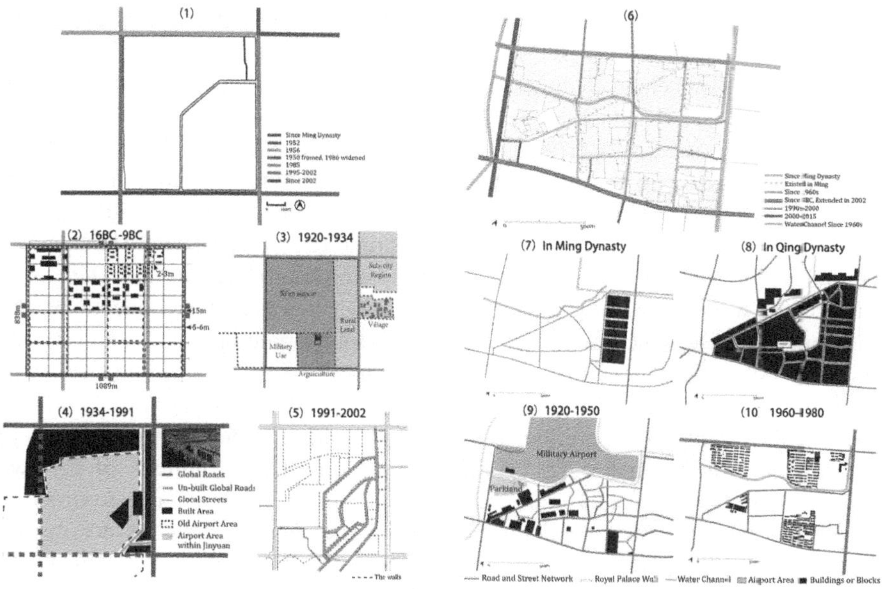

Fig. 3.11 Historical Transformation of the *Jinyuan* Superblock (on the *left* side) and *Daguangli* Superblock (on the *right* side) (Drawn by author according to Editorial Board of the District of *Lianhu* of *Xi'an* 2001, Shanxi Archaeology Institute 2008, Shi 1996, Editorial Board of local Chronicles of the City of Nanjing 2008, 2009, 2011)

in 1999 (Shaanxi Archaeology Institute 2008). *Liquan* was a walled Superblock (838 × 1032 m) next to the International Trade Market of the city. It accommodated about 10,000 people with a mix of temples, royal residences and pottery kilns (ibid. & He 2012). However, it was destroyed and gradually occupied by several villages as part of a suburb of *Chang'an* (the name of *Xi'an* in the *Tang* Dynasty). Later, the area became a walled military training field during the *Qing* Dynasty, and it was further converted into a walled airport from the 1920s to 1991 (see Fig. 3.11(4)). Because of the noise issue, the airport was moved and gradually changed into a number of walled residential and industrial compounds (Fig. 3.11(5)). By 2002, the expansion of south Global roads finally completed the Superblock, and four Glocal roads (orange lines) were formed and continue to serve this Superblock (Editorial Board of the District of *Lianhu* of *Xi'an* 2001).

Daguangli Superblock has been formed since the 1960s, and it is still under construction (see Fig. 3.11(6)). Several historical Global connections that can be traced back to South *Tang* Dynasty were extended to this area in *Ming* Dynasty. They provided links to other gates and formed a basic traffic skeleton for this area. Because of the proximity to the royal palace on the north side of the Superblock, this area was built with various government compounds and the residences for government officials at the time. Later in *Qing* Dynasty, it retained the same structure but with more subdivisions and the built area was expanded to the two sides of the diagonal street, which linked two city gates (see Fig. 3.11(7)). Between the 1920s and 50s, the area went through even more subdivisions in the southeast half, and the area on the north side of the water channel was used as a military airport. During this period, the shape of the Superblock was gradually formed with Global connections on three sides (see Fig. 3.11(8)). The resulting structure within the Superblock was not changed dramatically until the 1960s when many of the irregular streets were removed and the water channel and diagonal streets were modified with more regularity (see Fig. 3.11(9)). More importantly, the area was fully enclosed by four Global roads: two were extended and two remained narrow until the 1990s as indicated in Fig. 3.11(10). Some new gated communities were built in this period but were not walled at first: an extensive wall structure emerged as more housing for communities was constructed in this area. The arterial roads around *Daguangli* Superblock were historically connected to the city gates as global connections, and the wall and gate structure is still dominating the Superblock and exercising its power of controlling movement at a different scale.

The transformation processes of the two Supergrids and Superblock structures indicate the use of the Supergrid and the wall and gate structure as strong structural determinants in the making of *Xi'an* and *Nanjing*. The evolutions of the two Superblocks also indicate the consist and pervasive use of the wall and gate structure to various degrees. More importantly, despite the two Superblocks are mostly formed under the modern period; the Supergrids of the ancient cities still play a fundamental role in the process as indicated in Fig. 3.10.

3.6.5 *The Superblock Structure of* Jinyuan *and* Daguangli *Superblocks*

The investigations of the physical structure of *Jinyuan* and *Daguangli* Superblocks focus on the street networks and activity distribution. The street networks are categorized into Global Road, Glocal, Local and Internal streets.[5] Because the two Superblocks both have an extensive wall structure with many gates as indicated before, when taking the gates into consideration, they can be further classified into two types: Local and Internal gates.[6] Functional activities are classified into Consumption, Production, Service, and Residence.[7]

[5] **Global Road:** Mainly refers to the wide arterial roads that form the Supergrid; **Glocal Street:** Mainly refers to the streets that connect neighboring Superblocks; **Local Street:** Mainly refers to the streets within a superblock that create direct access to global roads at the edges, but do not go across the global roads into its neighboring Superblocks; **Internal Street:** Mainly refers to the local streets that do not directly connect to the edges of a Superblock providing internal links within Superblocks.

[6] **Local Gate:** Mainly refers to the gates within a superblock that create direct access to global roads at the edges; **Internal Street:** Mainly refers to the gates that do not directly connect to the edges of a Superblock providing internal links within Superblocks.

[7] **Consumption Activity:** represents a number of wealth-using activities that are related to the use of any commodity or service in cities; **Production Activity:** represents a number of wealth-creating activities (both physical and intellectual); **Service Activity**: represents a number of activities that are related to the assistance of consumption, production or living. It is mainly about the non-

3.6.5.1 Street Network

The street networks of *Jinyuan* and *Daguangli* Superblock present two types of structure and hierarchical importance of different street types. One particular feature is the large number of gates locating along the Global and Glocal streets, forming Global-Local gates and Glocal-Internal gates. Correspondingly, the numbers of Local gates of *Jinyuan* and the Internal gates of *Daguangli* are the largest. Similarly, the analysis of the junctions of both Superblocks exhibits clearly that the Global-Local type of T-section in *Jinyuan* and Glocal-Internal T-section in *Daguangli* are the largest (see Fig. 3.12). These analyses indicate that the gates along the Global

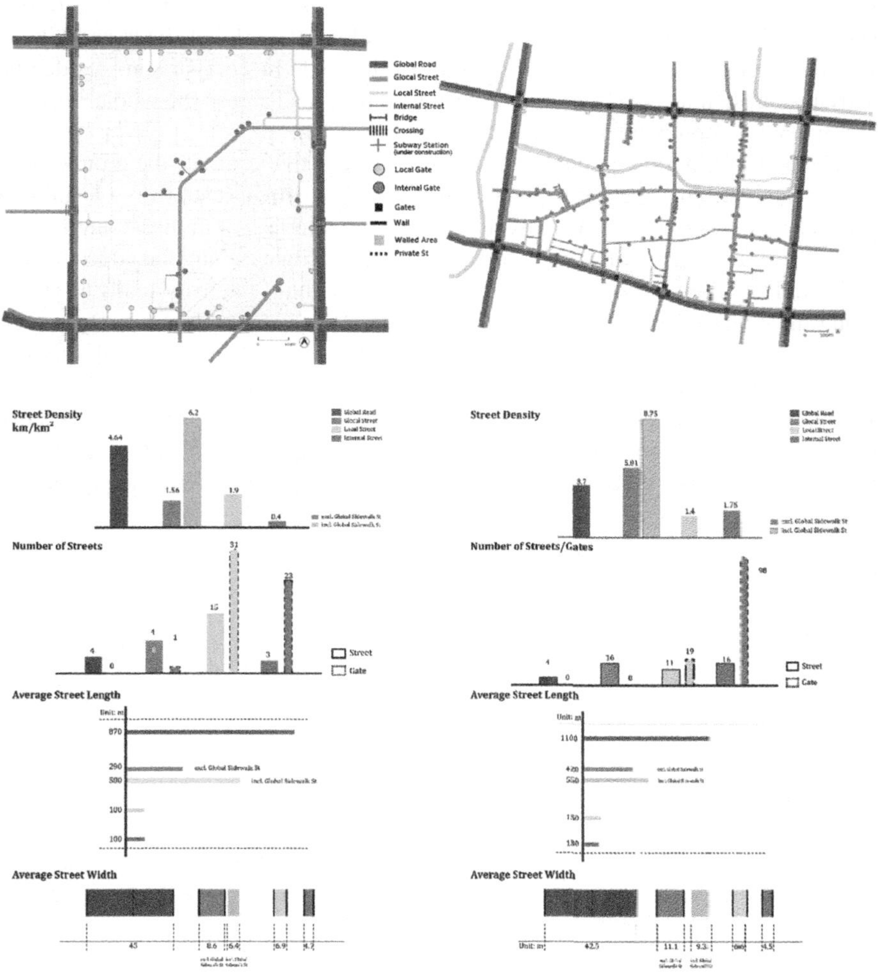

Fig. 3.12 Street network of Jinyuan (left) and Daguangli (right) Superblock (by author)

---

profit oriented social service; **Residence:** represents a number of activities that are related to living and residing, including all kinds of apartment and houses.

roads in *Jinyuan* and Internal gates along the Glocal street in *Daguangli* are the major structure that controls the movement within the two Superblocks. This indicates that walls and gates play an important role in the physical structure of the two Superblocks. On the other hand, Global roads and Local streets in *Jinyuan* have larger street densities and quantities than *Daguangli*. The average street lengths and widths in both Superblock are very similar, but the lengths of Glocal streets in Daguangli are shorter but wider than the same type in *Jinyuan* Superblock.

3.6.5.2 Functional Activity

Distributions of non-residential activities in *Jinyuan* and *Daguangli* Superblock show similar patterns, and they tend to gather along the Global roads and Glocal streets where gates of each walled area are located. In comparison, residential activities present no clear pattern, but they tend to spread across the whole sites and surrounded by non-residential activities (see Fig. 3.13).

The mix of function on the horizontal level (the activity mix of the ground floor use of each building measured by a 50 m grid) and vertical level (mix of building uses on different floors) also exhibits similar patterns. Both horizontally and vertically, the intensity of activity mixes is generally higher along the edges than in the middle of both sites. About a quarter of high activity intensity tends to gather along the edges of the sub-areas in *Jinyuan* (23.7%) or street blocks in *Daguangli* (25%) where the gates are located. The areas with low to medium levels of intensity (indicated in gray and pink) are also about the same in both Superblocks (76.2% of Jinyuan, and 75% of *Daguangli*). Of these, 51.8% of the area in Jinyuan and 45.6% of the area in *Daguangli* Superblock are for single use. Over 67% of the buildings have only one type of activity on the vertical level in *Jinyuan*, but the percentage

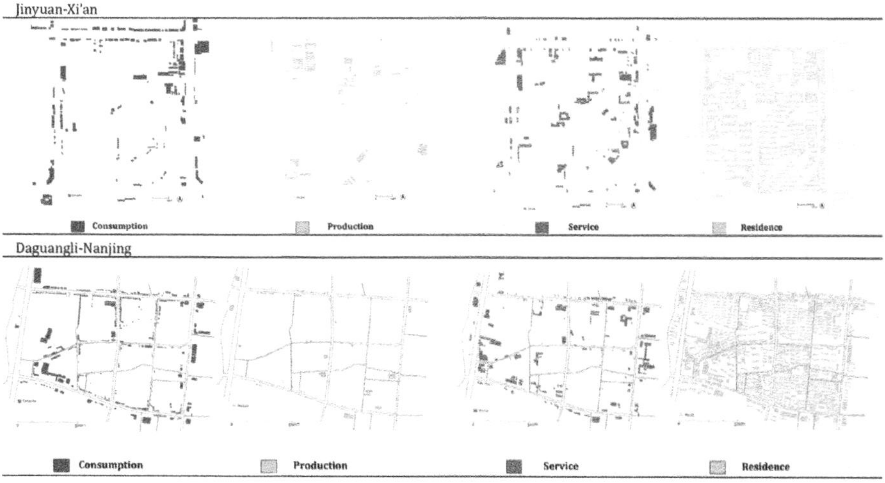

Fig. 3.13 Distribution of functional activity of *Jinyuan* and *Daguangli* Superblock (by author)

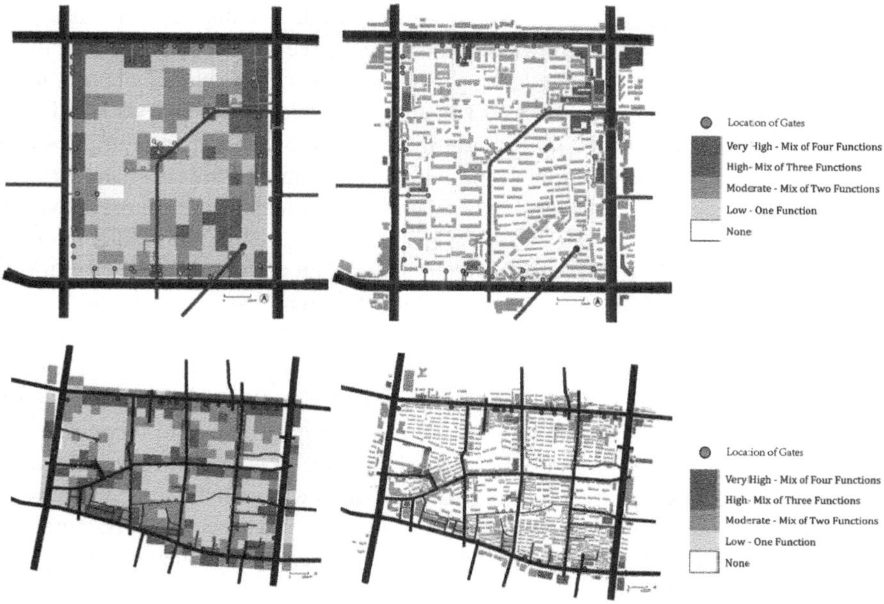

Fig. 3.14 Distribution of activity intensity and location of gates of Jinyuan (top two) and Daguangli (bottom two) Superblock (by author)

drops to 46% in *Daguangli* on the vertical level. High intensity (33% in *Jinyuan* and 47% in *Daguangli*) still mostly occurs along the edges where the gates of walled areas are located (see Fig. 3.14).

3.7 Conclusion

The investigation of the Supergrid and Superblock structure indicates a long-lasting tradition of the application of the structure in Chinese cities. Despite the strong western influence, many traditional design principles are inherited and continued to be used in modern times, especially creating a Supergrid as the basic skeleton and the use of Superblock in combination with wall and gate structure. It is especially evident in the case study of *Nanjing* where the Western small (not Super) grid structure was replaced by a Supergrid network, which fits the Chinese principles of spatial demarcation. The two Superblocks in China indicate the same principle of spatial organization, with the pervasive use of wall and gate structure as the major component in controlling and restructuring the street network. The *Jinyuan* in *Xi'an* and *Daguangli* in *Nanjing* represent two types of Superblocks in more central and edge parts of the Supergrid respectively. Both are surrounded by Global roads that enjoy some level of equality within the global network. Both indicate high levels of self-containment within the superblocks.

While *Jinyuan* has a large number of gates locating along the Global roads, *Daguangli* has more gates sitting along the Glocal streets. As a consequence, more activities are gathered along Global roads in the former Superblock, whereas most are concentrated along Glocal streets in the latter one. Thus, the wall and gate structure is the major factor influencing the distribution of activities in the street network because it is the locations of the gates that determine where most movements and activities take place. Local and Internal gates attract and generate most activities, followed by the junctions of Global roads and Glocal streets. The results strongly indicate the dominant role of the wall and gate structure as the main generator of activities, especially when they are mapped together with the street network.

Instead of taking for granted that the western street block is the solution as other existing studies suggested, this chapter offers a different voice from a spatial perspective. The findings provide an important understanding of the contemporary Chinese urban structure for both the academics and government to reconsider the guidelines. The Supergrid and Superblocks structure does not have to be totally replaced by the western street blocks structure, and the demolishing of all the walls around residential areas may not be a final solution as it has been deeply ingrained in the planning practice of Chinese cities as discussed in this chapter. Hopefully, more studies on Superblocks can inspire better solutions.

References

Bray, D., 2005. *Social Space and Governance in Urban China: The Danwei System from Origins to Reform*, California: Stanford University Press.

Calthorpe Associates, 2011. *Chenggong New Town, International Masterplan*, Available at http:// www.calthorpe.com/chenggong

Chen, W. & Gazzola, L., 2013. Comparative Study on the City Walls of Nanjing and Rome, Dongnan University Press

Chen, L. 2016a. Discussion on the theory and practice of urban block scale, *Shanxi Architecture*, 42 (16), 11-12

Chen, T. 2017. The Evolution of the Scale of Contemporary Urban Blocks—The Game between Modernism and Humanism, China Urban Science Research Association, Hainan Provincial Planning Commission, Haikou Municipal People's Government. Urban Development and Planning Conference Proceedings, Chinese Cities Scientific Research Association, Hainan Provincial Planning Commission, Haikou Municipal People's Government: Beijing Bonti Conference Co., Ltd.

Chen, X. F., 2018. A Comparative Study of the Supergrid and Superblock Structures in Chinese and Japanese Cities. *The International Journal of the Constructed Environment,* 9**,** 45-63.

Chen, S. Y., 2016b. Exploration of the Controlled Detail Planning Based on Street Block Concept—Taking Nanhai Area of Maoming, Guangdong Province as an Example, China Urban Planning Society, Shenyang Municipal People's Government. Planning 60 Years: Achievements and Challenges - 2016 China Urban Planning Annual Conference Proceedings (06 Urban Design and Detailed Planning), China Urban Planning Society, Shenyang Municipal People's Government: China Urban Planning Society, 14.

Chu, W.W., 2017. Strategies and Analysis on the Factors Affecting the Introduction of Traffic Microcirculation in the Planning of Urban Renewal Units in Shenzhen, Shenzhen University, Doctor Thesis

Deng, F. 2013. 'Gated Community and Housing Segregation: Empirical Evidence from Chongqing, China', Resource, Land Technology, Chongqing, SSRN Electronic Journal, Available at http://papers.ssrn.com/sol3/papers.cfm?abstract_id=2416663

Dou, Q. 2010. 'From Micro-district to Gated Community: The Evolution of the Morphology of Residential Housing in Beijing.' Multiple perspectives: Theory and Practice, Architecture Journal, Vol.3, 7-11, available at http://www.cqvip.com/QK/94827X/201003/33512771.html

Editorial Board of the District of Lianhu of Xi'an, 2001. *District Records of Lianhu*, Sanqin Publishing House, Xi'an.

Editorial Board of local Chronicles of the City of Nanjing, 2009. The Chronicles of the City of Nanjing: The City construction of Urban and Rural Area, China Local Records Publishing: Beijing

Editorial Board of local Chronicles of the City of Nanjing, 2008. *The Chronicles of the City Planning of Nanjing*, Vol. One & Two. Jiangsu People's Publishing House

Editorial Board of Local Chronicles of the City of Nanjing, 2011. The Chronicle of Baixia District of Nanjing 1986-2005, Vol. One & Two, China Local Records Publishing: Beijing, 80-290, 635-652

Feng, D. Breitung, W. Zhu, H. 2011. Space of fragmentation: A study of the linkages between gated communities and their neighborhoods in Guangzhou, China, *Geology, 30*(1), 61-70, Available at http://www.dlyj.ac.cn/CN/10.11821/yj2011010006

Friedmann, J. 2005. China's urban transition, Minnesota: University of Minnesota Press.

Gong T, Wang, F.X., Xie, D. 2017. The Analysis of Spatial Density of Traditional Business District – Taking the city of Cangzhou as an example, *Housing and Real Estate*, 18, 3-4.

Hartog, H. D. (ed.) 2010. *Shanghai New Towns: Searching for community and identity in a sprawling metropolis,* Rotterdam: 010 Publisher, Rotterdam.He, C. R. (2012). *The Former Capital City of Xi'an (Gu Du Xi'an)*, Qinghua University Press

Hu, Y. J., Yu, T. & Luo, X. L., 2011. Research on Urban Fragmentation of China under the Reconstruction of Public Space during Transformation Period, *Study of Urban Spatial Reconstruction*, 10-13, Available at http://www.mur.cn/nsfiles/8054FCE0B476E5388AFD80F087E40FD1-2011-03-10.pdf

Hu, S.R. 2010. The study of west district of Hangzhou Residential gated neighborhood: looking at cities from an urban design perspective, *Master Thesis Architecture Faulty, Zhejiang University,* Available at: http://wenku.baidu.com/view/47b24af 77c1cfad6195fa759.html?re=view###

Huang, Y., 2006. Collectivism, Political Control, and Gating in Chinese Cities, *Urban Geography 27*(6): 507-525. Available at: http://www.tandfonline.com.ezproxy1.library.usyd.edu.au/doi/pdf/10.2747/0272-3638.27.6.507

Jiang, Y., 2017. From closed unit to block system – A practical study based on road network connectivity, China Urban Planning Society, Dongguan Municipal People's Government. Sustainable Development Rational Planning - 2017 China Urban Planning Annual Conference Proceedings (07 Urban Design), China Urban Planning Association, Dongguan Municipal People's Government: China Urban Planning Association, 11.

Kan, H. Y., Forsyth, A. & Rowe, P. 2017. Redesigning China's superblock neighbourhoods: policies, opportunities and challenges. *Journal of Urban Design,* 22, 757-777.

Li, Y., 2017. Research on the design and planning of transportation under the street block system Environment, *Chang'an University,* Doctor Thesis.

Li, Z. T. & Yu Y., 2006. Open Block' Concept and its inspiration in urban planning of China's residential housing development, *Urban Planner, 2* (31).

Liao, P., 2004. Cancer of urban life: Problems of Gated Communities in China and Their Solutions (Cheng shi sheng huo de ai zheng – Feng bi shi xiao qu de wen ti ji dui ce). *Modern Architecture,* 5, 46-49.

Liang, P., & Su, H.M. 2011. The study of the Characteristics of the history of the enclosed superblock and grid pattern of Chinese cities, *City exploitation, The University of Xi'an Education University*, *4*(3), available at http://wenku.baidu.com/view/0060593409l2a2161479290b.html?re=view

Lu, Y. T., 2017. Discussion on the countermeasures of implementing the "block system" in the stock area of Hangzhou, China Urban Planning Association, Dongguan Municipal People's Government. Sustainable Development Rational Planning - 2017 China Urban Planning Annual Conference Proceedings (02 Urban Renewal), China Urban Planning Society, Dongguan Municipal People's Government: China Urban Planning Association, 13.

Lu, Y., Stamenovic, B. 2004. The Spatial Concept of Chinese Architecture. *Built Spaces: The Cultural Shaping of Architectural and Urban Spaces*, 9.

Lu, D., 2006a(a). *Remaking Chinese Urban Form: Modernity, Scarcity and Space, 1949-2005*, Routledge.

Lu, D., 2006b (b). Travelling urban form: the neighborhood unit in China. *Planning Perspectives*, 21, 369-392.

Liu, X., 1936. Li Lu Kao, Foodstuffs. 3, 575

Lu, J., Rowe, P. G., & Zhang, J., 2001. *Modern Urban Housing in China, 1840-2000*, Prestel.

Liu, Y., Li, Z. G., 2010. A Review of Studies of Gated Community since 1990s. *Human Geography*, 3, 10-15.

Liu, Y., & Li, Z. Y., 2009. 'From Gated Community to traditional and sustainable Chinese neighborhood', Modern Architecture, vol. 2, 20-24, Available at http://www.cqvip.com/QK/91309X/200902/29895053.html

Ministry of Housing and Urban-Rural Development of the People's Republic of China. 2016. *State Council Issues: 'Key Suggestions to Further Reinforce the Management of Urban Planning and Construction' Available at* http://www.mohurd.gov.cn/zxydt/201602/t20160222_226701.html

Mote, F. W., (ed.) 1977. *The transformation of Nanking, 1350-1400, in Skinner, G.W. The City in Late Imperial China*, Stanford, California: Stanford University Press, 20-49

Nieminen, S., 2012. *Chinese Superblock: A study and design tools for sustainable urban housing development in China.* Master Degree, Tampere University of Technology.

Peng, J., 2017, Study on one-way traffic organization under the condition of small block and dense road network—Taking the Honghua Airport area of Nanjing as an example, China Urban Planning Association, Dongguan Municipal People's Government. Sustainable Development Rational Planning - 2017 China Urban Planning Proceedings of the Annual Conference (06 Urban Transport Planning), China Urban Planning Association, Dongguan Municipal People's Government: China Urban Planning Association, 11.

Ren, C., 2008. An Analysis on the Evolution, Value and Applicability of High-density Gridiron Street Pattern and Streets of China: A Consideration on Grand—Road and Super-Block Phenomenon. *Urban Planning Forum*, 174, 53-61.

Ren, Y. Y., 2009. The Development of Xi'an Urban Planning Thought in Modern Times——Taking the Archives of the Republic of China from 1927 to 1947 as an Example, *Journal of Shaanxi Normal University* (Philosophy and Social Sciences Edition), 38(05), 105-112.

Rowe, P., Guan, C.H., 2016. Striking Balances between China's Urban Communities, Blocks and Their Layouts *Time+Architecture*, no. 6, 29-34.

Shane, D. G., 2014. *Block, Superblock and Mega-block, A short history* [Online]. Italia World Magazine. Available at http://www.arcduecitta.it/world/2014/01/block-superblock-and-megablock-a-short-history/

Shanxi Archaeology Institute, 2008. *The Kiln Site of Tricolor- Glazed Pottery at Liquanfang in Chang'an Capital City of Tang Dynasty*, Antique Publishing House

Shen, F., Li, L., Zhai. H., 2016. Road Network Design and Planning and of "Narrow Street and Small Block" Mode —Taking the planning of the core area of Chenggong new district in Kunming as an example, Urban Planning, 4(5) , 43-52.

Shi, N. H., 1996. *The Historical Atlas of Xi'an*, Xi'an Cartographic Publishing House

Steinhardt, N. C., 1999. *Chinese Imperial City Planning*, Hawaii, University of Hawaii Press, 10-60

Song, W. X., & Zhu, X. G., 2009. The gated communities in China: The negative impact of the social differentiation, *The Urban planner*, (11), 82-86, Available at: http://www.cqvip.com/qk/81000x/200911/32418185.html

Song, W. X., 2010. The Study Progress of Gated Communities in China, *Urban Planning, Nanjing University, 4*. Available at: http://www.cnki.com.cn/Article/CJFDTotal-CXGH201004009.htm

Tan, J. M., Liu, J. C., Xu, Y.H., 2017, Discussion on proper road network density based on street block system, *Shanxi Architecture*, 43(08), 1-3.

Wang, Y. H., 2003. *New residential Neighborhood,* Nanjing: Southeast University, 25-36, Available at: http://wenku.baidu.com/view/36b07a04e87101f69e3195fc.html

Wang, Y. H., 2010. The gated communities in Chinese cities: Problems and Countermeasures, *Study of Modern cities, The Southeast University,* Available at: http://www.mur.cn/MUR-2010-3-d96404f15a401232a648c9158bea5729.html

Wang, D., 1992. The Historical Importance of the Planning of the Luoyang City in Beiwei Dynasty (Beiwei luoyang gui hua ji qi li shi di wei). *Huazhong Architecture,* 10, 47-56.

Wallenwein, F., 2013. *The Housing Model xiaoqu: The Expression of an Increasing Polarization of the Urban Population in Chinese Cities?* Master Thesis from Heidelberg University, Available at: http://archiv.ub.uni-heidelberg.de/volltextserver/17143/1/The Housing Model xiaoqu_Wallenwein.pdf

Wang, J. Y., 2016. Exploration of the Implementation Path of Street Block in Urban CBD based on Neighborhood Development——Taking the Urban Design of the Central Business District of Bali Lake in Jiujiang City as an Example [A]. China Urban Planning Society, Shenyang Municipal People's Government. Planning 60 Years: Achievements and Challenges— 2016 China Urban Planning Annual Conference Proceedings (06 Urban Design and Detailed Planning) [C]. China Urban Planning Association, Shenyang Municipal People's Government: China Urban Planning Society,15.

Wei, W., Qin, L.F., 2011. The Analysis of Chinese Gated Communities, Zhejiang University Architectural Journal, 5-8, Available at: http://wenku.baidu.com/view/535c9b85cc22bcd126ff0cc5.html

Win, T. N., 2014, Historical Mapping Method: New Research Method for Research Methodology, *Journal of Perspectives on Development Policy in the Greater Mekong Region (ICMR Journal)*, Volume 2, Number 2, July-December. Available at: http://icmr.crru.ac.th/Journal/Journal%204/5%20Historical%20Mapping%20Method.pdf

Wu, L., 1999. *Rehabilitating the Old City of Beijing, A Project in the Ju'er Hutong Neighborhood*, University of British Columbia Press.

Wu, F., 2005. Rediscovering the 'Gate' under Market Transition: From Work-unit Compounds to Commodity Housing Enclaves, *Housing Studies*, *20*(2), 235-254. http://citeseerx.ist.psu.edu/viewdoc/download?doi=10.1.1.202.8763&rep=rep1&type=pdf

Wu, Y., 2017. Research on Traffic Organization under "Small Block, Narrow Street', China Urban Planning Association, Dongguan Municipal People's Government. Sustainable Development Rational Planning - 2017 China Urban Planning Annual Conference Proceedings (06 Urban Transportation Planning), China Urban Planning Association, Dongguan Municipal People's Government: China Urban Planning Association, 10. Mou, Q. Li, Y. H., 2017. Research on the method of Planning and Layout of Street Block Network, *Transportation Science and Engineering*, 33(03),82-86

Wu, Y. Q., 2018. Research on the design of the renovation of gated residential quarters in Zhengzhou, *North China University of Technology*, Doctor Thesis

Xiao, X. X., 2016. The spatial logic of the survival of urban Space from the street block structure, *Residential Area*,4,86-89

Xu, X., 2007. *Urban Road and Transportation Planning,* Beijing, China Construction Industry Press, 202-205

Xu, M., Yang, Z., 2007. The Theoretical Speculation of Gated Community in China, in Jingyu Zhou (Ed.) Sustainable and Harmonious: Housing and Environment of Human Settlement in Massive Urbanization. *Beijing: China City Press,* 839-846,

Xu, M. & Yang, Z., 2008a. Design history of China's gated cities and neighborhoods: prototype and evolution, *Urban Design International*, *14*(2), 99-117.

Xu, M., & Yang, Z., 2008b. Rethinking the multi-disciplinary studies of gated communities: urban design perspectives. *Urban Planning International*, *22*(4), 24-28

Xu, M., & Yang, Z., 2009, Theoretical debate on gated communities: genesis, controversies, the way forward, Urban Design International, *13*(4), 213-216

Xu, M., & Yang, Z., 2010. Superblocks plus Gated Community: The Death of Public Space, *Architectural Journal,* 3, 12-15

Xu, M., 2009. Gated communities in China: Urban design concerns, *Phd Thesis Cardiff University,* Available at: http://search.proquest.com.ezproxy1.library.usyd.edu.au/docview/1369556528/previewPDF/DA01A5AAC75F48CAPQ/1?accountid=14757

Xu, Z., Song, W. X., Zhu, X. G., Wu, Q. Y., 2009. The formation of Gated Community and its social impact, *Urban Issue, 168*(7), Available at: http://www.cnki.net/KCMS/detail/detail.aspx?filename=CSWT200907001&dbname=cjfqtotal&d bcode=CJFQ&v=MTIzODI4L3poWVU3enNPVDNpUXJSY3pGckNVUkwrZlllWnJGeXZtVzczSkpqN2Nlckc0SHRqTXFJOUZaWVI2RGc=

Yang, C. Y., & Min, S., 2008. Theoretical Framework and Chinese Model of Gated Communities, *Huazhong Architecture Journal, 26*(11), 139-142. Available at: http://www.cqvip.com/QK/91309X/200902/29895053.html

Yao, Z. J., 1992. An Investigation of the Ruins of the City of Ping of Beiwei Dynasty (Wei du ping cheng yi zhi shi tan). *Journal of Chinese Historical Geography*, 215-234.

Zhang, L., Nong, H. P., Xiong, N. J., 2016. Study on the Planning of the Implementation of the Controlling Road Network in the Block System—Taking the Detailed Controlled Planning of the YA-04 Unit of the Friendship Area of the Downtown Area of Nanning as an Example, *Planner*, 32, 38-41

Zhang, W., 2017. Evaluation of Jinan CBD Planning and Design Based on the Street Block Development, China Urban Planning Association, Dongguan Municipal People's Government. Sustainable Development Rational Planning - 2017 China Urban Planning Annual Conference Proceedings (07 Urban Design) [C]. China Urban Planning Association, Dongguan Municipal People's Government: China Urban Planning Association, 9.

Zhang, C.Y., Xie. K, 2017. Discussion on Urban Design Optimization Based on "Narrow Road Closed Network" Planning Model, *Urbanism*, 10, 114-117.

Zhang, J., Meng, Z. Z., 2011. A Comparative Study of The City of Ping and Luoyang in Beiwei Dynasty (Beiwei guo du ping cheng he luo yang cheng de dui bi). *Journal of Sanmenxia Polytechnic,* 10, 75-81.

Zhu, J., 2004. *Chinese Spatial Strategies Imperial Beijing 1420-1911,* Curzon, Routledge.

Zou, Y., & Bian, H.B., 2000. Comments on urban residential district pattern in China, Available at: http://wenku.baidu.com/view/36b07a04e87101f69e3195fc.html

Chapter 4
Adaptative Resilience: Traditional Wisdom of the Mulan Weir Water Conservancy System in Song-Yuan Period

Tianjie Zhang, Qiuyin Xu, and Shuangchen Chen

Abstract The Mulan Weir in southeast China, a World Heritage Irrigation Structure, has been serving for nearly one thousand years. It can adapt to natural disasters such as drought, floods and tides, ensuring regular agricultural irrigation. Through in-depth literature review, this chapter presents a framework for resilience analysis involving ecological, engineering, social and economic subsystems. Following this framework, the research reveals the traditional wisdom of adaptive resilience in planning, construction and management of the Mulan Weir system.

On the basis of archival research and fieldworks, the study finds out the ecosystem of the Mulan River basin was in a relative dynamic balance during Song-Yuan Period. The water conservancy system included three levels, and consisted of weirs, sluice gates and culverts. It subtly improved the local irrigation conditions and was flexible enough to adapt to different disaster scenarios. What's more, local social organizations and folk beliefs promoted self-organization and ensured the construction and maintenance of engineering facilities. The economic strategies further provided long-term momentum.

The research uncovers that the adaptive resilience of this outstanding ancient project depended not only on advanced engineering design, but also on the interaction and application of different subsystems. Active social organizations, engaging cultural atmosphere and flexible economic strategies all contributed to the adaptive resilience. They guaranteed the daily and also long-term maintenance of engineering facilities. Even in the face of disasters, these characteristics ensured faster recovery of the entire system.

Keywords Water conservancy system · Social-ecological system · Resilience · Adaptation · Self-organization

---

T. Zhang (✉) · Q. Xu
School of Architecture, Tianjin University, Tianjin, China
e-mail: arch_tj@126.com

S. Chen
China Academy of Urban Planning & Design, Beijing, China

L. Bian et al. (eds.), *Chinese Urban Planning and Construction*,
Strategies for Sustainability, https://doi.org/10.1007/978-3-030-65562-4_4

4.1 Introduction

In the field of natural disaster management, enhancing resilience in social-ecological systems is considered as a practical path to deal with future uncertainties and risks. Both engineering and non-engineering measures need to be taken seriously (Berkes, 2007; Schneidergruber et al. 2004). There are some resilient survival modes in human history. The technology used by the ancients was relatively simple and could adapt to the dynamic changes of the environment (Kaptijn 2018). As these modes have been tested by history, summarizing them can provide reliable support for modern construction. World Heritage Irrigation Structure—the Mulan Weir, located in Putian, Fujian Province, China, is a case worthy of attention.

Completed in 1083 A.D., the Mulan Weir is a middle scale irrigation project for water retention, tide prevention, flood discharge, sand removal and water diversion (Mohanan 2012). Until now, the weir has served for nearly a thousand years. Moreover, the list of World Heritage Irrigation Structures includes this masterpiece in 2014. Xinghua plain, where the Mulan Weir is located, is closed to the sea in the southeast and surrounded by mountains on other sides. Long-term river alluvial and sea tide toppings formed the marsh here. The Mulan River, with a total length of 105 km, is the longest river on the plain. It flows from west to east, dividing the plain into two parts (Nanyang plain and Beiyang plain). In terms of, Putian, where the weir is located, has a tropical maritime monsoon climate. The rainfall here is abundant with the significant spatio-temporal. Therefore, this region often faces natural disasters such as drought, floods and tides. By the completion of the Mulan Weir, water for agricultural production and domestic life was supplied steadily, completely changing the plain to the main production area of grain and sugarcane in Putian (Fig. 4.1). Hence, this chapter re-examines this thousand-year-old project and tries to reveal its traditional resilience wisdom.

4.2 Method and Materials

The research methods include literature review, on-site fieldworks and archival studies. After reviewing relative monographs and journal papers, this chapter builds up a theoretical framework for multi-dimensional analyses on the adaptive resilience of the Mulan Weir water conservancy system. The materials used include historical documents, local gazetteers, monographs, newspapers, field surveys, interviews, etc. They describe the history of the Mulan Weir, principles of water conservancy, and ecological changes in Song and Yuan Dynasties. Local gazetteers have recorded the details of the construction and later maintenance. Field surveys observe the spatial layout and effect of the entire system. Besides, local reputed historians, anthropologists and archaeologists are interviewed to comprehend the local clan power and water conservancy organizations.

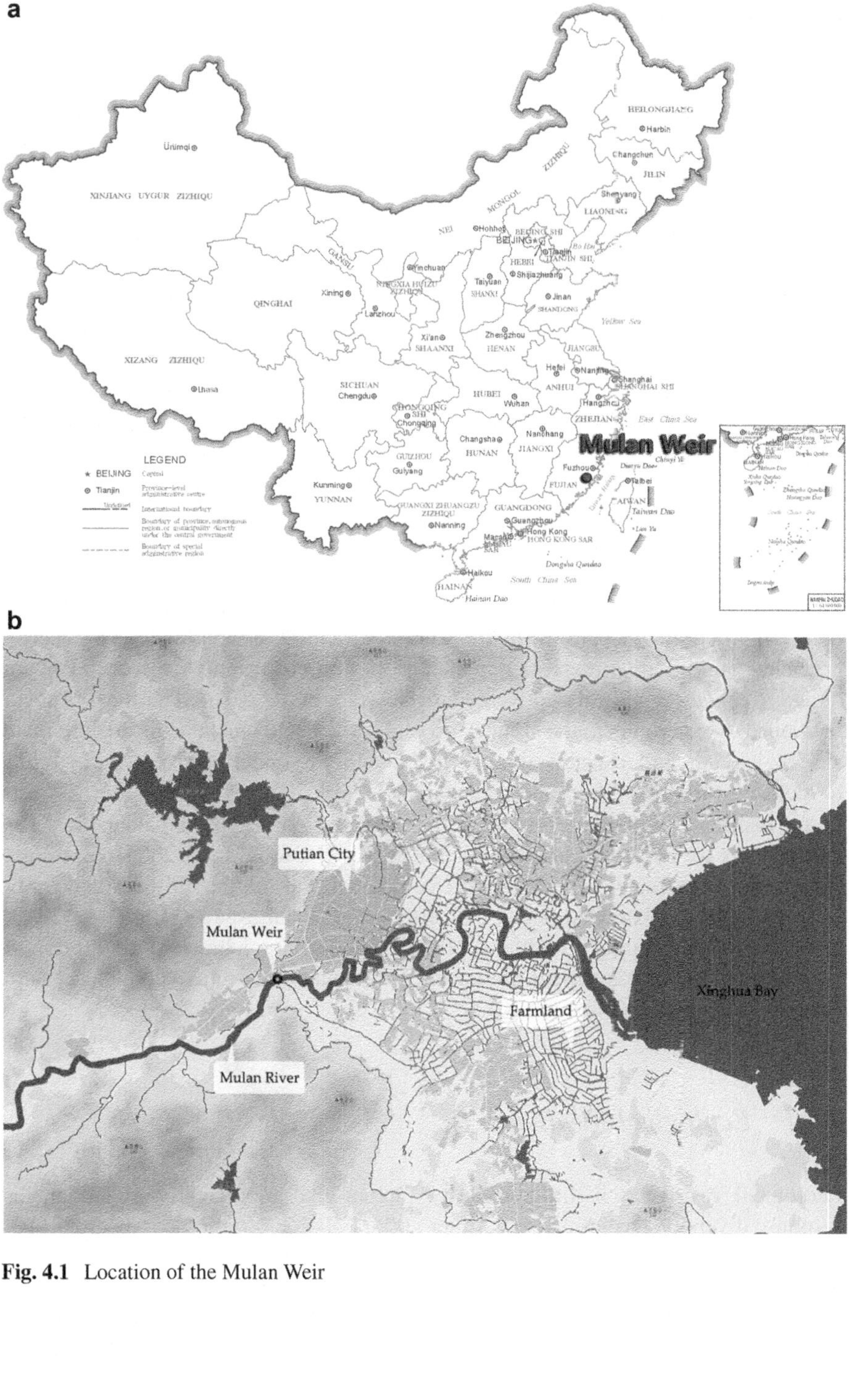

Fig. 4.1 Location of the Mulan Weir

4.3 Literature Review

The management of natural disasters involves the ecological environment, human society and built environment. Hence, it is a coupled system between human and nature, namely a social-ecological system (Adger et al. 2005; Berkes 2007). In this system, the meaning of resilience extends from "keeping the same or maintaining the same system", concepts from the fields of engineering and ecology (Holling 1973, 1996; Walker et al. 2004), to "being able to adapt to internal and external shocks, retaining its structure and processes" (Ernstson et al. 2010). Researchers studying disaster resilience try to consider this kind of intersecting systems more from a human-in-context perspective (Cutter et al. 2008; Ungar 2018). Human systems, environmental systems and the built environment interact to produce antecedent conditions, which contain both inherent vulnerabilities as well as inherent resilience (Susan et al. 2014). Some scholars believe that the realization of resilience comes from the processes of physical and social components as well as the interaction between them (Desouza & Flanery 2013; Godschalk 2015).

In general, the combination of different subsystems produced the resilience of the social-ecological system (Tompkins & Adger 2004). It can be broadly divided into four main areas (Chen et al. 2017). The first concept is ecological resilience, which refers to the extent to which the ecological environment can resolve changes before it is reorganized and formed into new structures. Secondly, the engineering resilience of infrastructure system, which is, the ability to predict, absorb, adapt and quickly recover from a disruptive event (Hosseini et al. 2016). The third is social resilience, which refers to the ability of social entities and mechanisms to anticipate, absorb, and adapt to disasters along with the ability to undertake recovery activities to reduce future disruptions and their impact (Kwok et al. 2016). The fourth is economic resilience, which refers to the ability and adaptive response that enables firms and regions to avoid maximum potential losses (Rose & Liao 2005). Currently, there have been many discussions on ecological and engineering resilience and remains a distinct knowledge gap concerning the social aspects (Davidson 2010). Besides that, developing a multidisciplinary theory that integrates different dimensions into a unified conceptual framework is the major theoretical challenge facing many scholars today (Jabareen 2013).

4.3.1 Ecological Resilience

Ecological resilience is influenced by factors like biodiversity, redundancies, response diversity, spatiality, etc. (Adger 2006; Folke 2006). Some indicators, such as wetland area, erosion rate, impermeable surface, biodiversity, can be used to measure ecological resilience (Cutter et al. 2008). The condition of the environment and the way human beings treat natural resources determine ecological resilience. So, the researchers strongly advocated the concept of sustainability. Unsustainable practices, such as large-scale deforestation and loss of coastal wetlands, may amplify the destructive power of natural disasters.

4.3.2 Engineering Resilience

As suggested by probability theory, a failure event with a zero probability of failure could still occur in engineering practice (Yodo & Wang 2016). Concepts of engineering resilience are used to address failure more systematically. There are two main aspects of engineering resilience. For one thing, the physical properties of the system, such as the quantity, strength, etc. (Cutter et al. 2008). For another, the dependence and interdependence on other infrastructure. Like a disruption in one sector cascades into impacts on other sectors, a high degree of dependence or interdependence reduces resilience(Chang et al. 2007). In all, the properties of resilient infrastructure—robustness, redundancy, resourcefulness, and rapidity—reduce the probability of failures (Bruneau et al. 2003). Minimization of failure, limitation of effects, administrative controls/procedures, flexibility, controllability, and early detection are factors that can enhance engineering resilience (Dinh et al. 2012).

4.3.3 Social Resilience

Social resilience is a concept that can move forward theory and policy on the disaster. Improving social resilience has been highlighted as a critical factor to enhance social-ecological resilience (Aldunce et al. 2015; Aslam Saja et al. 2018). The "5S" framework summarized by Aslam Saja provides a reference for us to analyse and evaluate the social resilience in disaster management. According to this "5S" framework, social resilience mainly includes social structure, social capital, social mechanism, social equity and diversity, and social belief (Aslam Saja et al. 2018). In terms of social structure, one fundamental component of social resilience is resilient households, which can be comprehensively measured through socio-economic status and attainment levels in health and education (Qasim et al. 2015). Greater population mobility can enhance social resilience and improve the effectiveness of disaster response (Adger et al. 2005). All forms of local social capital play a key role in shaping local adaptability (Pelling 2003). As it relates to social mechanisms, the process, in which stakeholders work together to achieve self-empowerment and foster strong relationships, is essential for achieving greater social resilience. Strategies and plans to promote such participation and cooperation are essential elements (Norris et al. 2008). Concerning social equity and diversity, diverse skills and labour are needed in pre- and post-disaster situations. Adequately trained personnel are an essential indicator to evaluate social resilience (Cutter 2015; Parsons et al. 2016). Besides, culture is another separate dimension, and adaptability of many regions are strongly grounded in culture and faith (Kwok et al. 2016; Ostadtaghizadeh et al. 2016).

Many scholars have proposed the concept of "adaptive governance" for the improvement of social resilience (Chaffin et al. 2014; Cosens et al. 2018). At present, natural disasters are not only related to destructive events but also related to gradual changes (Wardekker et al. 2010). Self-organized actions on an appropriate scale are needed to flexibly adjust the impacts of these disasters (Wang & Blackmore 2009). External interventions seeking to enhance resilience need to act on two levels. Firstly, fostering positive local self-organization to confront vulnerability. Furthermore, secondly, supporting institutional frameworks that will not constrain local self-organization (Pelling 2003). Building the attributes of social resilience, such as knowledge, skills and learning, community networks, people-place connections and engaged governance, provides the focus for management actions (Maclean et al. 2013).

4.3.4 Economic Resilience

Economic resilience is determined by industrial structure, labour market, financial arrangements and governance arrangements (Martin & Sunley 2014). The specialization of the industry affords resistance to shocks, while the diversified industrial structure supposedly promotes rapid and sustained recovery from shocks (Evans & Karecha 2014). In terms of the labour market, the ability of a region to attract and retain a skilled and creative workforce is probably one of the critical determinants of economic resilience and growth (Glaeser 2005). In terms of financial arrangements, national financial environment, loan conditions and attitudes of financial institutions, alternative sources of finance and local state support have a profound impact on economic resilience. As for governance arrangements, developing collective and forward-looking strategic positions on economies by local authority is of critical importance (Martin & Sunley 2014).

4.4 Ecological and Engineering Resilience of the Mulan Weir Water Conservancy System

4.4.1 Ecological Resilience

During the Song-Yuan Period (960–1368 A.D.), the environment (including climate, hydrology, vegetation, soil, etc.) of the Mulan River basin maintained a state of relative dynamic balance. On the one hand, though the ecosystem itself was very fragile, it had a specific capacity of self-recovery. The Mulan River basin was adjacent to the sea, with considerable annual precipitation which varied greatly in different time and areas. Such character resulted in disasters, such as periodic tide, flood, drought. In the Song-Yuan period, the Mulan River watercourse had expanded about 10 km to the sea. Compared with the straight river course, the natural meandering river had the functions of expanding water storage capacity, slowing down the impact of sea tide, and had higher redundancy.

On the other hand, the artificial reconstruction of the ecological environment was still within a reasonable range. During the Shaoxi years (1190–1194) in the Song Dynasty, the Xinghua county (now Putian, Fujian Province) had 72,353 households with a total population of about 360,000 (the current resident population is 2.9 million) (Lin 2012). That is 86 people per square kilometre. Even though the artificial agricultural reclamation was enhanced with the increase of population, on the whole, the ecosystem still had excellent functions of self-recovery and dynamic balance because of the low development intensity (Zhang 2015). According to the records, in Fujian areas, "mountains and forests are precipitous, stretching for thousands of miles" until the late Northern Song Dynasty, which indicated that the primaeval forest of most mountainous areas and some hilly areas were well preserved (Wang & Zhang 1999). The water of rivers and lakes had not been pollutional yet. At the same time, a particular understanding of the relationship between water resources, vegetation and soil erosion had been developed. Putian ancestors had noted that the reduction of vegetation in the upper reaches would lead to a large amount of soil erosion and increase the risks of floods. Thus, in the upper reaches of the Mulan River, trees were planted to conserve water and soil. And reclamation of mountainous areas was carried out in form of terraced fields.

4.4.2 Engineering Resilience

The Xinghua plain had experienced a long period of sediment deposition and artificial reclamation. The process could be roughly divided into four stages (Zheng, 2018). In the early days, mountains linked with the sea. Saltwater spread all over, making limited flat land unfit for farming (Fig. 4.2a). In the Tang Dynasty, people began to build ponds and reclaimed land from the sea (Fig. 4.2b). In the Song Dynasty, the residence dug ditches and built weirs, thus initially forming the pattern of the Xinghua plain (Fig. 4.2c). In the Yuan Dynasty, the ditch system was further enhanced, and significant water systems were connected (Fig. 4.2d). The construction of the Mulan Weir was accomplished in 1083. The functional components and overall layout are the core of the engineering resilience of the Mulan Weir protection system.

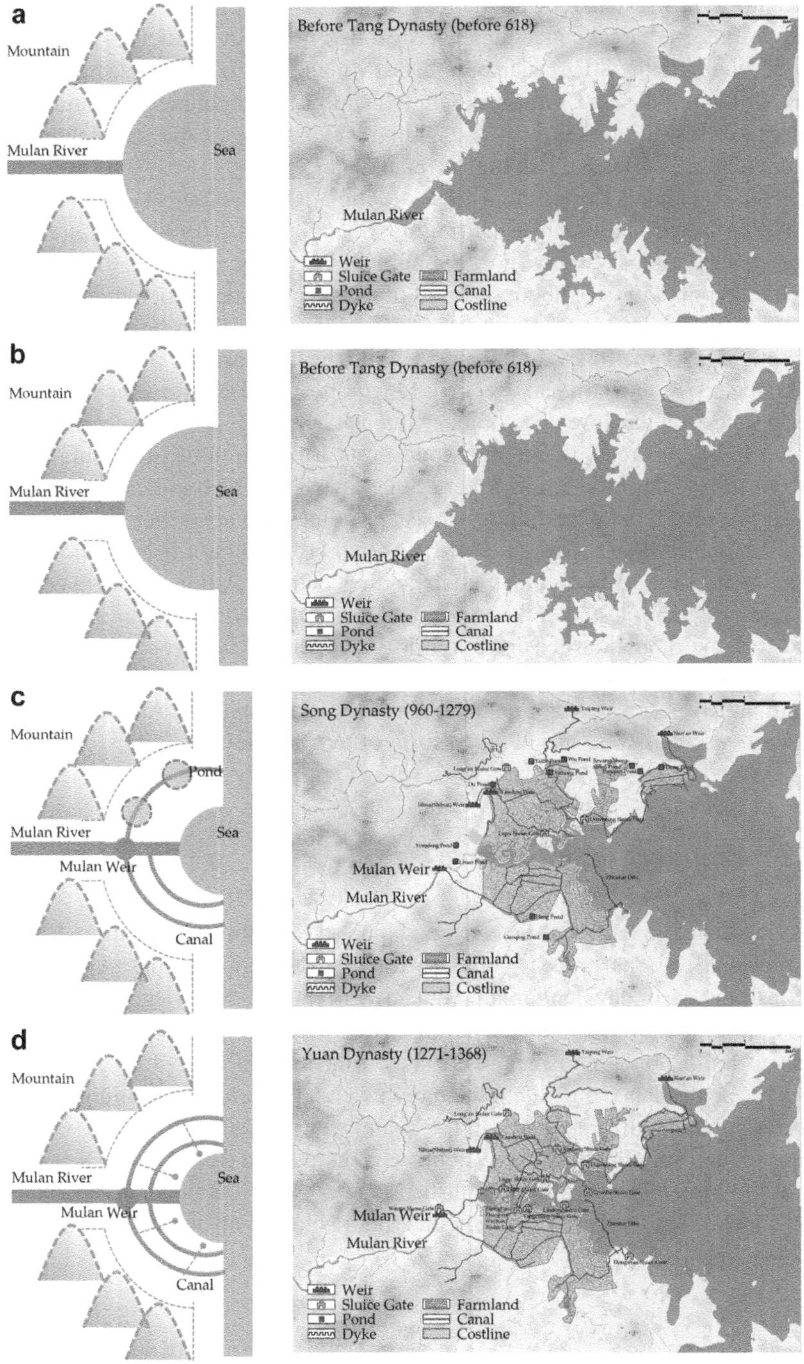

Fig. 4.2 The historical changes of the Mulan River basin: (**a**) Before Tang Dynasty (before 618); (**b**) Tang Dynasty (618–907); (**c**) Song Dynasty (960–1279); (**d**) Yuan Dynasty (1271–1368)

Fig. 4.3 Some basic components: (**a**) Weir; (**b**) Sluice gate; (**c**) Culvert

4.4.2.1 Major Functional Components

As far as the Mulan Weir Water Conservancy System was concerned, the weir was the core component. Also, there were sluice gates, culverts, ditches, ponds, dikes, etc. All of these components worked together.

1. The weir (Fig. 4.3a) was the backwater building above the river. It had the functions of both water storage and water diversion. Altogether there were 32 wooden gates in the Mulan Weir. The residents could open and close the gates as barriers against the sea tide.
2. The sluice gate (Fig. 4.3b) was the building set on a riverbed or the bank of a river or lake. It had the functions of moderating water, discharging flood, resisting the tide, sluicing sand, and controlling navigation.
3. The culvert (Fig. 4.3c) was overflow structure buried under the soil (mainly dike). The culvert was more straightforward than the sluice gate. It was the terminal of the water conservancy system which could make up for the functions that the sluice gate did not possess.
4. The ditch was utilized to connect water conservancy facilities, and there were canals and ditches.
5. The pond was to hold and collect water.
6. The dike was built along the coast. It was utilized to intercept streams inside and resist the sea tide.

It could be observed that there were various water conservancy components with flexible and changeable functions, thus guaranteeing the applicability in the disaster.

4.4.2.2 General Layout

During the Song and Yuan dynasties, the Mulan Weir Water Conservancy System has formed a relatively complete three-level water control system, which includes weirs, sluice gates and culverts (Fig. 4.4). The Mulan Weir was built on the site that featured by broad water surface, slow flow and stable matrix in Mulan River. The "raft foundation" technology used ensures the firmness and durability of the project (Mao 1986). Protective dike, more than 500-m long, was constructed in the south and north side of the weir. On the south and north banks, Huilan Bridge and Wanjin

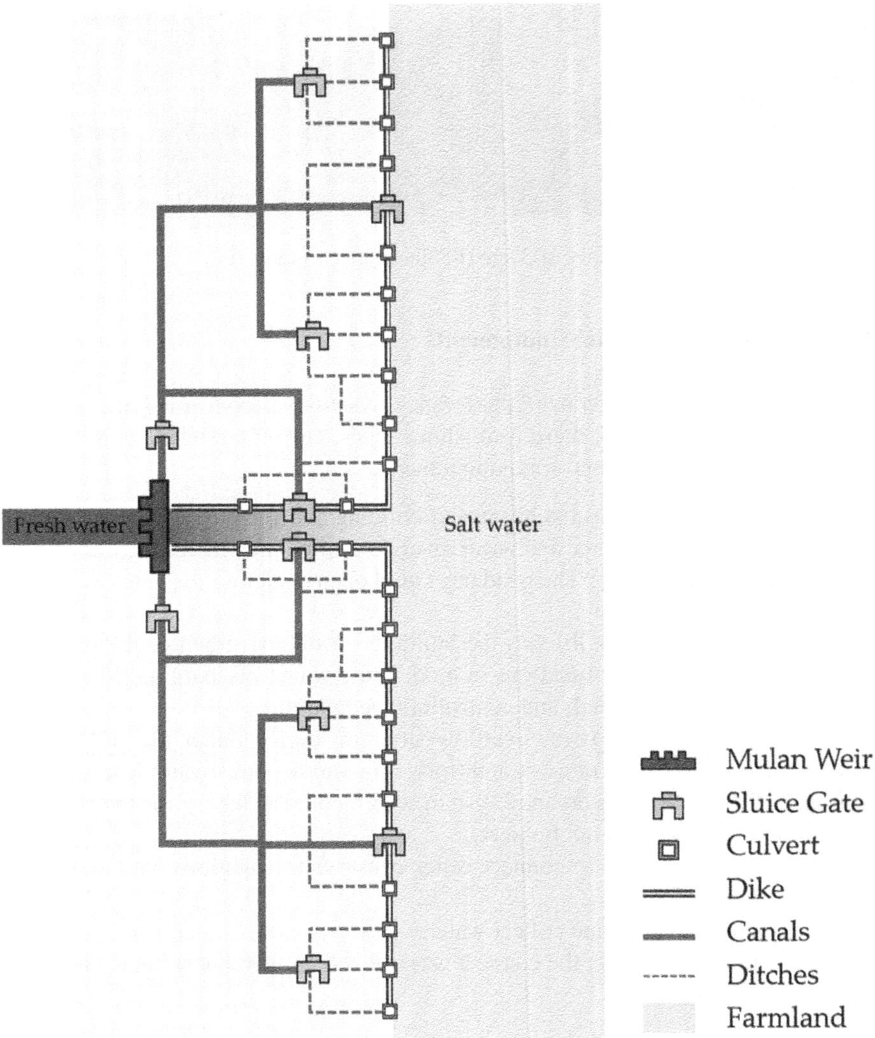

Fig. 4.4 Schematic diagram of the three-level water conservancy system

Sluice Gate were built respectively to divert water into the ditches on both sides of the plain. According to *Puyang Mulan Water Conservancy Records*, after the completion of the Mulan Weir, the local wealthy families donated fields to make ditches. Huilan Bridge was built on the south bank of the weir and ditches were dug straight down the bank to divert water. There were 7 canals and 109 ditches, which were divided into upper, middle and lower sections. Over ten thousand hectares of land was irrigated, and the annual output for army provision was thirty-seven thousand hu (1300 ton). Since then, canals and ditches were built successively, developing vast and crisscrossing canal networks. The total capacity of the canals and ditches exceeded 30 million cubic metres (Putian County Cultural Museum in Fujian Province 1978). The main canals were more than 120 kilometres long, and along with them, there were more than 300 sluice gates and culverts. As *the Mulan Water Conservancy Records* in the Song Dynasty goes, "Water conservancy system is like a person's body, the weir is the throat, and the canals and ditches are the intestines and stomach. Water enters the body through the throat and is received by the intestines and stomach. However, if there was no rump to discharge, the body will swell" (Fang Song Dynasty (n.d.)). Therefore, the end of the canals and ditches was connected to the Mulan River and the sea, where dikes were built for partition and sluice gates and culverts for discharging water. There were a large number of culverts, of which a small part was built in the internal dikes (to the Mulan River) and the vast majority in the outer dikes (to the sea).

4.4.2.3 Operation Principles

In addition to meeting regular agricultural irrigation, the Mulan Weir Water Conservancy System could also reduce drought, drain flood and resist the sea tide. Freshwater from the upper reaches of the Mulan River is separated by a ratio of three to seven in the Mulan Weir. Through Huilan Bridge and Wanjin Sluice Gate respectively, 30% of the water was diverted into the Beiyang Plain and 70% into the Nanyang Plain (Zheng 2004). Then through the ditches and sluice gates, freshwater finally flowed into the cultivated land. The sluice gates and culverts near the shoreline remained closed to hold streams, enabling the saline soil to be gradually cultivated (Fig. 4.5a). In the case of drought (Fig. 4.5b), the gates on the weir as well as

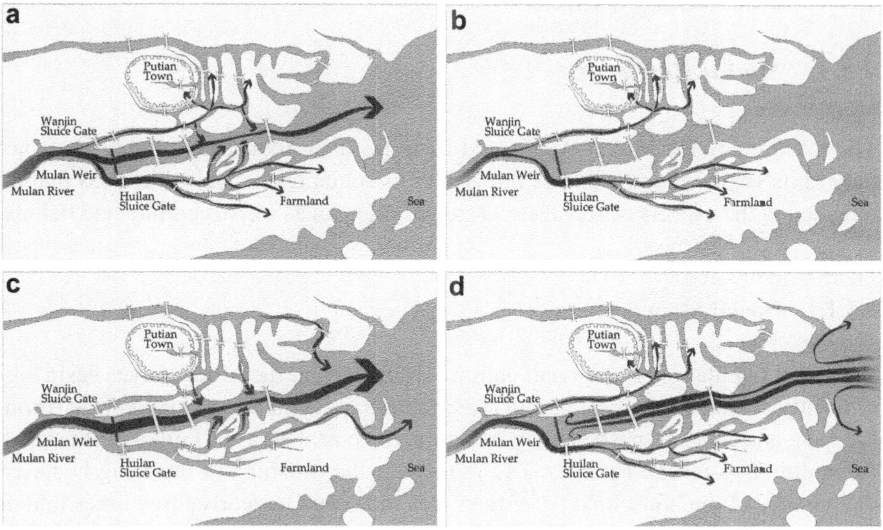

Fig. 4.5 Diagram of the operation of the Mulan Weir Water Conservancy System: (**a**) Normal irrigation; (**b**) During droughts; (**c**) During flood; (**d**) During the tidal attack

the culverts and sluice gates in the coastal area could be closed to hold freshwater in ditches and ponds.

In the case of flood (Fig. 4.5c), the change in water potential could be observed according to *Shuize* (a gauge to measure the water level) on the weir. If it was a small flood, water could directly fly over the weir into the sea. In case of a massive flood, it would be necessary to open the gates of the weir. The upstream water level could lower 0.39 m if all the gates were open (Putian County Cultural Museum in Fujian Province 1978). The sluice gates and culverts at the dikes could also open to help drain the flood in the hinterland.

When the sea tide attacked (Fig. 4.5d), the dikes on both sides of the weir were equivalent to mitigation zones, which could reduce the direct impact of the wave on the weir. The precise height design of the Mulan Weir ensured that the tide could not flow over the weir. At this time, the sluice gates and culverts close to the shore kept closed to prevent the seawater from flowing back into the cultivated farmlands.

Persisting in trying, experience in large and small water disasters improved people's engineering technology and management skills. The water conservancy technology gradually evolved, from resisting water mechanically to diverting water flexibly, from scattered technology to systematic technology. Only with the cooperation of the weir, sluice gates, culverts, ditches, ponds and dikes could the crisis be effectively resolved. At the same time, the redundancy of the construction ensured that, even if the disaster destroyed some structure, other structure could continue to be utilized. Thus, the entire hinterland would not be subject to the threat.

4.5 Social and Economic Resilience of the Mulan Weir Water Conservancy System

4.5.1 Social Resilience

The reason why the Mulan Weir could be utilized for thousands of years lays not only in its engineering resilience but also in its social resilience, which was mainly reflected in the aspects of social structure, mechanism as well as culture and beliefs.

4.5.1.1 Social Structure

The rapid population growth and cultural exchanges in the Mulan River basin significantly promoted the progress of water conservancy in Song-Yuan Period. From the Tang and Song Dynasties, the population of coastal areas of Fujian proliferated. According to *Sanshan Records*, 130 years after the founding of the Song Dynasty, the number of the households of Fujian area increased to nearly three times that of the founding of the Song Dynasty. Moreover, the population of the Mulan River basin was also increasing. The growing population included not only the local population but also some families newly moved in due to the war in north China. It could be said that various cultural circles were mingled (Clark, 2010). The more advanced technology and knowledge of water conservancy from the north were disseminated and even further developed. The construction of water conservancy facilities became extremely active. Three hundred twenty-four water conservancy projects were built in the Fujian area, exceeding the scale in the most prosperous Jiangsu and Zhejiang areas in the same period. Many of the remaining water conservancy projects, including the Mulan Weir, were built during this period (Hong & Zheng 1985).

Clans were the leading organizations in water conservancy projects and guaranteed abundant social capital. Large water conservancy projects were often constructed by several prominent families jointly. Also, each large family carried out land partition and reclamation after the completion of the projects. The community of the clan was further strengthened under the influence of Confucianism in the Song Dynasty. Without exception, every individual was included in the clan genealogy. Sacrificial activities were regularly held, so that descendants inherited the beliefs and spirits from ancestors. Meanwhile, the clans also had complete family education system, including the clan schools and family instruction, which not only cultivated family talents but also set up normative and restrictive behaviours. What's more, there was a system of public property to support the development of the clans, which explains why the clans could participate in the construction of the abundant water conservancy projects.

4.5.1.2 Social Mechanism

During the Song and Yuan Dynasties, especially after the political reform of Anshi Wang (1069–1085 A.D.), the central government attached great importance to water conservancy and inspired civil forces to participate. A series of water laws and regulations formed a strict water control system, from the central government to the local areas, from officials to the civilians. *Eight Regulations of Water Conservancy Interest* (1024 A.D.) covered the management and technical aspects of water conservancy: the beneficiary farmers would equally share the human resources required for the project; officials who failed to fulfil their duties of controlling water would be severely punished; the local elites were inspired to build small facilities on their own; the standards for construction dimensions, approaches as well as the calibration criteria of the water conservancy facilities were provided (Zhang 2010). In *Farmland Interest Treaty* (1069 A.D.), officials and civilians were encouraged to submit to the government-specific proposals for building farmland water conservancy; talents would be employed and rewarded; county officials who introduced new laws to build water conservancy projects would be rewarded according to their merits; lending policies were established to support water conservancy construction, and a large number of officials were sent to various places to inspect and supervise. The government "paid high salaries to attract participation in water conservancy construction. Thus civilians participated with pleasure" (Qi 2017). "The government offered capital, and the civilians offered manpower, and the two parties both gained benefits". The local elites, as the master of civil discourse and the main leader of the civil affairs, and ordinary civilians, as the direct participants and chief undertakers of the civil affairs, all participated in the initiation, fundraising, project management and even construction of the water conservancy.

In this context, the construction and management of the water conservancy projects in the Mulan River basin received joint participation of local governments and local people. A pattern of specialized maintenance work took shape, in which officials were in charge of supervision and civilians were responsible for the specific operation.

1. Generally, there were 8 to 10 gate-keepers in each sluice gate, responsible for guarding, opening and closing the sluice gates. In the core sluice gates, such as the Yangcheng Sluice Gate, there were also odd-job men from the local government. Gate-keepers and odd-job men lived near the sluice gate at night to guard the sluice gates.
2. As to the culverts, there was no definite literature on whether there was specialized personnel to guard. It was recorded in the existing material that, if the culvert built by the villagers destroyed the whole structure, the local officials would demolish it. It indicated that the beneficiary villagers themselves mainly managed the culverts, but the local government had some supervision. Nor did it exclude that some particularly necessary culverts were managed directly by the government.
3. The dikes were divided into several parts according to Jia (a unit of civil administration), and there were ten households each Jia specializing in the management of dikes. Also, a certain amount of land revenue was set aside for the maintenance of dikes. If the dikes broke, the local officials would organize the repair work (Chen 2015).

The maintenance of the Mulan Weir was the most complex. The "System of Dedicated Fields" was formed to solve problems like funds and the division of duties. After Siniang Qian and Congshi Lin failing to build the weir, official Hong Li was dispatched by the imperial court to construct the weir. His was not only assisted by Zhi Feng, a monk with knowledge of water conservancy, but also by 14 large local clans. What is more, after the completion of the weir, these 14 clans contributed to building a bridge through which water was diverted to flow eastward. Also, they dig more than 4995 mu of their farmlands as ditches (Xie 1981). After that, large ponds were gradually phased out. Li, together with the 14 clans, organized human resources to fill up the other five ponds except for Guoqing pond. A total of 2665 mu of land was reclaimed, among which 1325 mu were dedicated to supporting the management and maintenance of the Mulan Weir and the related facilities. These farmlands of 1325 mu were collectively referred to as Dedicated Fields, which were divided explicitly into weir construction fields, tobacco fields, foremen fields, coolie fields, ect. Proceeds from these fields are retained for exclusive use only. Shisheng Zhan, a local official, instituted this custom in *Regulations*

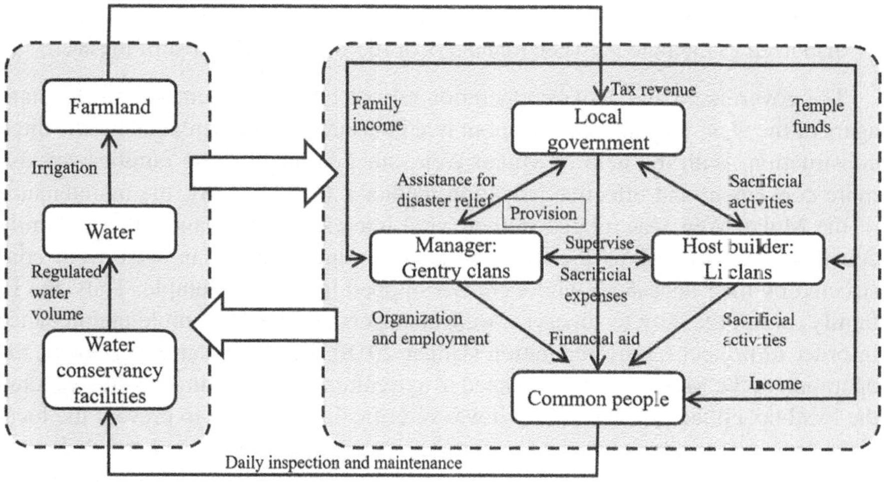

Fig. 4.6 Manage the mechanism of the Mulan Weir

of the Weir Management in 1119 (He & Hui 2018). Later generations adjusted according to the situation, and it was still utilizes until the early Qing Dynasty.

The managers of the weir, common people, the host of the weir and the local government were the main participants in the management of the Mulan Weir (Fig. 4.6).

1. As to the managers of the weir, there were a chief manager and a deputy chief manager and were rotated among the local 14 gentry families year by year. The managers of the weir were in charge of running the dedicated fields and managing the proceeds of the fields. They organized the local civilians to conduct routine inspection and repair of the water conservancy facilities regularly. During disasters, they assisted the local government in relief.
2. The routine management personnel included one foreman, two sailors and eight to nine coolies. They were responsible for patrol supervision, the opening and closing of the sluice gates during the flood, the routine maintenance and the opening and closing of the sluice gates respectively. The local farmers offered human resources and got paid from the proceeds of the dedicated fields. In the disaster, they served as the leading force participating in disaster relief (Lei & Zheng 2015).
3. The Li family, who presided over the construction of the Mulan Weir, was in full charge of the host of the weir. They were in charge of the incense of the temple and the folk sacrifices in spring and autumn, managing the proceeds of the tobacco fields, and supervising the work of the weir managers.
4. The local government was responsible for taking custody of the proceeds of the weir construction fields to apply to the overhaul of the Mulan Weir, checking out the accounts of the managers of the weir periodically, sending officials to supervise the progress of the water conservancy. During disasters, it organized disaster

relief and assigned specific tasks to managers of the weir. The local government also need to mediate between managers of the weir and the host of the weir.

The government and civil's cooperation rely on common interests. For the managers of the weir, they abandoned about five thousand mu of farmlands for the ditch construction. With the help of Mulan Weir, other farms of these families can use more convenient and effective irrigation methods. Consequently, the maintenance of the Mulan Weir was related to their vital interests. For instance, the Li family considered themselves as the host of the weir, since the Mulan Weir were constructed by their ancestors, who were worshipped in the weir temple. Thus, the Li family owned the duty to supervise weir managers and family temple maintenance in order to protect family reputation (Ungar 2018). For the local government, the operation of the Mulan Weir was related to agricultural production, and thus affected the local tax collection. Therefore, it was of critical importance to prevent the local clans from encroaching on the interests of water conservancy and the dedicated fields of the Mulan Weir. Many large governmental water conservancy projects during the same period were financed by the government to construct through corves or employing civilian workers. Compared with that, this pattern, in which the officials were in charge of the supervision while the civilians conducted the construction, gained high efficiency and little corruption. Therefore, it has dramatically advanced in that era.

4.5.1.3 Social Culture and Beliefs

In the Mulan River basin, local culture praised the spirit of heroes in water conservancy construction was formed, enhancing people's awareness of protecting water conservancy facilities. Primary cultural forms consisted of monumental buildings, sacrificial activities, biographies and inscriptions, folk arts, etc. (Table 4.1) (He & Hui 2018). The weir construction heroes became the object of universal commemoration and praise. The most representative one was the "Qian Concubine Culture" which commemorated the first heroine in weir construction. During the Song and Yuan Dynasties, Siniang Qian was shaped by the Putian people into a local goddess as famous as Mastu (goddess of the sea). A large number of inscriptions, biographies and even little ditties sang the praise of the heroes of weir construction. In such a cultural atmosphere, every local official would tour the Mulan Weir, visit the relics of the ancestors and read the inscriptions on both banks. Thus, they attached great importance to the maintenance work of the Mulan Weir, and coordinated and supervised the work of the managers and the host of the weir. The local people were grateful to the heroes of weir construction, and they followed the "System of Dedicated Fields", actively offered power during the disasters and completed the maintenance work.

Meanwhile, related biographies and inscriptions contained much information about the process of construction and technical measures. For instance, the process and the technology of constructing the Mulan Weir were recorded in Puyang Mulan

Table 4.1 Local water conservancy culture and beliefs

Form	Main content
Monumental building	Xieying Temple, Huiji Temple, Xiangshan Palace, Chongxun Temple; Sigong Pavilion, Persuasion Pavilion, Jiansi Pavilion, Meifeng Pavilion, Renmei Pavilion, Zhenghe Pavilion, Shengxian Pavilion
Sacrificial activities	Offering sacrifices in the Spring and Autumn Festival; Salute ceremony; Mourning.
Biographies and inscriptions,	Puyang Mulan Water Conservancy Records (presided over by Zhan Shisheng in Song Dynasty), Mulan Water Conservancy Records (Fang Tianruo, Song Dynasty), Rebuilding the Mulan Weir (Zheng Qiao, Song Dynasty), Rebuilding the North and South Shores of the Mulan Weir (Chen Shichu, Song Dynasty), Rebuilding the Mulan Weir (Liu Lirong, Yuan Dynasty), Record of Wanjin Sluice Gate (Jin Ruling, Yuan Dynasty), Repair of Nanyang Seawall in Yuan Dynasty (Yu Qianyi, Yuan Dynasty); Think of Qian Siniang (Wu Shugao, Song Dynasty), Biography of the Elder Li (Lin Daying, Song Dynasty);
Folk arts	"Bang drum bang" (now known as "li song")(Li 2013).

Water Conservancy Records. "The upper and lower weir was constructed first to block the stream and the sea, then dug the bottom of the river. An area of three 'zhang' and five 'chi' ('zhang' and 'chi' were both units of length in ancient China; one 'zhang' was equal to ten 'chi', and three 'chi' was equal to one meter) in-depth and thirty-five 'chi' in both length and width was filled with stones as the base". A wealth of information about the construction of the Mulan Weir was retained. The literature and historical information enabled the descendants to fully understand the material and approaches adopted in the original construction of the Mulan Weir, the problems encountered in the construction process, the causes of all the previous damages as well as the improvement conducted in the repairs. All this information provided significant reference value for the subsequent maintenance and protection work related to the Mulan Weir.

4.5.1.4 Summary

The resilient attributes of the Mulan Weir, including knowledge and skills, community networks, human–land relationship, management participation, were strengthened through a variety of means. They are as follows.

The first was the support of the national policies. On the one hand, macro-policy strongly encourages non-governmental forces to participate in the design and construction of water conservancy projects with particular rewards and forms a collaborative approach to participate in governance (Knight et al. 2002). On the other hand, scientific engineering technology has been widely promoted, and the learning of relevant knowledge and skills has enhanced the ability of individuals and groups to solve local needs and problems (Cuthill & Fien 2005).

The second was leadership. The local officials directed the construction and overhaul of the Mulan Weir. For instance, in the twenty-eighth year during the Shaoxing years in the Song Dynasty (1158), "water deviated its original path, flowing straight to Chongyuan (name of a county) from the north to the east". Yuansu Feng, the county magistrate, organized the repair work of the Mulan Weir, and "worked day and night for 90 days" under numerous difficulties and dangers. In the eighth year in the Yuan Dynasty (1218 A.D.), Darughachi Bahadiya (an official appellation in the Yuan Dynasty) organized civilian workers and "added fold stones, and set up five permanent pillars". In the twenty-eighth year in the Yuan Dynasty (1291 A.D.), Xiaosi Zhang, an investigation official, organized the repair work of the Nanyang seawall. Local officials had authority in the locality, and the elites were the core of the family network. Close social networks enable vital decision-makers to form alliances and take collective action in times of crisis (Carl 2003).

Thirdly, systematic management. During the Song and Yuan Dynasties, large and small civil water conservancy projects established actively, while the problem of water usage frequently led to conflicts among villages and prominent families. Some private construction behaviours damaged the overall functions of the hydraulic system. The local governments intervened in the construction to some extent, integrated all water resources and prohibited improper construction behaviour. Thus, the civilians gained the capability of self-organization. Moreover, they were also required to follow some norms. Considering the interdependence between human and environment (Dale et al. 2008), integrated and holistic management improved the ability to monitor and respond to environmental feedback(Cuthill & Fien 2005). Every disaster experience created opportunities for the system to adjust its internal structure, process and build relevant knowledge, and gradually formed diversified coping strategies (Folke 2006; Smit & Wandel 2006).

Fourthly, effective supervision mechanism. The local civilians themselves conducted the routine operations of the water conservancy projects, and the local government would carry out some supervision. For another example, the managers of the weir were in sole charge of the routine inspection and maintenance of the Mulan Weir. Nonetheless, they needed to be supervised by the local government and the host of the weir. Proceeds of the weir construction fields should be stored in the local government in time and withdrawn when necessary. The real participation of relevant stakeholders in public and community sectors in the whole process can effectively solve the problem.

4.5.2 Economic Resilience

The economic resilience of the Mulan Weir Water Conservancy System was reflected in diversified sources of funding for construction and maintenance. In addition to government funding, various forms were adopted to solve the problem of insufficient funding, such as sharing equally among households, encouraging wealthy households to contribute funds and low-interest or interest-free loans from the government (Ma 2005). During the construction of the Mulan Weir, there was Siniang Qian, who donated and raised one hundred thousand taels of silver (about 4.3 million dollars) and promoted the very first attempt to construct the Mulan Weir. Then, 14 large local families contributed funds of seven hundred thousand min (about 30 million dollars), which enabled the accomplishment of the Mulan Weir and Huilan bridge. In order to provide sufficient financial support for the future maintenance of the Mulan Weir, Dedicated Fields, including weir construction fields, tobacco fields, foremen fields, coolie fields, were designated to support its routine maintenance. The annual rental income was 1325 dan grain (dan is a unit of weight, one dan is about 70 kg). According to the records in *Regulations of the Weir Management*, the annual income of the weir construction fields was 330.4 dan. Among them, 105 dan belonged to the managers of the weir, 12 dan was regarded as the payment for the coolies. The rest income, about 213.4 dan and 613 guans (a string of coins, the total was about 35 thousand dollars) were under the custody of the government, which would be exclusively used for significant project maintenance in the future. Besides, after its completion, the Mulan Weir not only provided water for agricultural production and family life but also promoted enormously the development of shipping, aquaculture and other industries. Thus, it was enriching the types of local economic revenues, as well as providing indirect financial support for engineering construction and management maintenance.

4.6 Conclusion & Discussion

The resilience of Mulan Weir Water Conservancy System lies in the interaction and adaption of ecological, engineering, social and economic subsystems. The ecosystem was the foundation, provided necessary development space and physical basis for social life and economic production. However, the ecosystem itself was fragile. Reasonable and ingenious engineering system construction maintained the dynamic balance of the ecosystem and increases the adaptability to natural disasters. The fundamental components of Mulan Weir Water Conservancy System were set along with the topography with advanced construction technology to ensure the anti-disturbance. Each component was simple to operate and versatile. All components coordinate with each other and adjust flexibly in different disaster scenarios to meet different needs. In the process of ecological environment transformation, the ancestors of the Mulan River basin did not carry out destructive construction. The original forest vegetation in the upper reaches of the river was maintained, and the water network in the lower reaches became more abundant than the previous. Overall, the resilience of the ecological environment was strengthened, and the ability of the social-ecological system to adapt to uncertainties became strong.

In the social system, the social structure characterized by cultural integration and high clan power, and the social mechanism that encouraged the participation of civilian forces provided knowledge, skills, financial resources and human resources for the construction of the engineering system. Local self-organization of engineering system management had emerged for common agricultural interest in such social context. The local culture centring on the praise of water conservancy heroes strengthened the moral constraint of local self-organization. The elites of clans were in charge of the daily use and maintenance of water conservancy projects. They organized ordinary people to carry out the specific work of gate opening and closing, inspection and maintenance. Meaningfully, the local self-organizing behaviour was carried out in the macro-control and supervision of the government, which prevented encroachment on the overall interests and corruption within the self-organization.

The economic system was the driving force for the construction of the engineering system and the self-organization of social system. The construction of water conservancy projects improved the production conditions, created plentiful economic benefits and diversified industrial types. Reasonable fund arrangement and liberal loan policy provided financial support for further construction and maintenance, forming a virtuous circle of regional development. Safeguarding one's economic interests were the basis for the development of self-organization of water conservancy. The arrangements of funds represented by "System of Dedicated Fields" reasonably determined the distribution of interests in the social system. The funds not only included the individual labour remuneration but also included special funds for repairing engineering facilities during future catastrophes and supporting related cultural facilities and activities. The self-organization, both economically and socially, contributed to the adaptive resilience of the Mulan Weir and guaranteed its operation and maintenance throughout years.

Acknowledgments This research was funded by the National Natural Science Foundation of China (No. 51778403).

References

Adger, W. N., 2006. Vulnerability. *Global Environmental Change,* 16(3), 268-281.

Adger, W. N. et al., 2005. Social-ecological resilience to coastal disasters. *Science,* 309(5737), 1036-1039.

Aldunce, P. et al., 2015. Resilience for disaster risk management in a changing climate: Practitioners' frames and practices. *Global Environmental Change,* 30, 1-11.

Aslam Saja, A. M. et al., 2018. An inclusive and adaptive framework for measuring social resilience to disasters. *International Journal of Disaster Risk Reduction,* 28, 862-873.

Berkes, F., 2007. Understanding uncertainty and reducing vulnerability: Lessons from resilience thinking. *Nat Hazards,* 41, 283–295.

Bruneau, M. et al., 2003. A framework to quantitatively assess and enhance the seismic resilience of communities. *Earthquake Spectra,* 19(4), 733-752.

Carl, F., 2003. Freshwater for resilience: A shift in thinking. *Philosophical Transactions: Biological Sciences,* 358(1440), 2027-2036.

Chaffin, B. C., Gosnell, H. & Cosens, B. A., 2014. A decade of adaptive governance scholarship: Synthesis and future directions. *Ecology and Society,* 19(3), 56.

Chang, S. E. et al., 2007. Infrastructure failure interdependencies in extreme events: Power outage consequences in the 1998 ice storm. *Natural Hazards,* 41(2), 337-358.

Chen, C., 2015. Records of water conservancy in Putian In Editorial Committee of Chinese Water History Classics (Ed.), *Chinese history of water conservancy (Taihu lake and southeast area)* (Vol. 2). Beijing: China Water Power Press.

Chen, L., Zhu, X. & Sun, J., 2017. The basic concept, mechanism and planning ideas of resilient cities. *Modern Urban Research,* 09, 18-24.

Clark, H. R., 2010. Portrait of a community: Society, culture, and the structures of kinship in the mulan river valley (Fujian) from the late tang through the song (review). *China Review International*, 17.

Cosens, B. A., Gunderson, L. & Chaffin, B. C., 2018. Introduction to the special feature practicing panarchy: Assessing legal flexibility, ecological resilience, and adaptive governance in regional water systems experiencing rapid environmental change. *Ecology and Society,* 23(1), 4.

Cuthill, M. & Fien, J., 2005. Capacity building: Facilitating citizen participation in local governance. *Australian Journal of Public Administration,* 64(4), 63-80.

Cutter, S. L., 2015. The landscape of disaster resilience indicators in the USA. *Natural Hazards,* 80(2), 741-758.

Cutter, S. L. et al., 2008. A place-based model for understanding community resilience to natural disasters. *Global Environmental Change,* 18(4), 598-606.

Dale, A., Ling, C. & Newman, L., 2008. Does place matter? Sustainable community development in three Canadian communities. *Ethics, Place & Environment,* 11(3), 267-281.

Davidson, D. J., 2010. The applicability of the concept of resilience to social systems: Some sources of optimism and nagging doubts. *Society & Natural Resources,* 23(12), 1135-1149.

Desouza, K. C. & Flanery, T. H., 2013. Designing, planning, and managing resilient cities: A conceptual framework. *Cities,* 35, 89-99.

Dinh, L. T. T. et al., 2012. Resilience engineering of industrial processes: Principles and contributing factors. *Journal of Loss Prevention in the Process Industries,* 25(2), 233-241.

Ernstson, H. et al., 2010. Urban transitions: On urban resilience and human-dominated ecosystems. *Ambio,* 39(8), 531-545.

Evans, R. & Karecha, J., 2014. Staying on top: Why is Munich so resilient and successful? *European Planning Studies,* 22(6), 1259-1279.

Fang T., Song Dynasty (n.d.). Mulan water conservancy records (Mulan Shuiliji).

Folke, C., 2006. Resilience: The emergence of a perspective for social–ecological systems analyses. *Global Environmental Change,* 16(3), 253-267.

Glaeser, E., 2005. Reinventing Boston: 1630–2003. *Journal of Economic Geography,* 5(2), 119-153.

Godschalk, D. R., 2015. Urban hazard mitigation: Creating resilient cities. *Urban Planning International,* 4(3), 136-143.

He, Y. & Hui, F., 2018. The features and influences of system of dedicated fields during song and ming dynasties. *Agricultural History of China,* 4.

Holling, C. S., 1973. Resilience and stability of ecological systems. *Annual Review of Ecology Systematics,* 4(4), 1-23.

Holling, C. S., 1996. Engineering resilience versus ecological resilience. *Engineering Within Ecological Constraints*, 31-43.

Hong, Z. & Zheng, X., 1985. The development of agricultural economy in fujian coastal areas in the song dynasty *The Journal of Chinese Social and Economic History,* 4.

Hosseini, S., Barker, K. & Ramirez-Marquez, J. E., 2016. A review of definitions and measures of system resilience. *Reliability Engineering & System Safety,* 145, 47-61.

Jabareen, Y., 2013. Planning the resilient city: Concepts and strategies for coping with climate change and environmental risk. *Cities,* 31(2), 220-229.

Kaptijn, E., 2018. Learning from ancient water management: Archeology's role in modern-day climate change adaptations. *Wiley Interdisciplinary Reviews: Water,* 5(1).

Knight, B., Chigudu, h. & Tandon, r., 2002. *Reviving democracy: Citizens at the heart of governance*. London: Earthscan.

Kwok, A. H. et al., 2016. What is 'social resilience'? Perspectives of disaster researchers, emergency management practitioners, and policymakers in New Zealand. *International Journal of Disaster Risk Reduction,* 19, 197-211.

Lei, Y. & Zheng, Y., 2015. Summary of Mulan Weir (Mulanbeiji Jieyao). In Editorial Committee of Chinese Water History Classics (Ed.), *Chinese history of water conservancy (Taihu lake and southeast area)* (Vol. 2, pp. 475-561). Beijing: China Water Power Press.

Li, K., 2013. The story and culture behind Mulan Weir. *China Water Conservancy News.*

Lin, J., 2012. On the change of the registered population of Xinghua army in song dynasty. Retrieved from http://www.ptwhw.com/?post=4709

Ma, Y., 2005. On labor and fund in the construction of water conservancy during the period of song shen-zong. *Agricultural History of China,* 24(2), 59-66.

Maclean, K., Cuthill, M. & Ross, H., 2013. Six attributes of social resilience. *Journal of Environmental Planning and Management,* 57(1), 144-156.

Mao, Y. (Ed.) 1986. *Technical history of ancient bridges in China*: Beijing Publishing House.

Martin, R. & Sunley, P., 2014. On the notion of regional economic resilience: Conceptualization and explanation. *Journal of Economic Geography,* 15(1), 1-42.

Mohanan, M., 2012 Mulanbei water conservancy project. *Register of icid world heritage irrigation structures (whis).* Retrieved from http://www.icid.org/HisMore.php?ID=2

Norris, F. H. et al., 2008. Community resilience as a metaphor, theory, set of capacities, and strategy for disaster readiness. *Am J Community Psychol,* 41(1-2), 127-150.

Ostadtaghizadeh, A. et al., 2016. Community disaster resilience: A qualitative study on Iranian concepts and indicators. *Natural Hazards.*

Parsons, M. et al., 2016. Top-down assessment of disaster resilience: A conceptual framework using coping and adaptive capacities. *International Journal of Disaster Risk Reduction,* 19, 1-11.

Pelling, M., 2003. *The vulnerability of cities: Natural disaster and social resilience.* London

Putian County Cultural Museum in Fujian Province, 1978. Mulan Weir water conservancy project in the northern song dynasty. *Cultural relics,* 01, 82-87.

Qasim, S. et al., 2016. Community resilience to flood hazards in Khyber Pukhthunkhwa province of Pakistan. *International Journal of Disaster Risk Reduction,* 18, 100-106.

Qi, C., 2017. The folk force of song dynasty and its contribution to the local construction. *The Central Plains Culture Research.*

Rose, A. & Liao, S.-Y., 2005. Modeling regional economic resilience to disasters: A computable general equilibrium analysis of water service disruptions. *Journal of Regional Science,* 45(1), 75–112.

Schneidergruber, M., Cierna, M. & Jones, T., 2004. Living with floods: Achieving ecologically sustainable flood management in Europe. *POLICY BRIEFING.*

Smit, B. & Wandel, J., 2006. Adaptation, adaptive capacity and vulnerability. *Global Environmental Change,* 16(3), 282-292.

Susan, L. et al., 2014. The geographies of community disaster resilience. *Global Environmental Change,* 29(29), 65-77.

Tompkins, E. L. & Adger, W. N., 2004. Does adaptive management of natural resources enhance resilience to climate change? *Ecology and Society,* 9(2).

Ungar, M., 2018. Systemic resilience: Principles and processes for a science of change in contexts of adversity. *Ecology and Society,* 23(4), 34.

Walker, B. et al., 2004. Resilience, adaptability and transformability in social– ecological systems. *Ecology and Society,* 9(2), 5.

Wang, C.-H. & Blackmore, J. M., 2009. Resilience concepts for water resource systems. *Journal of Water Resources Planning and Management.*

Wang, Y. & Zhang, Q., 1999. *China's 5000-year eco-culture.* Wuhan: Central China Normal University Press.

Wardekker, J. A. et al., 2010. Operationalizing a resilience approach to adapting an urban delta to uncertain climate changes. *Technological Forecasting and Social Change,* 77(6), 987-998.

Xie, L., 1981. Stele recording the number of acres of ditches reclaimed by fourteen meritorious officials who abandoned farmland. Retrieved from http://blog.sina.com.cn/s/blog_912ed3b40102wi3c.html

Yodo, N. & Wang, P., 2016. Engineering resilience quantification and system design implications: A literature survey. *Journal of Mechanical Design,* 138(11).

Zhang, B., 2010. *Studying of laws of irrigation and water conservancy in northern song dynasty.* Zhengzhou University, Zhengzhou.

Zhang, Q., 2015. *The history of eco-environment changes in the north and south song dynasty* (Vol. 1). Beijing: Zhong Hua Book Company.

Zheng, L., 2004. *History of water conservancy*. Beijing: China Water Power Press.

Zheng, Z., 2018. Clan and religion in Putian plain——analysis of the inscriptions in xinghua, fujian. Retrieved from http://blog.sina.com.cn/s/blog_5178a6be0102wz42.html#_ftn5

Part II
Transformation of Traditional Values into Urban Planning

Chapter 5
From Concentration to Decentralization: The Spatial Development of Beijing and the Beijing–Tianjin–Hebei Capital Region

Yan Tang and Xiangyi Meng

Abstract The capital region of Beijing–Tianjin–Hebei (BTH) has developed unbalancedly over the past four decades, since Beijing and Tianjin have absorbed most of the regional investments, resources and talents for being the capital or the national central cities. And the emerging "mega-city diseases" of Beijing caused by high densities of urban functions and populations, such as heavy air pollution, water resource shortage, serious traffic jams and continuous environmental deterioration, have pushed the state government to continue exploring a more sustainable approach to promote the eco-friendly development of Beijing, as well as the wider BTH region. The national strategies of decentralizing the non-capital functions of Beijing and promoting the coordinative development of BTH region have been launched. Among them, the decision to construct Tongzhou, the Sub-City Center of Beijing, and the Xiong'an New Area, a new magnetic development center in Hebei province, are the two most significant actions. Based on literature review and policy analysis, this chapter first summarizes the evolution trend from "concentration" to "decentralization" of Beijing and the BTH region during the past decades, and then illustrates the main recent decentralization measures aiming at controlling the population and functions of Beijing from four spatial levels: the BTH region, the Beijing administrative region, the Beijing central city and the community level. The prospect of the decentralization strategy is further discussed in the conclusion and discussion.

Y. Tang
School of Architecture, Tsinghua University, Beijing, China

X. Meng (✉)
Department of Urban Planning, School of Architecture, Tsinghua University, Beijing, P. R. China
e-mail: mengxy14@tsinghua.org.cn

L. Bian et al. (eds.), *Chinese Urban Planning and Construction*,
Strategies for Sustainability, https://doi.org/10.1007/978-3-030-65562-4_5

Keywords Beijing–Tianjin–Hebei capital region · Concentration · Decentralization · Beijing master plan · Xiong'an new area · Sub-city center of Beijing

5.1 Introduction

China's rapid urbanization has created the "Chinese Miracle" through the past four decades. The mega-city regions, centered on mega-cities, have become the main driving force of the high-speed development of China's economy and urbanization. Among them, the Pearl River Delta with Shenzhen and Guangzhou as the core, the Yangtze River Delta with Shanghai as the core, and the Beijing–Tianjin–Hebei Capital Region with Beijing and Tianjin as the cores, are the three most significant mega-city regions in China (Wu et al. 1996; Wu 2009). In the early 1980s, the special reform and opening-up policies gave rise to the Shenzhen Special Economic Zone, which has developed into the leading economic and innovation center of the Pearl River Delta, using the policy support from the central government and its superior geographical location next to Hong Kong (Yeung et al. 2009). The 1990s witnessed the successful construction of Shanghai Pudong New Area with the influx of foreign capital and transnational enterprises, following which Shanghai has gradually become China's top economic and financial center and the Yangtze River Delta has stepped onto a fast track of regional integration (Wu 1998; Marton and Wu 2006). In the 2000s, the mega-event of the 29th Summer Olympic Games greatly promoted the update on urban construction and infrastructure in Beijing, as well as Beijing's global image and city brand (Zhang and Zhao 2009; Wu et al. 2012a, 2012b).

The BTH region has special national strategic significance, due to its unique political standing as the capital region. Located in the northern coast area and being one of the three major economic growth poles of China (Li 2019), the BTH region constitutes three provincial administrative units: the capital city of Beijing, the provincial-level city of Tianjin, and Hebei province (with 11 cities). As of 2017, the population of the entire region was 112.48 million, accounting for 8.1% of China's total population, and the Gross Domestic Product (GDP) was 8058.045 billion yuan, accounting for 9.77% of the national economic aggregate (National Bureau of Statistics of China 2018).

Since the founding of the People's Republic of China 70 years ago, the BTH region has witnessed a remarkable development from the "mono-centric" structure to a more "polycentric" regional structure, and from "concentration" to "decentralization." The government's policy-making power has played a crucial role in guiding and triggering this trend. Especially in the recent years, a number of large-scale, national level projects with far-reaching impacts have been launched, which reorganized the spatial pattern of both Beijing and the BTH region. These are marked by the construction of Tongzhou, the Sub-City Center of Beijing, and the Xiong'an New Area, a new magnetic development center in the Hebei province.

Therefore, through reviewing historical planning documents and policies released in the past decades, this chapter focuses on analyzing the spatial strategies, transformation of public policies and implementation actions planning of Beijing and the BTH capital region, hence to summarize their development track and characteristics. It not only reveals the evolution trend of Beijing from "concentration" to "decentralization," but also illustrates exquisitely the main decentralization measures aimed at reducing the population and functions of Beijing from four spatial levels: the BTH region, the Beijing administrative region, the Beijing central city and the community level. The prospect of the decentralization strategy, including both challenges and missions, is further discussed in the conclusion and discussion. The case study of Beijing and the BTH region can provide profound Chinese experiences, eastern wisdom and references to other world cities and global city regions.

5.2 Method and Materials

In this chapter, the method of literature, policies and planning documents review was adopted to study the spatial developing process of Beijing and the BTH region. The sources of research materials include journal papers, academic books, online policy documents from the public sector, open urban planning documents, local chronicles and historical archives. The chapter carried out multi-dimensional reviews on the seven editions of Beijing master plans, the literatures about coordinated development process of the BTH region, and the latest decentralization actions of Beijing municipal government on population and functions. The online databases of the ScienceDirect, Google Scholar, and Chinese Knowledge Network (CNKI.net) were used to carry out the literature research by searching the keywords "Beijing–Tianjin–Hebei region," "Beijing–Tianjin–Hebei coordinated development," "Beijing master plan," "Xiong'an New Area," and "the Sub-City Center of Beijing," as well as some other relevant keywords, such as "Beijing–Tianjin–Tangshan," "Sub-Center of Tongzhou" and "mega-city region."

Information from local chronicles with historical event records is always comprehensive and authoritative, especially in terms of urban policy studies. Local chronicles reviewed in this chapter include the *Beijing Local Chronicles-Urban Planning Volume, Pictures on Beijing Urban Planning:1945–2005*, and *Papers on Beijing Urban Planning*, which were compiled by the Beijing Municipal Planning Commission, Beijing Urban Planning and Design Institute and other front-line planning institutions in Beijing. Most master plan figures cited in this chapter are adapted by the authors from these local chronicles.

Policy documents, planning outcomes and news reports reviewed are mainly sourced from the official web pages of the Central People's Government of the People's Republic of China (http://www.gov.cn/index.htm), the State Council Information Office of the People's Republic of China (http://www.scio.gov.cn), the

People's Government of Beijing Municipality (http://www.beijing.gov.cn/) and the Beijing Municipal Commission of Planning and Natural Resources (http://ghzrzyw.beijing.gov.cn/). In addition, the chapter also reviewed related policy documents from the local municipal and regional planning agencies to summarize the latest measures and trend of population and function dispersing process in the BTH region, with the core information listed and analyzed by graphing.

By reviewing the aforementioned categories of literature, this chapter attempts to analyze the spatial development of Beijing and the BTH region around "concentration" and "decentralization" from the following aspects: (1) the transformation of Beijing master plans and the related urban development, as well as their rising influences on the BTH region over the past few decades; (2) latest planning actions and mega-projects in Beijing and the BTH region, and their potential impacts on the future development of the region.

5.3 Literature Review on the Development of Beijing and the BTH Region

From the early ideas of Le Corbusier's "Radiant City" and Frank Wright's "Broadacre City" to Peter Hall's study on *The Polycentric Metropolis: Learning from Mega-city Regions in Europe*, the issue of concentration and decentralization remains to be the center of debate on the sustainable development of cities and city regions.

Over the past decades, the BTH region has developed in a significantly unbalanced way, since Beijing, as the core city and national capital, has absorbed most advantaged resources and talents from the BTH region. Capital flow, headquarters of enterprises, retail business, advantaged clinical and educational resources have quickly flowed into Beijing, and the urban population of Beijing has increased dramatically from less than five million in 1949 to over 20 million in 2019. Guided by the city master plans, the development of Beijing has significantly affected the spatial development of the entire BTH region. Researches on these mainly focus on the historical background (Qian 1981; Wang 2005a), overall development trend, main planning points (Liu 1984; Ke 1993; Wang 2005b; Shi et al. 2017a), and planning implementation (Shi and Zhao 2005; Wu et al. 2012b; He et al. 2014; Shi et al. 2017b). Li and Han (2013) summarized the main points of each edition of Beijing master plans from 1949 to 2004, and divided Beijing's urban development into 4 periods: the preliminary development toward socialist urban cities, the transformation from "city of consumption" to "city of industrial production," the transformation from planned economy to market economy, and the new exploration under the market economy period. They believed the spatial policies identified by Beijing master plans reflected the attitude of the state and related national macro-policies on urban development. Dong (2006) divided Beijing's urban planning history since

1949 into four periods: initial formation (1949–1953), twisted improvement (1957–1982), adaption for market mechanism (1990s), and contradiction and balance (2000s to present). Liu et al. (2009) and Wang (2017) traced the history to the ancient feudal dynasty, when Beijing was first established as the capital over 600 years ago, and analyzed the evolution of urban pattern and municipal management in Beijing thereafter. However, most of these studies are focused on Beijing's master plans before 2005, and important policies such as the establishment of Beijing Sub-City Center and recent decentralization policies on the population and functions have not been sequentially explored.

The unbalanced "2 + 11" development model of the BTH region has led to a distinct spatial structure of "two strong centers (Beijing and Tianjin) – a weak hinterland" (Tang et al. 2017) (Fig. 5.1). Due to China's highly centralized political system, the 13 cities were accustomed to dialogues with their upper-level leaders, while ignoring or even refusing dialogues with their neighbors, forming a situation that has long hindered the BTH's sustainable integrated regional development (Tang et al. 2017). It was not until mid-1980s that the concept of BTH region began to gain recognitions from both the academia and the government. However, as early as in 1950s, experts like Liang Sicheng, Chen Zhanxiang, Hua Lanhong and Wu Liangyong had actively promoted the city master planning for Beijing and other BTH cities, among which a regional concept of Beijing–Tianjin–Tangshan has been proposed in late 1960s. In late 1970s, Beijing, Tianjin, and Hebei started to consider more on their core-periphery relationship in planning. In early 1980s, several research projects on the BTH region were carried out, including the *study on territorial plan of the Beijing–Tianjin–Tangshan region* and the *study on economic plan of the Bohai coastal region* (Lu 1985; Wu et al. 2002; Mao 2004). In 1986, Xu Shuli

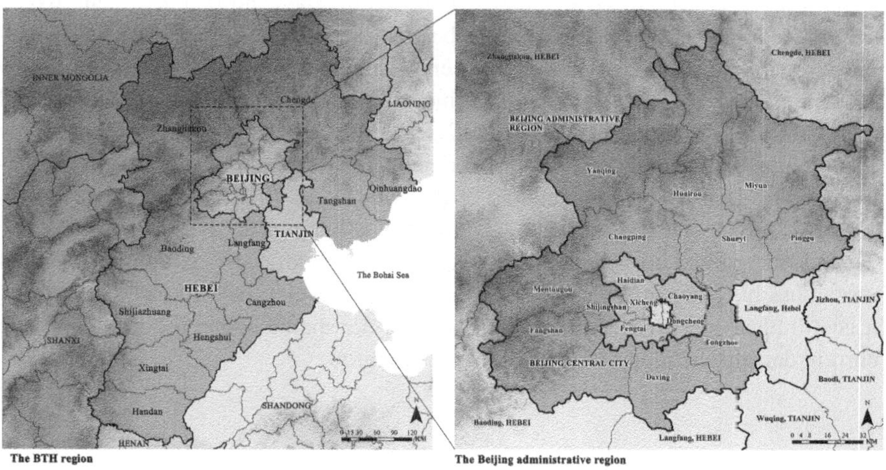

Fig. 5.1 The BTH capital region and the Beijing administrative region

published an article titled *Development of Horizontal Economic Integration in Beijing–Tianjin–Hebei Region*, which proposed both the BTH terminology and the economic integration potential of the Beijing–Tianjin–Hebei region (Xu 1986). Even though the importance of regional coordinated development of the BTH had been gradually recognized and accepted from then on, the official cooperation and exchanges of the three subparts were still relatively rare (Cui et al. 2019).

From 2000 to 2014, researches on the BTH became much more fruitful, especially on industrial and economic integration of the region (Lu 2002; Lv and Xie 2007; Sun et al. 2008; Zhu 2009; Sun and Ding 2012) and coordinated development methods of the urban agglomeration (Yang 2005; Li and Chen 2009; Xiao et al. 2011). To solve the more severe regional problems, such as heavy air pollution, vicious competition, resource pressure, and lack of regional cooperation mechanism, Wu Liangyong proposed a connotation of "Greater Beijing" to promote the cross-border joint development in the BTH, and suggested Beijing to learn from other global cities like London, Paris, and Tokyo to relocate some of the functions into its peripheral areas. Based on his thinking, the Tsinghua University team published three series of reports on the urban and rural development of the BTH region, advocating the coordination layout of town network, transportation network, ecological network and cultural network (Wu 2000; Wu et al. 2002; Wu and Mao 2004; Wu et al. 2006; Wu et al. 2012a, 2012b; Wu et al. 2013). It has provided significant theoretical support to the decentralization strategy of Beijing for dispersing its population and non-capital functions. The governments of the three subparts also started to actively explore cooperation mechanisms through seminars, inter-governmental conferences and development forums (Cui et al. 2019). Some inter-governmental cooperation in the fields of economy, public services, infrastructure development and environmental improvement has also been launched.

In 2015, "*The Outline of Collaborative Development of Beijing, Tianjin and Hebei Province*" was approved and implemented by the central government, which symbolized that the Beijing–Tianjin–Hebei coordinated development became a designated national strategy. Later on, the central government issued a series of policy documents and spatial planning to promote the environmental improvement and regional transport integration (Mao 2017; Cui et al. 2019). So far, researches on the BTH region have mainly focused on regional industrial synergy and market integration (Zhang et al. 2014; Sun and Yao 2015; Wang et al. 2018), collaborative governance on environmental and ecological issues (Zhang et al. 2015; Wei and Zhao 2016), and regional infrastructure connectivity (Sun et al. 2016). Although abundant researches have made certain progress around the development of the BTH region, it is still necessary to increase the research on the spatial strategies, especially the current decentralization policies, of the BTH region.

5.4 From Concentration to Decentralization: Spatial Policies and Development Trend Defined by Beijing Master Plans

City master plan is always the core planning policy document in China (Li and Han 2013). Since the founding of the P.R. China, Beijing has approved and implemented seven editions of master plans respectively in 1953, 1958, 1973, 1983, 1993, 2005, and 2017 (Shi et al. 2019). From the seven Beijing master plans, it can be seen that the planning goals, city functional orientation, city population scale, spatial structure strategy, satellite town ideas, industrial layout, etc., have reflected an obvious spatial development trend from "concentration" to "decentralization," which has driven the BTH region to present a same change trend in its spatial structure. In fact, the central city of Beijing has developed in an urban structure of "ring-radial pattern," reflecting typical centralized features. Therefore, when "concentration" has been the mainstream in Beijing for its capital attraction, "decentralization" has been a constant mitigation measure and guided development trend from the government to reach a better urban future (Fig. 5.2).

5.4.1 *Urban Expansion Based on the Historic City: 1949–1957*

From 1949 to 1957, Beijing was developed as the capital city of the newly-found China after the Anti-Japanese War and the Chinese Civil War. The urban population and industrial scale of Beijing expanded non-stop in a trend of "concentration." The city established the Urban Planning Commission and invited experts from home and abroad to make a new version of city master plan. In 1953, Beijing Municipal Government completed the first city master plan named *Draft Plan for Beijing's reconstruction and expansion*. The plan defined Beijing as the "national political center," "national cultural center," "advanced industrial base," and "national S&T (Science & Technology) center," which supposed to accommodate five million population and expand to 600 square kilometers. The proposed city functions of the master plan highly corresponded to the socialist national ideology, and was beneficial for solving the problems of unemployment and industrial decay at that time. Aiming at saving construction cost, the plan decided to locate the administrative center in the historic city of Beijing and started a concentrated approach of urban development (Fig. 5.3). However, when the draft edition was submitted to the

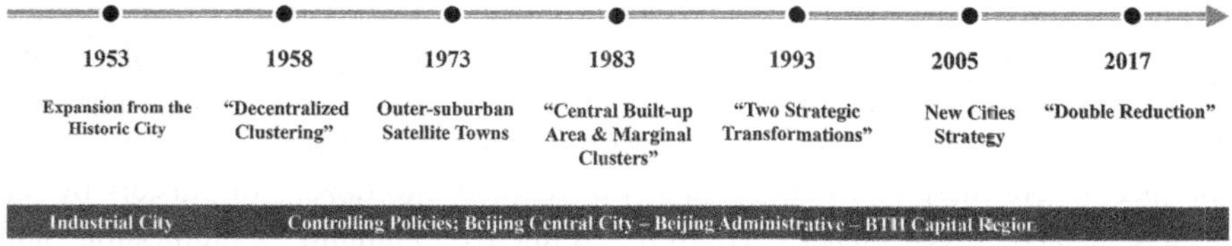

Fig. 5.2 The trend from "concentration" to "decentralization" of the Beijing master plans

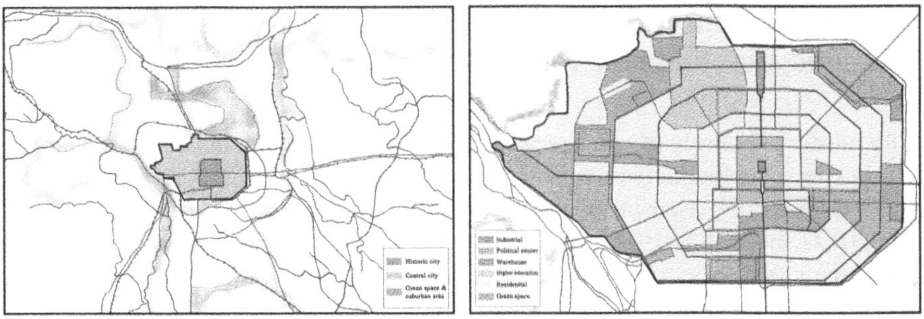

Fig. 5.3 Draft Plan for Beijing's reconstruction and expansion, 1954 (Left: the city-region scale; Right: the Beijing central city). Source: Authors' adaption from Beijing Local Chronicles Committee (2009)

central government, the State Development Planning Commission, which was in charge of the national economic and planning strategies, claimed that Beijing should not be defined as the "advanced industrial base" and the urban population should be controlled to four million. In consequence, the master plan was not approved by the central government, but urban construction was still carried out under its guidance (Beijing Municipal Planning Commission and Chronicles Committee at Beijing Urban Planning and Design Institute 2004; Dong 2006; Beijing Local Chronicles Committee 2009).

5.4.2 Political Movements and the Beginning of Decentralization: 1958–1982

From 1958 to 1982, influenced by western planning theories and practices, the idea of distributing functions to the outer satellite towns was emerging in Beijing. In 1957, Beijing completed the "*Beijing preliminary construction master plan,*" in which the city was defined as the "national political center," "national cultural and educational center," "modern industrial base," and "national S&T center." The plan firstly put forward the spatial strategy of forming a "Maternal bond" urban structure, which advocated to plan and build over forty satellite towns in peripheral areas to undertake the industrial production in the central city, and to avoid the excessive concertation of population and industry. In August 1958, the central government decided to launch the political movements of "Great Leap Forward" and forming rural People's Communes. Under such circumstance, Beijing revised the plan to the *Draft of the Beijing Master Plan Notes* (Fig. 5.4). It proposed a "Decentralized Clustering" spatial structure with a predicted population of ten million people for Beijing, where the urban built-up area was divided into dozens of clusters by the wedge-shaped green spaces (Beijing Municipal Planning Commission and

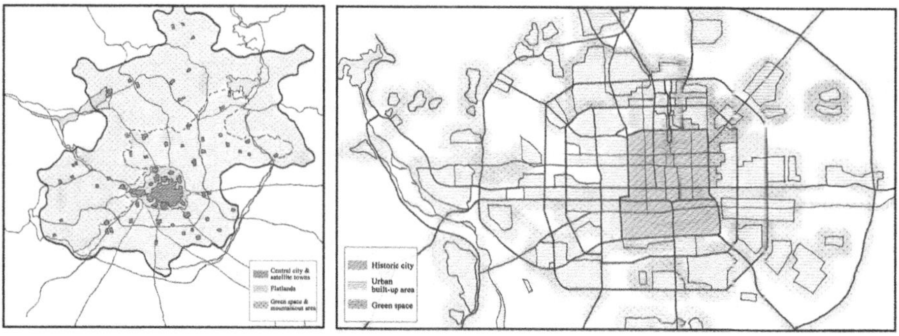

Fig. 5.4 Draft of the Beijing Master Plan Notes, Sep 1958 (Left: the Beijing administrative region; Right: the Beijing central city with the structure of "Decentralized Clustering"). Source: Authors' adaption from Beijing Local Chronicles Committee (2009)

Chronicles Committee at Beijing Urban Planning and Design Institute 2004; Dong 2006; Beijing Local Chronicles Committee 2009).

Due to the industrial and population dispersing needs brought by the political movements, problems of environmental pollution, water shortage and urban sprawl emerged in Beijing, which made the principle of "decentralization" the guideline for new urban construction. In 1964, Li Fuchun, Vice Premier of the State Council, submitted the *Report on Beijing Urban Construction Progress* to the central government, and proposed to locate new construction projects to the outer-suburban satellite towns. Later on, the 1973 revised Beijing master plan named *Report on Issues of Beijing Construction Master plan,* with a goal of 3.7–3.8 million urban population in 1980, proposed to relocate high-polluting industrial enterprises to the outer-suburbs, and to strengthen suburban satellite towns according to industrial distribution layout (Beijing Municipal Planning Commission and Chronicles Committee at Beijing Urban Planning and Design Institute 2004; Beijing Local Chronicles Committee 2009).

5.4.3 Development of the Outer-Suburban Areas and Construction of New Cities: 1983–2016

From 1983 to 2016, Beijing master plans continued the trend of "decentralization." In 1983, the approved *Beijing Urban Construction Master Plan* restricted the roles to only "political center" and "cultural center" for Beijing. It estimated that by the year of 2000, the population size of Beijing should be controlled to ten million, and that of the Beijing central city should be controlled to four million. A spatial layout of "Central Area & Marginal Clusters" was proposed on the basis of the former "Decentralized Clustering" layout (Fig. 5.5). The strategy of actively constructing

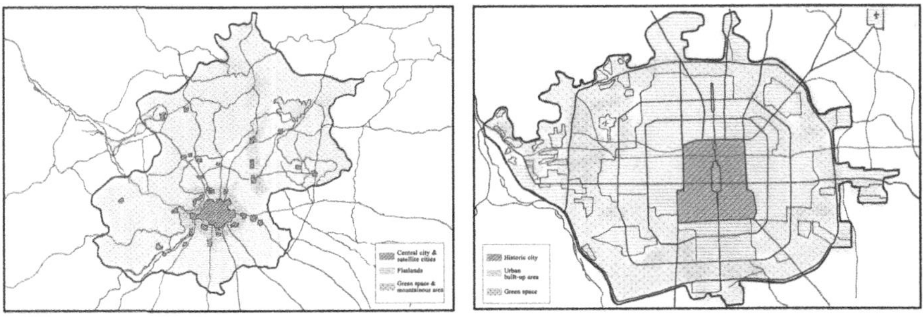

Fig. 5.5 Beijing Urban Construction Master plan, 1983 (Left: the Beijing administrative region; Right: the Beijing central city). Source: Authors' adaption from Beijing Local Chronicles Committee (2009)

the outer-suburban satellite cities continued in this period. (Qian 1981; Liu 1984; Beijing Municipal Planning Commission and Chronicles Committee at Beijing Urban Planning and Design Institute 2004; Beijing Local Chronicles Committee 2009).

The *Beijing Master Plan (1991–2010)* approved in 1993 pointed out that Beijing should be the "political center," "cultural center," "world-famous ancient capital," and "modern international city" in the national urban hierarchy system. It also predicted that by 2010, the permanent residents of Beijing would reach 12.5 million, and that of the Beijing central city would reach 6.5 million. The strategy of "Two Strategic Transformations" was put forward then, which emphasized that the construction would be directed from the central city to the outer-suburban areas, and the urban expansion would transform into urban renewal (Ke 1993; State Council 1994; Beijing Municipal Planning Commission and Chronicles Committee at Beijing Urban Planning and Design Institute 2004; Beijing Local Chronicles Committee 2009).

Approved in 2005, the *Beijing Master Plan (2004–2020)* maintained the same city functional orientation in the previous edition. It proposed that by 2020, the population size of Beijing should be controlled under 18 million, and that of the Beijing central city should be controlled to 8.5 million. More importantly, it proposed a spatial structure of "Two axes, Two bends, and Multiple centers" with an urban system of "Beijing Central City–New Cities–Towns" (Fig. 5.6). Therefore, eleven new cities, such as Tongzhou, Shunyi, Yizhuang, and Daxing, have been planned and developed in the suburban areas based on the previous satellite towns, in order to support the relocated functions and population from the central city (People's Government of Beijing Municipality 2005; Shi and Zhao 2005; Wang 2005a).

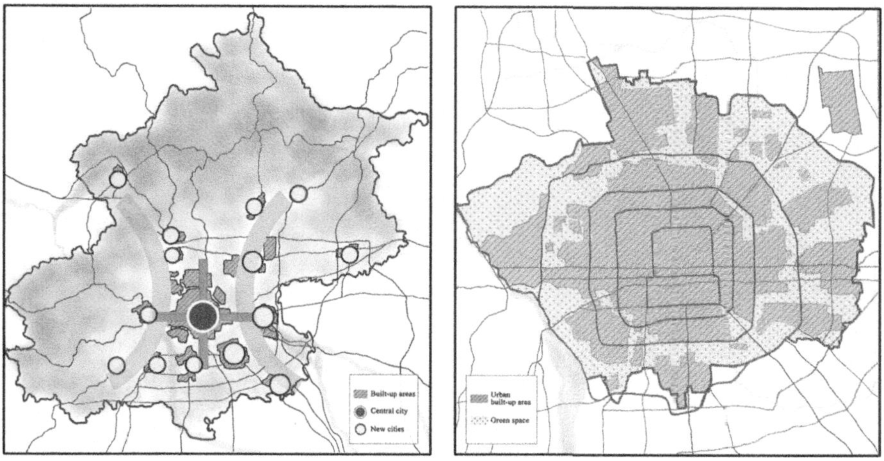

Fig. 5.6 Beijing Master plan 2004–2020, 2005 (Left: the Beijing administrative region; Right: the Beijing central city). Source: Authors' adaption from People's Government of Beijing Municipality (2005) and Beijing Municipal Planning Commission et al. (2006)

5.4.4 The "Double Reduction" and Regional Coordination Development: 2017–Present

In 2016, Beijing completed the *Beijing Master Plan (2016–2035)*, and it was approved by the central government in 2017 (CPC Central Committee & State Council 2017). The plan clarified Beijing as the national "political center," "cultural center," "center of international exchange," and "S&T innovation center." It also put forward a controlling principle of "Double Reduction" to reduce both the scale of the population and the urban-rural construction land. Specific reduction goals on the population and functions from 2020 to 2035 were also put forward. The plan proposed to limit the scale of population of Beijing to 23 million and that of the central city to 10.85 million by the year of 2020. The excessive population and functions of the central city would be dispersed into the five main new cities of Shunyi, Daxing, Yizhuang, Changping, and Fangshan, and the wider BTH region (CPC Central Committee and State Council 2017; CPC Beijing Committee & People's Government of Beijing Municipality 2019).

Most importantly, the plan has proposed to construct Tongzhou New City as the Sub-City Center of Beijing to support the decentralized non-capital functions from the central city. Thus, an urban system of "Beijing Central City–Sub-City Center/Tongzhou–New Cities–Towns" has formed in Beijing (Fig. 5.7). The connotation of "non-capital function" that will be dispersed from Beijing to suburban areas or other cities is clearly defined in the plan; specifically the amenities and industries that do not match Beijing's "four centers" roles, such as the primary manufacturing industries, regional wholesale trading markets, logistics and warehouse facilities, and polluting industries (CPC Beijing Committee & People's Government of Beijing

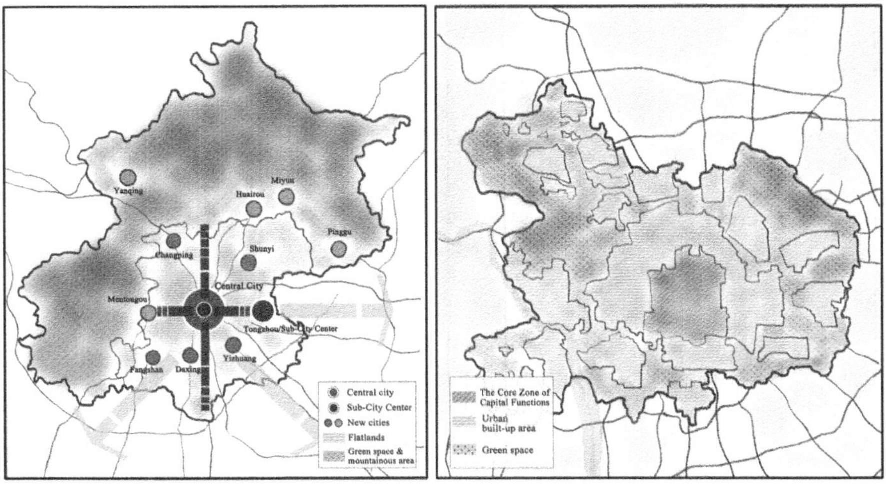

Fig. 5.7 Beijing Master Plan (2016–2035) (Left: the Beijing administrative region; Right: the Beijing central city). Source: Authors' adaption from CPC Beijing Committee & People's Government of Beijing Municipality (2019)

Municipality, 2019). The plan also proposed to disperse and relocate other functions, like public organizations, research institutes, headquarters of financial enterprises in the BTH region for regional sharing (Shi et al. 2017a; Shi et al. 2017b; CPC Beijing Committee & People's Government of Beijing Municipality, 2019; Shi et al. 2019).

From the abovementioned analysis, it can be seen that Beijing has experienced a spatial development from "concentration" to "decentralization," highlighted by its urban planning control on functions and population, the decentralized spatial structure and construction of satellite cities, etc. On one hand, the dominant function of Beijing is under change. While the functional orientations of "political center" and "cultural center" are consistent in all versions of city master plans, the roles of "economic center," "industrial base," and "mega-industrial city" have been abandoned since the 1980s and were eventually replaced by the "center of international exchange" and the "S&T innovation center." On the other hand, battling the "siphon effect," which has led to consistent concentration, all city master plans after 1958 have emphasized on controlling the scale of urban population. The implementation of the decentralization strategy from top-down could be clearly seen in the related development process. First, the "decentralized clustering structure" that was used to control Beijing's urban sprawl since 1958, has been consistent in all ensuing master plans. The scale of population and the area of construction land were controlled, and the strategy progressed to the recent trend of "Deducted Development." Second, the spatial scale for the decentralization and relocation has enlarged over the past decades, where the non-capital functions are first decentralized to the suburban "satellite towns," then to the "outer-suburban satellite cities," later to the "new cities," and finally to the whole BTH capital region.

5.5 Four-Level Actions: The Decentralization of Population and Functions in the BTH Region

Beijing has continuously suffered from "mega-city diseases" such as air pollution, traffic congestion and water shortage, due to the long-term, high-density concentration of population and functions. Meanwhile, Hebei province, as the surrounding hinterland of the capital Beijing, lags far behind in economic development and is in urgent need of adequate resources and advantaged policy opportunities. From the data of Beijing, Tianjin, and Hebei cities (Table 5.1), there has been an enormous gap in major social-economic indexes in the BTH region. Thus, the *Beijing Master Plan (2016–2035)* proposed the commission of "actively implementing the BTH coordinated development strategy with the focus on dispersing Beijing's non-capital functions, transforming current urban development mode, improving the urban governance mechanism, and effectively mitigating the mega-city disease" (CPC Beijing Committee & People's Government of Beijing Municipality, 2019).

The national strategies of decentralizing the non-capital function of Beijing and promoting the coordinative development of Beijing, Tianjin, and Hebei, have been launched in the BTH region since 2014. At present, the decentralization spatial

Table 5.1 Major social-economic data of Beijing, Tianjin, and cities in Hebei province in 2017

City	Permanent population (10,000 persons)	GDP (100 million yuan)	Total investment in fixed assets (100 million yuan)	Per capita disposable income (10,000 yuan)
Beijing	2170.7	28,014.9	8948.1	5.7
Tianjin	1556.9	18,549.2	11,288.9	3.7
Shijiazhuang	1088.0	6177.0	6353.2	2.5
Chengde	356.5	1465.5	1751.0	1.8
Zhangjiakou	443.3	1427.0	1655.3	2.0
Qinhuangdao	311.1	1500.3	890.5	2.2
Tangshan	789.7	6530.2	5365.3	2.8
Langfang	474.1	2881.0	2663.9	2.7
Baoding	1169.1	3449.7	2939.5	2.0
Cangzhou	755.5	3643.4	3759.1	2.1
Hengshui	446.0	1523.2	1322.7	1.8
Xingtai	735.2	2090.6	2169.6	1.8
Handan	951.1	3379.5	4029.1	2.1

Note: The total investment in fixed assets include investment in real estate development, infrastructure investment, construction, and installation investment

Source: Beijing Municipal Bureau of Statistics and NBS Survey Office in Beijing (2018), Tianjin Municipal Bureau of Statistics and Survey Office of the National Bureau of Statistics in Tianjin (2018), General Office of the people's Government of Hebei Province, Bureau of statistics of Hebei Province, and Hebei Academy of Social Sciences (2018) measures have been implemented comprehensively at different levels, including the BTH region level, the Beijing administrative region level, the Beijing central city level and the community level. At the BTH regional level, Xiong'an New Area, which has high hopes from China's top leadership, was built 150 kilometers south of the Beijing city center. At the level of Beijing administrative region, the Sub-City Center was established on the basis of the former Tongzhou New City, which was 25 kilometers east of the Tian'anmen Square. At Beijing central city level, a series of studies have been conducted on the potential of building a special Central Administrative District in its core zone. And at the community level, measures of "dispersing and renovating" designated space corresponding to the macro policies have been implemented.

5.5.1 BTH Regional Level: The Establishment of Xiong'an New Area in Hebei

In April 2017, China's central government launched the national strategy to plan and construct the Xiong'an New Area in Hebei province, which was officially described as the "Millennium City." Adopting the highest international construction standard and the advanced urban concepts (including the highest international standard, low-carbon principles, smart and digital urban administration, intelligent transportation network, resilient city, and utility tunnels), the Xiong'an New Area has been planned to take on the dispersed population and functions of Beijing and create a construction model showcasing Chinese planning wisdom to the world (Xinhua News Agency 2017; CPC Hebei Provincial Committee and People's Government of Hebei Province 2018; CPC Central Committee and State Council 2018).

Undoubtedly, the implementation of the national endeavor in the BTH region has caused intense concerns from the international academia, social media, urban planners and policy makers. The new area has been defined as "a strategy crucial for the coming millennium, and an issue be of great national significance" by the central government. It was positioned as a national new district on a par with the Shenzhen Special Economic Zone and the Shanghai Pudong New Area. The Xiong'an New Area covers around 2000 square kilometers, with a development zone of around 200 square kilometers (Xinhua News Agency 2017). On April 14, 2018, the CPC Central Committee and the State Council approved the *Outline of Hebei Xiong'an New Area Planning* (CPC Central Committee and State Council 2018), and on December 25, 2018, the State Council further approved the *Hebei Xiong'an New Area Master Plan (2018–2035)* (State Council 2019). These documents clearly pointed out that Xiong'an New Area is a special zone for undertaking Beijing's non-capital functions. Specifically, the functions which will be relocated into the new area include university branch campuses, scientific research centers, headquarters and branches of financial institutions, high-end service enterprises (like software, information

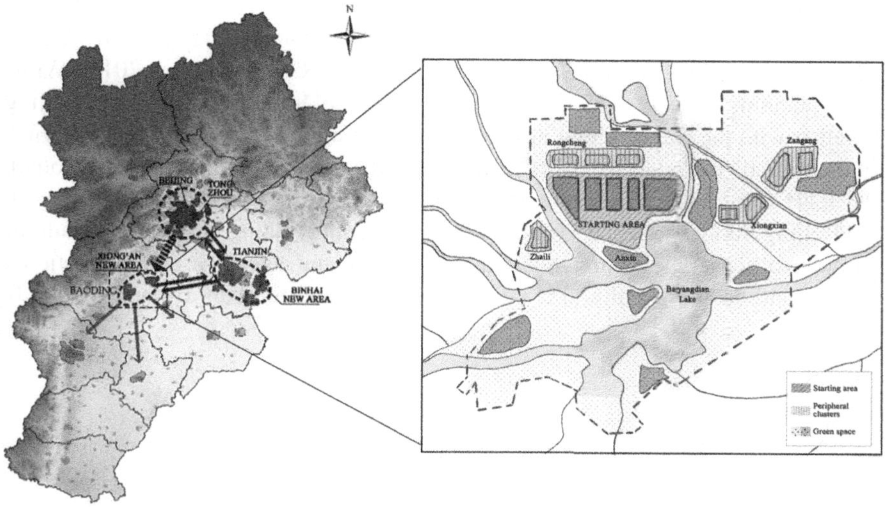

Fig. 5.8 Location and planned spatial structure of the Hebei Xiong'an New Area. Source: Authors' adaption from CPC Hebei Provincial Committee and People's Government of Hebei Province (2018)

services, design, creative, and consulting), and high technology industries (like modern logistics, headquarters of e-commerce enterprises, electronic information, biological medicine, and new materials) (CPC Central Committee and State Council 2018; State Council 2019), etc.

Without doubts, the newly-built Xiong'an New Area will become a new growth pole for economic development and scientific innovation in the BTH region. It is expected to transform the current "Duel Core" regional development pattern focusing on Beijing and Tianjin to a "Balanced Triangle Pattern" focusing on Beijing, Tianjin, and Hebei (Fig. 5.8). Thus, cities in Hebei province like Tangshan, Baoding, and Zhangjiakou can all benefit from the momentum of the coordinated regional development.

5.5.2 *Beijing Administrative Region Level: The Construction of Tongzhou/Sub-City Center*

At the Beijing administrative region level, after years of studies on its feasibility, Beijing officially launched the decision to establish the "Beijing Sub-City Center" and organized an international urban design consultation for the project. The new sub-center was allocated to the east of Beijing central city (the former Tongzhou New City) and planned to undertake Beijing's municipal administrative functions in the future.

On December 27, 2018, the CPC Central Committee and the State Council officially approved the *Detailed Regulatory Plan for Beijing Sub-City Center (2016–2035) (Block Level)* (CPC Central Committee & State Council 2019). With a total planning area of 155 square kilometers, the Sub-City Center of Beijing, together with the Hebei Xiong'an New Area, is positioned as one of the "two wings of the capital" to undertake the relocated Beijing non-capital functions and promote the BTH Coordinated Development Strategy (Fig. 5.9). Different from Xiong'an, the Beijing Sub-City Center is mainly engaged in undertaking the municipal administrative offices (all Beijing municipal governmental agencies and institutions), finance and business, cultural creativity industries and tourism dispersed from the

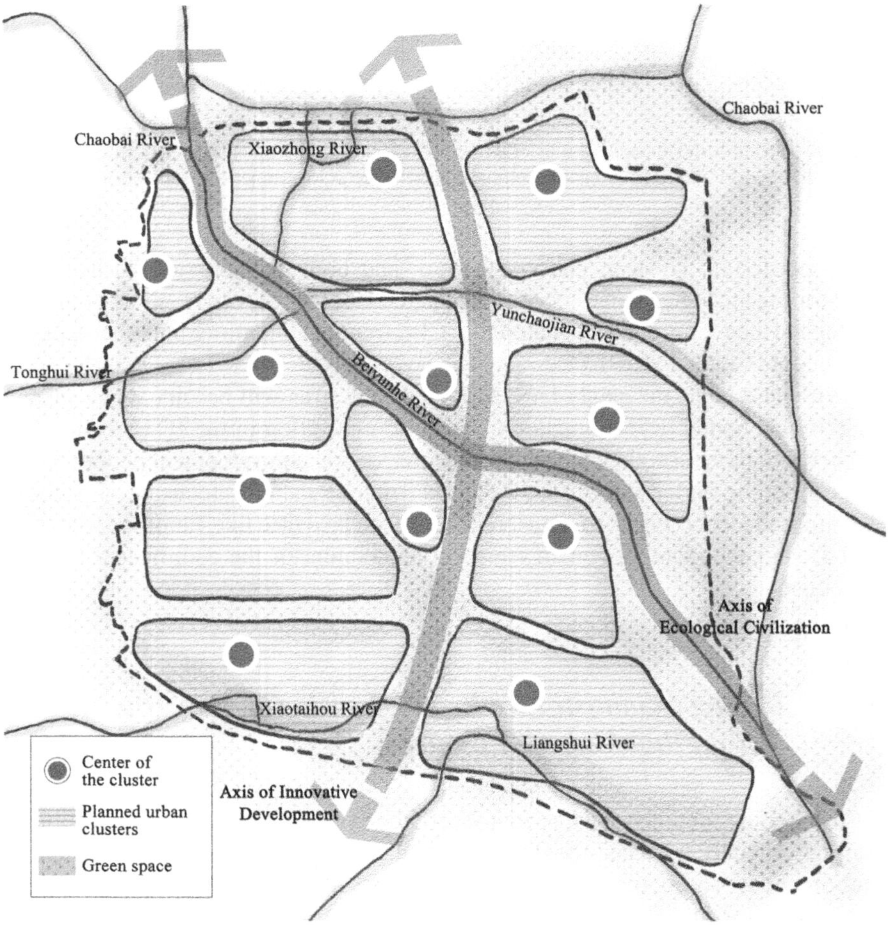

Fig. 5.9 Detailed Regulatory Plan for Beijing Sub-City Center. Source: Authors' adaption from Beijing Municipal Commission of Planning and Natural Resources (2019)

Beijing central city (Beijing Municipal Commission of Planning and Natural Resources 2019). On January 11, 2019, Beijing municipal administrative agencies officially started to move into the Sub-City Center (Beijing Municipal Commission of Planning and Natural Resources 2019; Beijing Daily 2019), marking the reorganizing process of Beijing's urban spatial structure. Then, the Beijing central city will be focused on serving the central administration and capital functions, while the sub-city center will focus more on the Beijing municipal affairs.

The released *Detailed Regulatory Plan for Beijing Sub-City Center* proposes to control the permanent population of the sub-city center to within 1.3 million by 2035, among which 400 to 500 thousand permanent residents will come from the Beijing central city. Meanwhile, the plan also puts forward the principles to strictly control the scale of urban–rural construction land, and the scale of overground and underground buildings. Like Xiong'an New Area, the principles of sustainable development, like low-carbon, green, urban resilience, digital, and smart, will infiltrate the whole construction process. The plan proposed to build a "small blocks, dense road network" transportation system so to reach a transit-oriented development in the Sub-Center, as well as building a municipal infrastructure network featured by utility tunnels (Beijing Municipal Commission of Planning and Natural Resources 2019).

5.5.3 Beijing Central City Level: An Idea of Building the Central Administrative District

At the Beijing central city level, drawing on the experience of Washington D.C., a study on establishing a Central Administrative District at the heart of the central city has been conducted, which probably would occupy 92.5 square kilometers area (CPC Beijing Committee & People's Government of Beijing Municipality, 2019). With the relocation of Beijing municipal administrative agencies to the Tongzhou/ Sub-City Center, the idea of transforming the core zone of Beijing central city into a demonstration area for central administrative functions and national political movements has the potential to be realized in the future. It is proposed to locate at the seat of Beijing historic city, dating back to Ming Dynasty, and constitute Dongcheng and Xicheng districts. The foreshadowing of this undetermined strategy can be seen in the *Beijing Master Plan (2016–2035)*, which emphasizes on providing necessary space for central government affairs, foreign exchange celebrations and major festival activities through decentralizing population and non-capital functions of the central city (CPC Beijing Committee & People's Government of Beijing Municipality, 2019).

5.5.4 *Community Level: The Dispersing, Renovating, and Improving of Designated Urban Space*

The implementation of Beijing's decentralization measures has a strong "top-down" attribute, determined by the different levels of government. Therefore, Beijing has also initiated the decentralization process at the community and block levels. In particular, since 2017, Beijing has taken comprehensive actions such as demolishing illegal buildings, stopping the informal commerce, and renovating the negative space, so as to realize informal function relocation, population dispersing and environmental improvement.

To be specific, according to the *Implementation Opinions of the People's Government of Beijing on Organizing and Conducting the Special Action of "Dispersing, Renovating and Improving" (2017–2020)* (People's Government of Beijing Municipality 2017), the core measures include ten actions: demolishing the unauthorized constructions; stopping informal commerce and restoring its space; improving the urban-rural joint areas; renewing old communities; mitigating urban risks and environmental improvement; relocating the primary manufacturing and polluted enterprises; relocating the wholesale markets; decentralizing the educational and clinical amenities; removing the informal residence; and the renovation of shanty communities in the historical areas.

In November 2017, a fire accident happened in an informal rental apartment in Daxing District and caused serious casualties, attracting the world's attention concerning the phenomenon of informal residence in China (Xinhua Net 2017). Afterwards, Beijing has reinforced the dispersing force and steps. However, the strict dispersing policies that influence so much on immigrant workers in Beijing, have also aroused some criticism and arguments from the public for considering inclusiveness, justness and equity of the city.

5.6 Conclusion and Discussion

This chapter focuses on the evolution process of spatial development of Beijing and the BTH region by reviewing related literature, planning documents and local chronicles. The overall trend, summarized as "from concentration to decentralization," reveals the strong intervention of Chinese government on urban and urban regional development.

After the founding of the P. R. China, Beijing developed with the concentration on its historic city in the early period. Then all the master plans after 1958 emphasized to strictly control the city's population and functions, and actively decentralize its urban functions to suburban satellite towns, making the increasing process of the concentration of population and industries in Beijing in step with a decentralization trend. After 1983, the city functional orientations as the "economic center" and "industrial base" were removed from the Beijing master plans, and the construction focus was shifting from the Beijing central city to the suburban areas of Beijing. In this case, the original outer-suburban satellite cities grew into new cities to undertake functions and population from the Beijing central city. The 2017 new city master plan decided to implement the strictest deduction control policies on population and construction land growth in Beijing, and clearly defined the "non-capital functions" of Beijing that should be decentralized from Beijing central city to its suburban area, the BTH region and even the entire northern China.

In recent years, the decentralization measures of the BTH region have been implemented at four spatial levels. At the BTH regional level, the Hebei Xiong'an New Area was set up to undertake non-capital functions such as research institutes and financial headquarters. The Tongzhou/Sub-City Center has been constructed to undertake the Beijing municipal administrative agencies. At the Beijing central city level, exploration and research on setting up the Central Administrative District are conducted. Finally, the community level has witnessed a "cleaning-up" campaign to deal with the informal buildings, residences, and commerce.

In the future, will the decentralized policies continue? What exact impact will these decentralization measures have on Beijing and the BTH region? How much population and non-capital functions of central Beijing can be attracted by the Xiong'an New Area and Sub-City Center of Beijing? Will the Central Administrative District be set up soon? Such things are still unpredictable. But some implications could be drawn from the ongoing development process.

Firstly, Beijing's spatial policies should properly handle the relationship between "national capital" and "municipality," and between "dispersing" and "obtaining." As the most significant city of the state, Beijing should "concentrate on something but not everything" (CPC Beijing Committee & People's Government of Beijing Municipality, 2019). The main reason for the excessive concentration of population and functions in Beijing is the highly imbalanced distribution of social-economic resources in the whole BTH region. What simmers under concentration and decentralization are the social-economic gaps between Beijing, Tianjin, and other cities in the BTH region. Thus, the fundamental solution to the "mega-city diseases" and "marginalized regional development" is to reach a more balanced urban and regional development by narrowing the gap of economic development, fundamental public services, employment, high-wage opportunities and so on. Therefore, establishing reasonable mechanism of ecological compensation and resource allocation, developing open and shared market, and providing fair policy supply are the main tasks for the coordinated development of the BTH region in the future.

Secondly, the process from "concentration" to "decentralization" in Beijing and the BTH region is a great window to catch a glimpse of the "top-down" administrative mechanism and political relations among the governments of central level, capital city level and local cities in China. The dispersing policies have started within the administrative boundary of Beijing municipality, and then expanded gradually to the whole BTH region and recently became a regional coordinated movement, which has greatly influenced the development of the BTH region for decades. The central government has obliged the official negotiations and cooperation among Beijing and other cities in the region. During this process, assuring and meeting the demand of Beijing has always been the top priority, which has restrained local petitions and development rights of other cities to some extent. On the other hand, the less-developed areas in BTH region are obtaining new opportunities by undertaking non-capital functions and the population transferred from Beijing. The functions like manufacturing industries, higher-educational institutions and headquarters of giant enterprises could bring capital flow and high-quality labor forces to the other cities in the BTH region, resulting in renewed economic growth in the economically lagging areas. In addition, regional issues like ecological protection and air-pollution treatment, have obtained evident success under the coordinated "decentralization" governance. People in the BTH region will also benefit from the coordinative measures in transportation and tourism development. However, there are still other aspects, such as regional industrial division, openness of united market, and balance of central, capital and local power, where significant progress remains to be made in the future. Thus, it is highly recommended that governments of Beijing, Tianjin, and other cities in Hebei should establish a mechanism of regular communication and cooperation, not only in order to end the vicious cycle of "marginalized regional development" between Beijing and other cities, but also to help improve the quality of life and well-beings of the citizens in the whole BTH region.

Finally, implemented through the vertical administration system in China, the decentralized spatial policies of Beijing are consistent in the contemporary era, working as a "top-down" spatial tool for reaching balanced development. The aforementioned mega-projects and ambitious plans are the most typical examples and results of such administration mode. However, we could also notice some other phenomena, such as involuntary relocation of the informal commerce and floating population, as the consequence of the "vertical administration." In the current society, where the public cherish so much the spirits of market and democracy, the transformations in administration mode from "administration" to "governance" and from "vertical" to "horizontal" have become increasingly important for the future development of the BTH region. Many questions remain to be answered in detail, for instance, how can Xiong'an New Area and the Sub-City Center of Beijing be developed as attractive places as wished? And how can the decentralization policies at community level achieve both dispersing goals and the protection over the people and the property? Hence, the planning decision-making process that can guarantee the individual and group interests should be pursued further. Apart from governmental actions, diversified market forces like social organizations, local autonomous agencies and the "grass-root forces" of individual citizens should also be well considered in improving urban governance. Only when the "top-down" policies well correspond to the "bottom-up" motivations, can the planning goals be reached gradually in the future.

Acknowledgments The authors are grateful to the National Natural Science Foundation of China (grant no. 51978363), the National Key Research and Development Program of China (grant no. 2018YFC0704600), and Beijing Social Science foundation (grant no. 19GLB034) for funding this study.

References

Beijing Daily, 2019. Beijing Municipal Administrative Agencies Officially Moved in the Tongzhou/ Sub-City Center. Available at: http://www.gov.cn/xinwen/2019-01/12/content_5357221.htm [Accessed October 21, 2019].

Beijing Local Chronicles Committee, 2009. *Beijing Local Chronicles - Urban Planning Volume*. Beijing: Beijing Publishing house.

Beijing Municipal Bureau of Statistics & NBS Survey Office in Beijing, 2018. *Beijing Statistical Yearbook 2018*. Beijing: China Statistics Press. Available at: http://data.cnki.net/Yearbook/Single/N2019010235 [Accessed November 22, 2019].

Beijing Municipal Commission of Planning and Natural Resources, 2019. Detailed Regulatory Plan for Beijing Sub-City Center (2016-2035) (Block Level). Available at: http://ghzrzyw.beijing.gov.cn/art/2019/1/4/art_5160_565183.html [Accessed October 19, 2019].

Beijing Municipal Planning Commission & Chronicles Committee at Beijing Urban Planning and Design Institute, 2004. *Papers on Beijing Urban Planning*. Internal unofficially-published documents.

Beijing Municipal Planning Commission et al, 2006. *Pictures on Beijing Urban Planning 1945-2005*. Internal unofficially-published documents.

CPC Beijing Committee & People's Government of Beijing Municipality, ed., 2019. *Beijing Master Plan (2016 – 2035)*. Beijing: China Architecture & Building Press.

CPC Central Committee & State Council, 2017. The CPC Central Committee and the State Council: Approval for the Beijing Master Plan (2016-2035). Available at: http://www.gov.cn/zhengce/2017-09/27/content_5227992.htm [Accessed October 19, 2019].

CPC Central Committee & State Council, 2018. The CPC Central Committee and the State Council: Approval for the Outline of Hebei Xiong'an New Area Planning. Available at: http://www.gov.cn/zhengce/2018-04/20/content_5284572.htm [Accessed October 19, 2019].

CPC Central Committee & State Council, 2019. Detailed Regulatory Plan for Beijing Sub-City Center (2016-2035) (Block Level). Available at: http://www.gov.cn/zhengce/2019-01/03/content_5354591.htm [Accessed October 19, 2019].

CPC Hebei Provincial Committee & People's Government of Hebei Province, ed., 2018. *The Outline of Hebei Xiong'an New Area Planning*. Beijing: People's Publishing House.

Cui, D., Wu, H. & Wu, D., 2019. A review and prospect of regional governance in the Beijing-Tianjin-Hebei region. *Progress in Geography*, 38(1), 1–14.

Dong, G., 2006. Glorious Development of the Fifty-Seven Years -- Beijing Urban Planning Development since the Founding of the People's Republic of China. *Beijing Planning Review*, 05, 13–16.

General Office of the people's Government of Hebei Province, Bureau of statistics of Hebei Province & Hebei Academy of Social Sciences, 2018. *Hebei Economic Yearbook 2018*. Beijing: China Statistics Press. Available at: http://data.cnki.net/Yearbook/Single/N2019040177 [Accessed November 22, 2019].

He, C. et al., 2014. Evolution of Beijing Master Plans and Compilation Innovation. *City Planning Review*, 38(10), 28–34.

Ke, H., 1993. Planning a Cross-Century Grand Master Plan to Build a Modern Capital – The Beijing Master Plan has been Revised. *City Planning Review*, 05, 1–5.

Li, D. & Han, G., 2013. Interrelationship between Urban Planning and Development of Beijing from 1949 - 2004. *Social Sciences of Beijing*, 05, 144–151.

Li, G. & Chen, X., 2009. Empirical Research on Influencing Factors on Population Growth of Beijing-Tianjin-Hebei Metropolitan Region. *Geographical Research*, 28(01), 191–202.

Li, G., ed., 2019. *Report on Coordinated Development of Beijing Tianjin and Hebei 2019*. Beijing: Science Press.

Liu, Q., 1984. Problems on the Implementation of the Beijing Master Plan. *Urban Problems*, 01, 1–5.

Liu, X. et al., 2009. *Planning, Construction and Management of Beijing under the National Capital Context -- 600 Years as the Imperial Capital of China and 60 Years as the Capital of the P. R. China*. Beijing: China Architecture & Building Press.

Lu, D., 1985. Regional Development and Spatial Structure of the Beijing - Tianjin - Tangshan Region. *Economic Geography*, 01, 37–43.

Lu, J., 2002. The Spatial Spread of the Beijing - Tianjin - Hebei Urban Economic Region. *Economic Geography*, 22(05), 574–578.

Lv, Z. & Xie, J., 2007. Developing Perspective and Strategic Conception of the Beijing, Tianjin and Hebei Region Economic Integration. *Economy and Management*, 21(08), 77–80.

Mao, H., 2017. Innovation of Mechanism and Regional Policy for Promoting Coordinated Development of the Beijing-Tianjin-Hebei. *Progress in Geography*, 36(01), 2–14.

Mao, Q., 2004. Review and Prospect of the Coordinated Development of Beijing-Tianjin-Hebei Region. *Beijing City Planning & Construction Review* , 04, 53–54.

Marton, A. M. & Wu, W., 2006. Spaces of globalisation: Institutional reforms and spatial economic development in the Pudong new area, Shanghai. *Habitat International*, 30(2), 213–229.

National Bureau of Statistics of China, 2018. *China Statistical Yearbook 2018*. Beijing: China Statistics Press. Available at: http://www.stats.gov.cn/tjsj/ndsj/2018/indexch.htm [Accessed October 16, 2019].

People's Government of Beijing Municipality, 2005. Beijing Master Plan (2004 – 2020). *Beijing City Planning & Construction Review*, 02, 5–51.

People's Government of Beijing Municipality, 2017. The Implementation Opinions of the People's Government of Beijing on Organizing and Conducting the Special Action of "Dispersing, Renovating and Improving" (2017-2020). Available at: http://www.beijing.gov.cn/zhengce/zfwj/zfwj2016/szfwj/201905/t20190522_59825.html [Accessed October 20, 2019].

Qian, L., 1981. Emancipating the Mind, Making Full Use of Advantages and Avoiding Disadvantages, and Accomplishing an Excellent Beijing Master Plan. *City Planning Review*, 01, 4–10.

Shi, W. et al., 2017a. Following the National Strategy, Responding to the Call of Age: Several Key Points in New Beijing City Master Plan. *City Planning Review*, 41(11), 17–22.

Shi, W. et al., 2019. Transformation and Exploration of the New Urban Master Plan of Beijing. *Urban and Rural Planning*, 01, 86–93.

Shi, W. & Zhao, F., 2005. Inheritance, Development and Innovation of Beijing Master Plan. *Beijing City Planning & Construction Review*, 02, 73–80.

Shi, X. et al., 2017b. Main Characteristics and Thinking of the New Urban Master Plan of Beijing. *Urban Planning Forum*, 06, 56–61.

State Council, 1994. Approval Documents of the State Council on the Beijing Master Plan. *Beijing Real Estate*, 01, 17.

State Council, 2019. The Approval of State Council for the Hebei Xiong'an New Area Master Plan (2018-2035). Available at: http://www.gov.cn/zhengce/content/2019-01/02/content_5354222.htm [Accessed October 21, 2019].

Sun, M. et al., 2016. Integrated Transportation Development for Beijing-Tianjin-Hebei Megalopolis: Problems & Countermeasures. *Urban Transport of China*, 14(03), 61–66.

Sun, J., Deng, H. & Ye, Z., 2008. The Discuss of the Beijing, Tianjin and Hebei Region Economic Integration and Cooperation Way. *Journal of Capital University of Economics and Business*, 02, 55–60.

Sun, J. & Ding, H., 2012. Research on the Stage of Beijing – Tianjin – Hebei Regional Integration. *Research on Economics and Management*, 07, 52–58.

Sun, J. & Yao, P., 2015. Industrial Relocation in Jing-Jin-Ji Region, Regional Specialization and Coordinated Development: An Analysis Based on the Framework of New Economic Geography. *Nankai Journal (Philosophy, Literature and Social Science Edition)*, 01, 81–89.

Tang, Y., Yang, D., Chen, K. & Zhu, H., 2017. The Beijing-Tianjin-Hebei Metropolitan Region: Regional Governance under a Highly Centralized Political System. in, David, Gómez-Álvarez., Eduardo, López-Moreno. & Robin, Rajack., eds. *Steering the Metropolis: Metropolitan*

Governance for Sustainable Urban Development, Mexican city of Guadalajara: University of Guadalajara, 341–350.
Tianjin Municipal Bureau of Statistics & Survey Office of the National Bureau of Statistics in Tianjin, 2018. *Tianjin Statistical Yearbook 2018*. Beijing: China Statistics Press. Available at: http://data.cnki.net/Yearbook/Single/N2019030264 [Accessed November 22, 2019].
Wang, J. et al., 2018. Transfer and Cooperation Mechanism of Manufacturing Industry in Beijing, Tianjin and Hebei under the Synergetic Perspective. *Economic Geography*, 38(07), 90–99.
Wang, K., 2005a. On the Background, Contents and Method of the New Comprehensive Plan of Beijing. *City Planning Review*, 29(04), 65–71.
Wang, K., 2005b. From "Liang-Chen Plan" to "Two-axis, Two-belt and Multi-center". *Beijing City Planning & Construction Review*, 01, 32–38.
Wang, Z., ed., 2017. *Beijing Sub-City Center*. Beijing: Beijing Science and Technology Press.
Wei, N. & Zhao, C., 2016. The research on collaborative governance of cross-regional air pollution based on the case of Beijing, Tianjing and Hebei region. *Hebei Academic Journal*, 36(01), 144–149.
Wu, L., 2000. Preliminary Studies on the Urban and Rural Spatial Development Planning of the Beijing-Tianjin-Northern Hebei Region. *City Planning Review*, 24(12), 9–15.
Wu, L. et al., 2002. *Research on the Rural and Urban Spatial Development Planning for the Greater Beijing Region (Beijing, Tianjin and Hebei)*. Beijing: Tsinghua University Press.
Wu, L. et al., 2006. *The Second Report on the Rural and Urban Spatial Development Planning Study for the Capital Region (Beijing, Tianjin and Hebei)*. Beijing: Tsinghua University Press.
Wu, L. et al., 2013. *The Third Report on the Rural and Urban Spatial Development Planning Study for the Capital Region (Beijing, Tianjin and Hebei)*. Beijing: Tsinghua University Press.
Wu, L., Mao, Q. & Zhang, J., 1996. Facing the 21st Century: the Future of Sustainable Development of the Chinese Mega-Cities: Case Study of Beijing, Shanghai and Guangzhou. *City Planning Review*, 04, 22–27.
Wu, L. & Mao, Q. 2004. Discuss of the Middle Cities Development in China and the Research of the Middle Cities in Beijing, Tianjin and Hebei Province. *Urban Studies*, 11(05), 49–53.
Wu, L., Wu, W. et al., 2012a. *Research on the "Beijing 2049" Spatial Development Strategy*. Beijing: Tsinghua University Press.
Wu, V., 1998. The Pudong Development Zone and China's economic reforms. *Planning Perspectives*, 13(2), 133–165.
Wu, W., 2009. Development of Mega-city Regions in China: Current Situation, Problems and Perspective. *Journal of Urban and Regional Planning*, 3, 84–103.
Wu, W. et al., 2012b. Strategic Thinking on Beijing's Urban Spatial Trends and Layout -- An Evaluation Study on the Implementation of the Beijing Master Plan 2004 ~ 2020. *Beijing Planning Review*, 01, 15–19.
Xiao, L., Huang, J. & Sun, G., 2011. Temporal-Spatial Characteristics of Evolution of the Urban System in Jingjinji Metropolitan Region. *Progress in Geography*, 30(02), 215–223.
Xinhua News Agency, 2017. The CPC Central Committee and the State Council decided to establish the Hebei Xiong'an Newe Area. Available at: http://www.gov.cn/xinwen/2017-04/01/content_5182824.htm [Accessed October 19, 2019].
Xinhua Net, 2017. A fire accident broke out in Xinjian Village, Xihongmen Township, Daxing District, Beijing, killing 19 people. Available at: http://www.xinhuanet.com//politics/2017-11/19/c_1121976989.htm [Accessed October 19, 2019].
Xu, S., 1986. Development of Horizontal Economic Integration in Beijing-Tianjin-Hebei Region. *Hebei Academic Journal*, 04, 96–98.
Yang, L., 2005. Metropolis Circle of Beijing, Tianjin and Hebei -- the Third Area of China's Increasing Economy. *Hebei Academic Journal*, 25(04), 57–64.
Yeung, Y., Lee, J. & Kee, G., 2009. China's special economic zones at 30. *Eurasian Geography and Economics*, 50(2), 222–240.
Zhang, L. & Zhao, S. X., 2009. City branding and the Olympic effect: A case study of Beijing. *Cities*, 26(5), 245–254.

Zhang, G. et al., 2014. Study on Co-ordination between Beijing-Tianjin-Hebei based on Industry Matching and Transfer. *Economy and Management*, 28(04), 14–20.
Zhang, Y. et al., 2015. Application of ecological conservation cooperation for Beijing-Tianjin-Hebei regional integration. *Resources Science*, 37(08), 1529–1535.
Zhu, E., 2009. Production Upgrading and Integration in the Process of the Integration of Beijing-Tianjin-Hebei. *Economic Geography*, 29(06), 881–886.

Chapter 6
From Tsingtau to Qingdao

Klaus R. Kunzmann and Erpeng Zhan

Abstract Qingdao is a thriving port city on the coast of the Yellow Sea in the Shandong province of the Peoples Republic of China with a population of around nine million. During 1897 to 1914 the city had been a German mandated territory, named Tsingtau. The city is an important industrial centre, one of the biggest ports in China, a centre of advanced seas and oceans science-technology research and development, a UNESCO city of film, a much favoured tourist city, and not to forget, the production site of the famous Tsingtau beer offered in Chinese restaurants all over the world. The "German" history of the city is still visible in Qingdao. In contrast to many other cities in China, the old city centre of Qingdao has been almost fully preserved. It is well protected, but is looking ahead to benefit from professional and economically viable urban regeneration policies. The essay will introduce into the "German" history of the city and the innovative land regulation, the German administration had launched to avoid excessive land speculation 100 years ago. Achievements of modern city development, such as the Sino-German Eco-development Park are presented. The essay will end with a brief account of challenges the city will face in the years ahead to conserve its historical heritage and meet the requirements of a modern Chinese metropolis.

K. R. Kunzmann
TU Dortmund, Dortmund, Germany

Bartlett School of Planning,
University College London, London, UK
e-mail: klaus.kunzmann@udo.edu

E. Zhan (✉)
Sino-German Eco-Park, Qingdao, China
e-mail: zhanerpeng@163.com

L. Bian et al. (eds.), *Chinese Urban Planning and Construction*,
Strategies for Sustainability, https://doi.org/10.1007/978-3-030-65562-4_6

Keywords Tsingtao · German architectural heritage · German land regulation · Urban regeneration · Eco-Park · Jimo City

6.1 Introduction

Qingdao is a thriving port city in the Shandong province of the Peoples Republic of China on the coast of the Yellow Sea with a population of around nine million in 2019. The city is an important industrial centre, one of the biggest ports in China, a centre of advanced seas and oceans science-technology research and development, a UNESCO city of film, a much favoured tourist city, and not to forget, the production site of the famous Tsingtau beer offered in Chinese restaurants all over the world. During 1897 to 1914 the city had been a German mandated territory, named Tsingtau. Numerous bird's-eye plans from this period depict the city and its growth Fig. 6.1). They were drawn by Chinese artists and taken by German naval personnel back to Germany as commemorative documents. Today, most of the bird's-eye views are in German public archives, libraries and museums or in private hands. In an impressive book published in two languages (German and Chinese) they are shown together with numerous images and photos of German buildings and ships (Hinz/Lind 1997, Kaster, 2017).

The "German" history of the city is still visible in Qingdao. In contrast to many other cities in China, the old city centre of Qingdao has been almost fully preserved. It is well protected, but lacks professional and economically viable urban regeneration policies. The essay will introduce into the "German" history of the city and the innovative land regulation that the German administration had launched to avoid excessive land speculation 100 years ago.

Achievements of modern city development, such as the Sino-German Eco-development Park, have shown that the concept of modernization in China has changed very much. In the years ahead Qingdao will have to cope with all the challenges of a modern Chinese metropolis, though meeting the requirements of sustainable development and conserving its historical heritage will remain important tasks.

Fig. 6.1 Bird's view plan of Tsingtao. Cover of book by Gert Kaster. Source. Gerc Kaster

6.2 Method and Materials

The authors combine European, particularly German experience, and local Chinese expertiose. The Chinese co-author benefiztted from working many years in the local administration as responsible chief planner. The academic literature, on which the paper is based, includes international, German and Chinese articles and books on the history of Qingdao, when it has been occupied by Germany during 1897 and 1914, and the dissertation of the second author, the Chinese co-author, which had been submitted to the Technical University of Hamburg in 2002. The first author, the German co-author has frequently visited China and Qingdao over the last 20 years, first in 1996 and last in 2019. During the visits to China, he has participated as an advisor to numerous urban development projects, among others in Haikou, Changchun and Zhengzhou. In 2017, the province of Jilin has awarded him with "The Certificate for Outstanding Foreign Experts" in appreciation for his contributions to the development of Changchun. Hence 20 years of work experience and desk-research on urban planning in China made him familiar with the related academic literature and the practice of urban development. During his visits to Qingdao he had been involved in local development discussions. As a member of the advisory committee to explore new uses for the old international airport district, he learnt much about local strategies to make Qingdao an international location. The second author has much insight into local political milieus and access to urban development documents. Both authors met frequently and discussed the challenges and successes of urban development and urban regeneration in the city.

6.3 From Tsingtao to Qingdao

During the last decade of the nineteenth century a German frigate cruised the Kiautschou Bay on the Eastern Coast of China, exploring the suitability of the area for acquiring a share of Chinese territory for a German naval base (Hinz & Lind, 1998; Wilhelmy, 1903) Germany was a colonial latecomer, and alarmed by the British establishing Hong Kong as a crown colony in 1842, Ferdinand von Richthofen, a renowned German geographer, urged his government to establish a mandate in China. The place, at that time called Tsingtao (now Qingdao), had been suggested by von Richthofen, who, based on experience during lengthy travels during 1862 and 1872, wrote the first comprehensive geographical encyclopaedia on China written by a Westerner (von Richthofen, 1912). A few years later, authorized by Wilhelm II—the German emperor—and his chancellor Otto von Bismarck, Admiral Alfred von Tirpitz successfully negotiated with the Chinese authorities a 99-year leasing contract for the territory. A feasibility study, commissioned thereafter, soon confirmed the suitability of the bay for a solid port installation. During an extremely short period of 12 years, a German settlement town, named Tsingtao was developed on the site with up-to date infrastructure and residences in German style (Cai and Li 2010; Herold 2006; Leuthner/Mühlhahn 2006; Mühlhahn 2007; Parger 2012). Twice a week a German freighter ship linked Hamburg and Tsingtao to bring German administrative staff and their families, businessmen and consumer products to Tsingtao. The "*Deutsch-Asiatische Bank*" even issued a local currency (Fig. 6.2). In 1913, one year before the Japanese army took over the German Colony, the city had a population of 53,312 Chinese, 2400, German soldiers 2069 European and Americans, and 205 Japanese.

Fig. 6.2 Tsingtao Dollar, issued by the Deutsch-Asiatische Bank in 1914. Source: Archive Kunzmann

The globally renowned beer brand "Tsingtao" is the outcome of the demand of thirsty German mariners for German beer. The transport via sea (around three weeks) took too long. Consequently, a local brewery had to be built. A project-finding mission positively assessed the local quality of water and malt, and since then and until today, Tsingtao beer is served in China restaurants around the world. The privately owned brewery is now mainly in Chinese hands, after Anheuser, an American company, sold its share to Ashai, a Japanese corporation, who in turn sold it to Chinese shareholders on the Hong Kong stock exchange in 2017.

The townscape of Tsingtau, developed within just a decade, during 1897 to 1914, reflected the architectural style and the identity of a middle size German town, with church towers and red roofs. This is not surprising, though the introduction of a very innovative land regulation (land regime) order is worth mentioning. For planners, the way in which the German occupiers have handled the intricate land issue in the colony is of particular interest. The German *Landordnung* (land regime) aimed at addressing the challenge of land speculation, which at that time had already affected urban development in other Chinese cities, primarily in Canton (now Guangzhou) and Shanghai. Having occupied the territory, the German government negotiated a 99-year leasing contract with the Chinese government for an area of 540 km^2. The contract was signed in 1898. Originally, the German military administration wanted to buy land from Chinese land owners, though due to the lack of funds, only the pre-emption rights were acquired, mainly in order to hamper speculative activities. In order to guide and control the development of the city, the naval office took over the responsibility for all planning and development activities. The compulsory purchase of land from the Chinese land owners was in the hands of Wilhelm Schrameier, a German interpreter. He drafted the very innovative *Landordnung* for the German

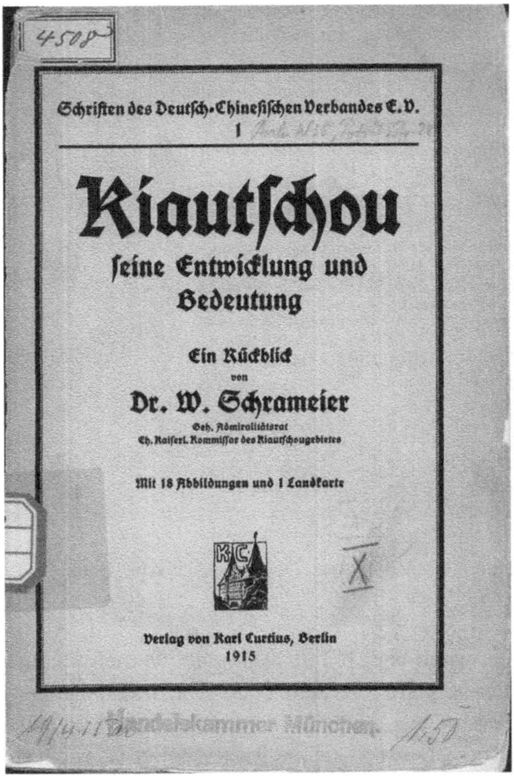
Schriften des Deutsch-Chinesischen Verbandes E.V.
1

Kiautschou
seine Entwicklung und Bedeutung

Ein Rückblick
von
Dr. W. Schrameier
Geh. Admiralitätsrat
Ch. Kaiserl. Kommissar des Kiautschougebietes

Mit 18 Abbildungen und 1 Landkarte

Verlag von Karl Curtius, Berlin
1915

Fig. 6.3 Nook cover Schrameier (1915) Kautschou, seine Entwicklung und Bedeutung. Ein Rückblick. Source: Archive Löhr

governor (Schrameier, 1915; Warner, 1996; Warner, 1999) (Fig. 6.3). The regulation aimed at controlling excessive land speculation by later buyers of land from the government. Landowners, who had bought land from the German administration and sold the land thereafter to third parties, had to return one third of the incremental land value. The land regulation also aimed to hinder the negative and socially unacceptable implications of future housing development. Chinese buyers had to accept that land had to be developed within three years after purchase or otherwise pay an annually increasing tax to the German administration to compensate them for infrastructure development.

The pre-emption right enabled the German planners of Qingdao to design their concept according to planning criteria without any intervention from landowners. Based on this plan, the administration leased the land step by step in the Western part of the leased area, in which the city could be developed. Thus, the planners could work without considering the interests and stakes of land owners. They were able to set up the plan for a new city under ideal conditions (Löhr, Fu, & Zhou, 2014). The city was built on hilly territory. Around the central government offices were European and Chinese businesses and residential quarters—mainly German-type villas. Three

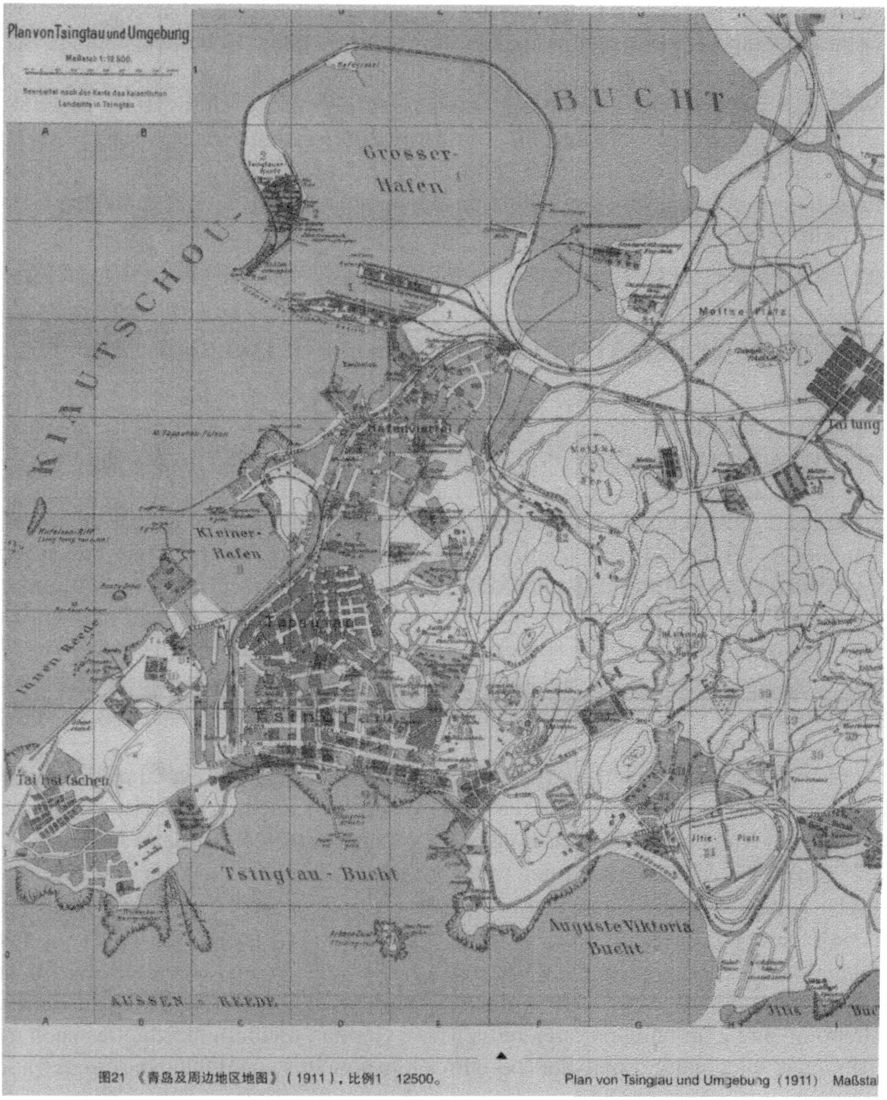

Fig. 6.4 Tsingtau 1911. Source: Archive Kaster

kilometres away from the port an industrial area, two settlements for workers and three military barracks for German soldiers were developed (Fig. 6.4).

Sun Yat Sen, the founder of the Kuomintang and first President of the Republic of China was very much impressed by the innovative land regulation of Wilhelm Schrameier. He ordered the translation of the document into Chinese and made Schrameier his personal advisor. Sun and Schrameier jointly developed a land programme, which was then enshrined in law in 1930. Due to the upcoming civil war, the law couldn't be enforced. It had, however a great impact on the land regime in Taiwan.

(Matzat, 1985, p. 32). The Qingdao land regime was much more advanced than most of the contemporary land regimes in Western countries today (Bohnsack, 2000, p. 18).

The German occupation ceased in 1914 with the beginning of the First World War, and 1919 Tsingtao came under Japanese occupation until 1922. In 2018, almost 100 years later, the technical infrastructure system (water, sewage) in the historical German town is still in good shape. In contrast to other cities in China, most houses built a century ago in the historical area of Qingdao are protected from demolition.

6.4 The German Architectural Heritage of Qingdao

Despite the visible economic decline of the historic centre of Qingdao, the approximately 3 km² "German" old town is a very popular place for citizens and tourists due to its location on the coast, its structural quality and its European architectural style and atmosphere. More than other major Chinese cities, Qingdao has preserved much of its original identity thanks to its German architectural heritage. This inner city district has withstood the pressure of investors interested in profit, who have destroyed local identities in most major Chinese cities. The loss of identity seems to be a special Chinese problem (Esherick 1999; Hassenpflug 2010). In contrast, local building laws and land use regulations protect the architectural heritage of Qingdao. The historic old town with an area of about 6 km² aims to be declared as a UNESCO cultural heritage site.

Over the last 10 years the city administration had drawn-up several plans for the protection and redevelopment of the old town. At the same time, the city—with the help of government bonds—has bought back numerous street districts (Jie-fang and Li-yuan). Nevertheless, the structural condition of the houses has become increasingly critical, as no convincing strategy for the economic use of the old town has yet been developed. Therefore, the "German" Old Town is still waiting for the cautious modernisation and development towards a lively, creative and economically flourishing, functionally mixed urban quarter, while avoiding to turn it into a tourist shopping and entertainment area only. Many German medium-sized cities such as Tübingen, Oldenburg, Landshut and Ulm show how this could happan (Fig. 6.5).

Fig. 6.5 Modern urban skyline of Qingdao. Source: Drawing by Klaus R.Kunzmann

6.5 Qingdao 2019

Qingdao is more than the exotic remnant of German colonial history. The "German" quarter is only a historical island in the 11.240 km^2 large archipelago of the city, in which around nine million inhabitants were registered in 2018 This made Qingdao the largest city in Shandong Province (2016: 99.47 million inhabitants) and the twelfth largest city in China (QIO 2018).

As early as 1984 Huangdao (a district of the city of Qingdao) became one of 14 Chinese economic zones on the coast that were authorized to trade with foreign countries. Since then, Qingdao has developed into a successful industrial city with one of the country's largest ports. The new port of Qianwan, a 12.7 km^2 free trade zone, is located directly opposite the "German" old town on the other side of the entrance to the bay. The port is the largest iron ore transhipment point in China and one of the starting points of the new maritime silk road from China to Europe. A terminal for cruise ships is under construction on the location of the old port of Qingdao. There, an attractive 4.3 km^2 district with hotels, a shopping centre and numerous museums, better connected to the city by public transport, will be developed. Via Jinan Qingdao is well connected into the dense network of high-speed trains in China. Eighteen daily connections to Beijing (3.6 h travel time for 819 km) have been available in 2019, and later four connections to Shanghai (about 3.5 h travel time for 754 km). Further expansion of the high-speed rail network will make Qingdao even more attractive from an economic point of view and give new impetus to spatial development in the urban region.

Public transport has high priority in the city. The general transport plan (Qingdao Urban Rail Transit Network Plan) provides for a total of 16 metro lines with a total length of 838 km to be built in 2050. The government's target is that 80% of the population in Qingdao will not live more than 800 m from one of the 400 stops of the metro network. Five lines are expected to be in operation by the end of 2020. As early as 2017, 95% of the population were served near their homes by bus lines (500 m radii). In addition, a tram network on separate routes is under construction. The first 8.8 km long section of the Qingdao tramway was opened in 2016. A novelty in Chinese tram history is the fact that the vehicles are equipped with fuel cells.

The international airport of Qingdao connects the city with 157 destinations in China and with 27 international destinations. From 2020 onwards the new international airport in Jiaodong, about 50 km from downtown Qingdao, near the city of Jiaozhou (Kiautschou), will go into operation. It will further improve the international accessibility of the site. The development of new industrial and economic special zones has also begun there. The development of the new airport and the associated expansion of the regional transport network are intended to narrow down the distances between the old and new districts of Qingdao.

Impressive tunnels under and bridges over the bay open up the city region for new development. The old airport, together with its 30 km^2 area, most of which is densely built-up, is to be developed into a "new urban living space". The city administration plans to build on the site a new regional shopping centre, an international conference centre and another cultural centre there.

Qingdao is an industrial city, an important port and a much visited tourist place in the province of Shandong. The city has an industrial focus on railway construction, household appliances (Haier, Hisense, Aucma, etc.), shipbuilding automotive engineering, pharmaceuticals and petrochemicals, port and marine technology, and above all high-tech railways. The first prototypes of 600 km fast maglev trains have been successfully tested. They will be introduced in 2021.

Qingdao is also a city of knowledge with over 20 universities. China Ocean University, one of China's most renowned universities, is particularly renowned for its research and development in the fields of physical oceanography, life sciences, ocean geology and the environment. In addition to the universities, Qingdao has numerous science and technology parks as well as 10 national research centres, including the National Laboratory for Marine Science and Technology.

Besides Beijing, Shanghai, Guangzhou, Shenzhen and Hangzhou, Qingdao is also one of the most attractive Chinese locations for national and international trade fairs and congresses, including the International Consumer Goods Fair and the International Marine Fair. Qingdao is one of 13 UNECSO Cities of Film besides Sydney and Rome (not Los Angeles and Berlin!). In 2018, a private company was founded in the city, the Oriental Movie Metropolis. The aim is to promote the national TV and film industry and to try to establish an Oriental Oscar comparable to the Oscar award in the USA.

In Chinese society, the industrial and port city on the Yellow Sea is also a popular destination for Asian tourists. Since 1991, the annual Oktoberfest has been particularly attractive, attracting Chinese visitors from China to the city's golden beaches. Since the 29th Olympic Games—Qingdao hosted the yacht competition for the 2008 Olympic Games in Beijing—Qingdao has also marketed itself as Sailing City of China, the city of sailing. Qingdao is the stage of numerous international regattas (Volvo Ocean Race, ISAF Sailing World Cup). For 2030, the city expects over 80 million tourists annually. Since 1991 an annual two-weekly international beer festival is held in the city attracting Chinese visitors to its golden beaches.

As a result of the relocation of the municipal city government in 1991 and the development measures to hold the Olympic sailing competitions in the east of the "German" old town, a new, modern, densely built-up district of around 9.5 km^2 with apartments, offices and shopping centres had been created. There the population density there is 21,000 inhabitants/ km^2. This development led to an increasing economic, social and cultural polarization between the old town and the new district.

The current urban development plan (Master Plan Qingdao 2011–2020) assumes that Qingdao will develop into a nationally significant city in which the port, the technology-oriented industry, the local knowledge industries, culture and tourism flourish. In this context, the city government is focusing strongly on environmental protection, improving social living conditions and reducing poverty. Many chemical and steel factories have been relocated from downtown Qingdao to rural areas. Rivers have been and are being cleaned throughout the city. Many residential areas with low quality housing (*Hüttenviertel*) were demolished. Most inhabitants were compensated and resettled in new high-rise buildings in adjacent neighbourhoods (Zhan 2002).

On the basis of the Qingdao *New-type Urbanisation Plan* (2014–2020), Qingdao is pursuing a strategy to change the regional settlement structure. The strategy aims to achieve polycentric decentralization with the development of small model cities and the preservation and upgrading of rural villages.

6.6 A Closer Look to Three Urban Development Projects

Three urban development projects in Qingdao are particularly worth mentioning: the tunnels and bridges over Jiaozhou Bay, the Sino-German Eco-Development Park west of Jiaozhou Bay and the municipal retro development project in the Jimo district.

Jiaozhou Bay tunnel underpass and bridge: The impressive and regionally significant tunnel underpass and the bridge over Jiaozhou Bay opened up the urban region for new developments. The tunnel underpass, completed in 2010, is about 6 km long. This has considerably shortened the journey time from Qingdao to Huangdao and changed the entire urban structure. Another new tunnel between Qingdao and Huangdao will be built after 2020. The Jiaozhou Bridge, which had been completed in 2011, is 42 km long and is thus the world's longest above highway bridge above water. It connects Qingdao with Huangdao, Jiaozhou and the old and new airport. The impressive bridge has already triggered important impulses for the economic development of the Qingdao urban region. In connection with the new airport in the east of Jiaozhou and the Free Trade Area Shandong on the west side of Jiaozhou Bay, several new model industrial parks are planned, which are well connected to the city with this bridge (Figs. 6.6 and 6.7).

The Qingdao Sino-German Eco-park: The functionally mixed park, designed by the German architects and engineers is an example of future-oriented sustainable urban development in China and recognized as such by being awarded the UN Sustainable Urbanization Prize at the 13th United Nations Sustainable Urban Development Forum. The park combines workspaces with residences. A number of support institutions in the park aim to attract German industries to Qingdao and promote German- Chinese co-operation. The German football club Bayern Munich manages a professional football school in the park, which is training a new generation of football players from all over China and which benefits from Germany's reputation for world-class football. A team of prominent German and Chinese advisors, including a former CEO of Siemens, is advising the project on its promotion in economic and political circles. The villagers, who lived in the area, were compensated with owner-occupied apartments in both multi-storey and high-rise buildings. Some innovative environmental features (e.g. passive house technology, allotment gardens) give the eco-park its unique profile (Fig. 6.8).

Jimo: A most unusual project is underway in Jimo, a city in the wider administrative territory of Qingdao (Fig. 6.9). The city with a population of just 1.3 million

Fig. 6.6 The new International Airport of Qingdao. Source: Qingdao Local Government

Fig. 6.7 The Jiaozhou Bridge. Source: Qingdao Local Government

is one of the most affluent middle-sized cities in China. Its local economy is based on "blue"—ocean related industries. The old inner city quarter shows a typical rectangular Chinese layout. With the exception of two small houses from the 1930s, the old town has been demolished and replaced by a totally new Chinese town in old style, with a boutique hotel, a Chinese garden, a museum, numerous shops and

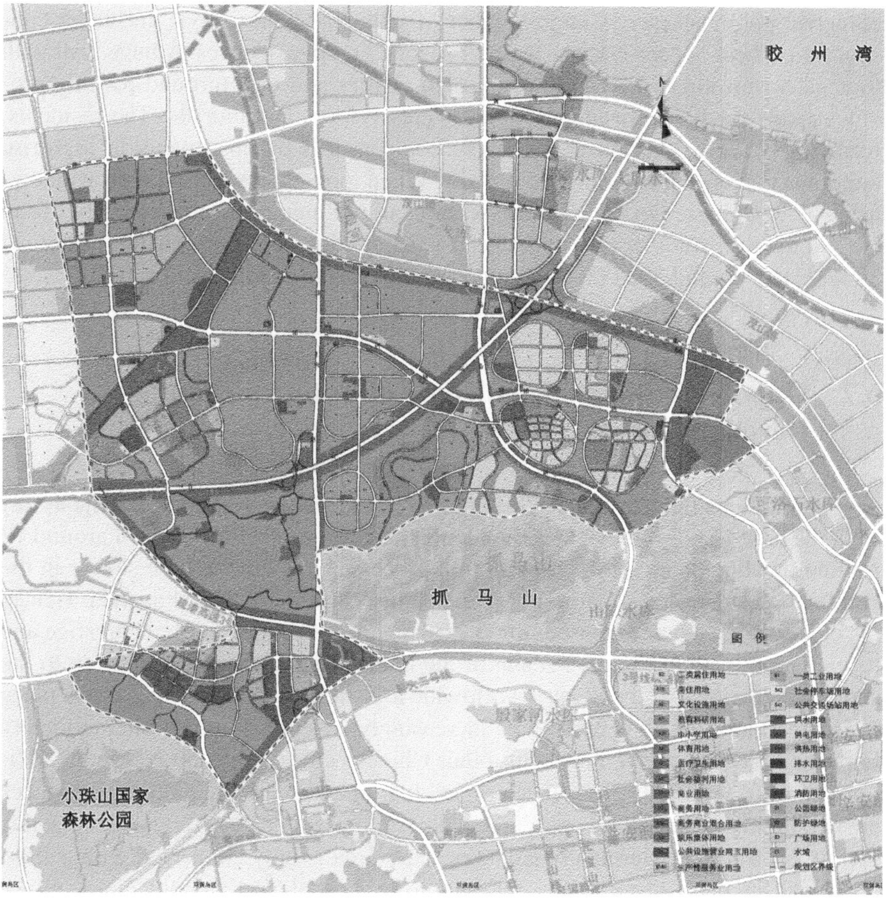

Fig. 6.8 The Sino-German Eco-development Park. Source: Archive Erpeng Zhan

Fig. 6.9 Jimo City Center (left). Source: Photo Klaus R.Kunzmann; New (old) city center Jimo (right). Source: Photo Klaus R. Kunzmann

attractive public spaces. A large number of villas in Chinese style are under construction. These are for lease as main or second home residences. A huge wall with four gates surrounds the whole new urban quarter and underground parking provides space for future residents, shop owners visitors and tourists. The residents, formerly living in the old centre, were compensated by apartments in high-rise buildings that are surrounding the old city. When looking down from their windows to the new Chinese Disneyland, they can recall old memories of their daily life in the old quarter. Another project in Jimo is an ambitious city expansion project of 80 km^2 with around 500.000 residents, branded *Vocational Education City*.

6.7 Opportunities and Challenges of Future Urban Development

Qingdao will continue to grow in the coming next decades, perhaps not quite as fast as in the past decades. It can be assumed that the city will be the home of around 12 million inhabitants in 2030. Urbanization processes in China will be slower in the coming years decades. Quality not just quantity is on the Political agenda. Hence, the creation of a better quality of life in city regions will enjoy political prestige and receive more priority on the agenda of the national government. According to the summit conference of the SCO, a worldwide logistics association, which took place in Qingdao in 2018, the city is to be increasingly developed into an internationally important city. In competition with other Chinese cities, Qingdao can score additional points with its preserved “German” heritage, as German culture and values are still highly valued in the opinion of Chinese society (Fig. 6.10). The fact that this identity as a former German mandate or lease area is only connected with a short phase of Qingdao’s history does not seem to be a problem. Like all Chinese cities, Qingdao faces many challenges in the coming years.

Some neighbourhoods of the German Quarter are in bad physical shape (Fig. 6.11). The local government has already acknowledged the high value of this quarter, though has not year developed a convincing strategy to revitalize the quarter. The argument is that finance hinders these approaches, although it seems to be more about political priority. A convincing concept for revitalizing the old city quarter and making it a thriving creative place and tourist target is not yet on the political agenda, although the city has started to explore ways and means to find a convincing concept. Until today urban revitalization by modernizing existing building stocks, carried out together with residents and shop owners, is still a challenge for local governments in China. Meeting the requirements of Central Government to increase GDP, they demolish obsolete structures and replace them with high-rise apartment houses. The old German quarter in Qingdao would be a pertinent case for carefully guided gentrification. This requires a slow revitalization process, managed by an independent local district manager intensively communicating with the few remaining residents and local shop owners, as well as with artists, future shop owners and start-ups in creative industries. Within a few years the quarter could be a thriving

Fig. 6.10 Christ Church, Tsingtau built during 1908-to1910. Source; Drawing by Klaus R. Kunzmann

inner city quarter, where a younger generation of Qingdao residents and tourists would enjoy strolling around, shopping and enjoying local cuisine. However, preserving identity in the historic "German" quarter of Qingdao is a challenge that requires "only" political will, careful planning and professional implementation of a convincing urban renewal concept. First demonstration projects are already underway. They will be finish during 2020 (Figs. 6.12 and 6.13).

Like all Chinese cities, Qingdao will have to cope with the consequences of economically and politically promoted automobility. This will not succeed with the use of electric vehicles and smart technologies (self-propelled cars) alone, not even through the further expansion of public transport, access and parking restrictions in

Fig. 6.11 The regeneration zone of the German Quarter. Source: Photo Wang Fang

Fig. 6.12 The German and Chinese skylines of Qingdao. Source: Photo Klaus R.Kunzmann

Fig. 6.13 The Port of Tsingtau. Source: Archive Kaster

individual city districts. It will only be possible to reduce the mobility of citizens through administrative measures that allow a stronger mix of living and working places and through behavioural change. Therefore, it is necessary to optimize the spatial structure and functional layout of the whole larger city around the Jiaozhou Bay. But given the economic conditions in China, Qingdao itself can do little to reduce the general level of market-driven automobility in the country.

It is difficult to create something like a sense of home and local distinctiveness in the modern districts of a modern city that enables the residents to identify with their living spaces and actively participate in their design. Important dimensions are attractive jobs, social security, locally accessible public facilities and affordable rents. Not urban expansion, but urban renewal or revitalisation is the great challenge for which politicians, planners and citizens in China must prepare themselves—not least, however, the universities that train planners for future repair challenges.

With great enthusiasm, Qingdao has developed the Eco-Park into a sustainable mixed-use district of around 30 km^2, which at least from a Chinese perspective meets some criteria of sustainability. With its initially formulated development goals—idyllic city, green development, beautiful life—the entire Eco-Park is now regarded in China as a model for sustainable urban planning. The planned metro connection between the Eco-Park, the city centre and the new airport, which will further contribute to the park's economic development. Transferring the comparatively successful model of sustainable urban development to other development projects in the city is a challenge that the city of Qingdao will have to face in the years ahead.

After completion of the new airport of Qingdao, it is necessary to develop a convincing, perhaps even more sustainable concept for the use of the old airport in Liuting and its urban surroundings. The relocation of the airport will force many companies in the logistics sector to change their location and invest in the new airport. This could become a lesson for the sustainable renewal of an existing urban neighbourhood. But this also requires political visions, professional strategic planning at regional level and committed urban development management. The relocation of the industrial port to Huangdao and Dongjia-kou and the construction of the Free Trade Area Shandong on the west side of Jiaozhou Bay between Jiaozhou and Huangdao will offer new opportunities to use the old industrial quarters in the city systematically for the sustainable improvement of work and quality of life in the inner city. But this also requires far-sighted political decisions.

Despite its more than 20 universities and colleges, Qingdao is still not a significant, internationally outstanding knowledge location in the eyes of Chinese society. In some fields (marine technology, shipbuilding and electrical household appliances, etc.) Qingdao appears in international ranking tables. However, the city's future competitiveness will primarily depend on how it manages to distinguish itself from competing cities in the country such as Suzhou, Hangzhou, Wuhan, Chongqing, Jinan or Nanjing. Factors of competition attracting qualified young labour are easily accessible and affordable cultural, educational, recreational and health services, and an attractive cityscape, managed by an efficient city administration.

6.8 Outlook

Qingdao is not a competitor to the five super city clusters (Great Bay Area City Cluster Shenzhen, Hong Kong and other cities), the Yangze River Delta City Cluster (among other cities Shanghai), the Yangze River Middle Reach City Cluster (Wuhan, Changsa and other cities, Jing_Jin-Li (Beijing, Tianjin) and the Chengdu-Chongqing City Cluster. Like other second tier cities, Qingdao will have to strive in the future to strengthen its own profile in the area of tension between top-down and bottom-up policies, to align urban development with the requirements of ecological sustainability and resilience, to balance between cultural, tourism and industrial development, and to secure the quality of life of its inhabitants within the framework of their local development strategies. This will probably only succeed if the national government makes even greater efforts to improve the framework conditions for sustainable spatial development at all administrative levels. Chinese cities still have much to learn in the fields of sustainable urban development and regional spatial planning. As learning is traditionally deeply rooted in the Chinese society, Qingdao will certainly succeed to shape its future.

References

Bohnsack, Gustav. (2000): Vor 100 Jahren: Landordnung von KIAUTSCHOU (Tsingtau), *Allgemeine Vermessungswesen Nachrichten (AVN)* , No 1/2000, 9-18.

Cai, Lui and Li Baihao (2010) A Study on History of Early Modern City Planning of Qingdao (1891-1949) *China City Planning Review* Vol. 19, No. 4 , 63-72.

Eshervik, Josef W. (1999) *Remaking the Chinese City. Modernity and National Identity, 1900-1950.* Honolulu: University of Hawai'i Press.

Herold, Heiko (2006) : *Deutsche Kolonial- und Wirtschaftspolitik in China 1840 bis 1914. Unter besonderer Berücksichtigung der Marinekolonie Kiautschou.* 2., verbesserte und erweiterte Auflage. Köln: Ozeanverlag Herold,.

Hinz, Hans-Martin and Christph Lind Hrsg. (1998) *Tsingtau; Ein Kapitel deutscher Kolonialgeschichte in China 1897 -1914.* Berlin: Deutsches Historisches Museum.

QIO (=Qingdao Information Office) (2018) *Qingdao 2018.* Municipals People's Government. Qingdao Publishing House.

Kaster, Gert und Stadtarchiv Qingdao (2017) *Die Vogelschaupläne von Tsingtau.* Kiel: Ludwig

Leuthner, Mechthild und Mühlhahn, Klaus Hg. (2006) *Kolonialkrieg in China. Die Niederschlagung der Boxerbewegung 1900-1901.* Berlin Ch. Links Verlag.

Mühlhahn, Klaus (2007) Deutschland's Platz an der Sonne? Die Kolonie "Kiautschou". In: Leutner, Mechthild und Klaus Mühlhahn , Hg. (2007) *Kolonialkrieg in China. Die Niederschlagung der Boxerbewegung 1900-1901.* Berlin: Links 43-48.

Löhr, Dieter, Shihe. Fu and Li Zhou (2014): The Qingdao Land Regime-Lessons Learned. Journal of Translation from Foreign Literature of Economics (JTFLE), Special Issue on Land Taxation, January. Online: http://se.xmu.edu.cn/jzyc/UploadFiles/ 2014371831770554751115776.pdf

Matzat, Wilhelm. (1985): *Die Tsingtauer Landordnung des Chinesenkommissars, Wilhelm Schrameier.* Online: http://www.tsingtau.org/wp-content/uploads/chinesenkommissar_schrameier_von_w_matzat.pdf

Mühlhahn, Klaus (2000) *Herrschaft und Widerstand in der "Musterkolonie" Kiautschou. Interaktionen zwischen China und Deutschland 1897–1914* (= *Studien zur internationalen Geschichte.* Bd. 8). Oldenbourg, München 2000 (Zugleich: Berlin, Freie Universität, Dissertation, 1998).

Parger, Hans Georg (2012) *Tsingtau/Qingdao. Deutsches Erbe in China.* Mit einem Vorwort vom Leiter des Chinesischen Historischen Museums in Tsingtau. Graz: Ares.

von Richthofen, Ferdinand (1912) *Atlas von China.* Berlin Dietrich Reimer.

Schrameier, Wilhlm (1915) *Aus Kiautschous Verwaltung. Die Land-, Steuer und Zollpolitik des Kiautschougebietes.* Jena: G. Fischer.(published in Shanghai in Chinese in 1923).

Warner, Thorsten (1996) *Die Planung und Entwicklung der deutschen Stadtgründung Qingdao (Tsingtau) in China. Der Umgang mit dem Fremden.* PhD Dissertation TU Hamburg-Harburg.

Warner, Thorsten (1999): Die Landordnung von Tsingtau, in: *Zeitschrift für Sozialökonomie*, Vol. 36, 15-18.

Wilhelmy, Emil (1903) *China: Land und Leute. Illustrierte Geschichte des Reiches und seiner Wirren.* Berlin: W. Hertlet.

Zhan, Erpeng (2002) *Entstehung, Wandlung und Sanierung der „Hüttenviertel" in Qingdao. Tsingtau; Städtebau und Veränderungen der Stadtstruktur als soziales Problem in den Küstenstädten der VT China.* Berlin: Logos Verlag.

Chapter 7
Three Stages of Urban Community Development and Regeneration Planning in Chongqing (2010-2020)

Ling Huang, Junhang Luo, and Xiang Peng

Abstract Since 2010, when the urbanization rate of Chongqing City had exceeded 50%, the city has shifted its focus from pure economic growth to living quality improvement. Urban planners began to focus more on community-based interventions within the city context that called for a better quality of life and a new and more people-oriented type of urbanization. Many theories and approaches have been developed and tested by urban planners to satisfy the residents' needs and to resolve social problems emerging in local community programs. This chapter characterizes and reviews multidisciplinary approaches of community planning in Chongqing City over the past 10 years. This process can be divided into three stages. At the germination stage (2010–2013), community development was mainly based on the formation of community neighborhood committees. Continuous declines of the quality of the physical environment in older communities had a tremendous negative impact on residents' daily life, which later developed into a major social problem. The planning aimed to improve the built environment of a community by identifying and recognizing its cultural assets and promoting its environmental renovation by applying a strategy of cultural renaissance. At the rising stage (2013–2016), planners aspired to incorporate innovative social governance into the community's regeneration program. The applied context was that social governance was part of China's national governance in 2014. Using the Asset-based Community Development (ABCD) approach, all community resources could be understood as community assets and could be divided into human assets, social assets, and physical assets of a community. The whole planning process was further divided into built environmental regeneration and community governance. The project finally yielded the following achievements: community action plans were developed, and broad-based community participation was achieved. The project showed that the

L. Huang (✉) · J. Luo (✉)
School of Architecture and Urban Planning, Key Laboratory of New Technology for Construction of Cities in Mountain Area, Chongqing University, Chongqing, China
e-mail: hwawa1025@vip.163.com; luojhang@cqu.edu.cn

X. Peng
Changsha Planning & Design Institute Co. Ltd, Changsha, China

L. Bian et al. (eds.), *Chinese Urban Planning and Construction*,
Strategies for Sustainability, https://doi.org/10.1007/978-3-030-65562-4_7

community planning of Chongqing City had expanded its content from the renewal of the physical environment toward comprehensive social planning. At the developing stage (since 2016), the demands and contradictions of community development have become increasingly complex and diverse. Furthermore, the ABCD approach has been applied to several community regeneration practices under various socio-economic conditions. By applying the ABCD approach, the goal of community diversification development can be achieved. The internal forces of a community can be further stimulated, and the relationship between community development and planning can be further strengthened. The significance of this chapter is the clear presentation of the community development process in Chongqing City over the last 10 years. This information enables both planners and policymakers to recognize the critical roles of communities in urban development as well as the roles of residents in community planning. Over the next decade of Chongqing's community development, it will be necessary to adopt ABCD to further highlight the locality and diversity of community planning. More importantly, the community governance framework requires further exploration and improvement. This forms a new opportunity and challenge for Chongqing City.

Keywords Asset-based Community Development (ABCD) · Urban community regeneration · Community planning · Chongqing City

7.1 Introduction

In 2016, the United Nations released the New Urban Agenda (NUA) at the Habitat III Conference. The NUA proposed the goal of "producing just, safe, healthy, accessible, affordable, resilient and sustainable cities and human settlement" and evoked the vision of "cities for all." In 2018, the United Nations Conference on Sustainable Development issued the Guidelines for Sustainable Cities and Communities (SUC). Criteria such as "inclusive community facilities and services," "safety," and "community management" were applied to evaluate community sustainability. According to global consensus, the community and the people in it exert essential roles in the urban development during the new era.

The 2017 document, "*Opinions on Strengthening and Improving Urban and Rural Community Governance*" was the first document on urban and rural community governance issued by the Party Central Committee and the State Council. Its aim was to "improve the urban and rural community governance system, and strive to build a harmonious, orderly, green, innovative, inclusive, and shared happy home for urban and rural communities." In February 2019, the Ministry of Housing and Urban-Rural Development issued the *Guidance on the Joint Creation of a Better Environment and a Happy Life in the Construction and Remediation of Urban and*

Rural Living Environment. People's livelihood orientation and community-based urban regeneration became critical starting points toward strengthening the quality of urban space. The need to coping with social transformation in the new era presents an opportunity for urban transformation and development from the perspective of sustainable community development. It can also achieve harmonious living, enhance urban competitiveness, and improve the city image.

In 2018, the "*Chongqing City Improvement Action Plan*" was introduced to increase the urban quality of life. This action plan presents 37 tasks, including "upgrading urban planning levels", "highlighting the characteristics of stereo cities," "promoting co-construction and co-governance and sharing," and "enhancing the quality of the city's humanities." It proposed to "use historic and cultural resources such as streets and lanes to connect historic blocks and traditional scenic areas" in old urban areas. Thus, it became a promising route of urban experience. Moreover, "accelerate the construction of a multi-level, full-coverage and humanized basic public service network, allowing the public to enjoy community-level public services in 10-minute walking and street-level public services in 20-minute walking". At the same time, this action plan stipulated to complete the transformation and upgrading of the old residential areas in the main urban area by 2020.

In 2018, the urbanization rate of Chongqing City was 65.5%, indicating that more than 20 million people lived in the city. Chongqing is composed of 38 districts and counties, with a per capita GDP of $9400. Yuzhong district is the birthplace of Chongqing City, has an urbanization rate of 100%, and a per capita GDP of $24,400. During the 13th Five-Year Plan (2016–2020), Yuzhong district shifted its focus from urban extension to urban regeneration, with a particular focus on the regeneration of the old community. This critical shift created vital spaces for the public, especially for low-income groups, and viewed the community as a place with assets that need to be activated. The creation of vital communities not only enhances the competitiveness of cities and the quality of urban life, it also brings social equity for citizens.

Guided by both global consensus and national policies, the content of community planning in Chongqing City has extended. The traditional grassroots-level administration of the past has developed a notable planning structure over the past decade. Achievements include improvements of e.g., the physical environment, social governance, public services, and cultural renaissance.

Instead of a needs-based approach to community development, asset-based community development (ABCD) encourages people to recognize the definitive assets of their community and start to identify ways to better utilize these assets. This chapter traces the history and evolution of the community planning of Chongqing City over the past 10 years. The connotations of community planning, to promote the quality of life through community development, are summarized.

7.2 Method

Since 2010, Planning has been part of developing the urban communities of Chongqing City. By introducing the "Asset-based" theory, this chapter explores a localization path, suitable for the progress of communities of Chongqing City. At the same time, the planning discipline itself has gradually been subject to innovation and improvement. As bystanders and witnesses, researchers analyze the evolution of community development and planning based on community assets, by applying a holistic and dialectic perspective. This chapter analyzes the macro policy background and the specific characteristics of problems of urban community development in Chongqing City. The characteristics and mechanisms of planning interventions and the transformation of the planner's role are analyzed based on community assets (Fig. 7.1).

To identify and analyze the developmental stages of Chongqing City community planning over the past decade, a method of theoretical deduction and empirical research is applied. Theoretical deduction refers to the formation of a relatively complete and detailed community development system. This system combes the evolution rules of Chongqing City's regional characteristics, policies, and relevant factors. This combination applies the premise of summarizing the context of community development and planning intervention both in China and internationally. This empirical research explores the process and transformation of planning interventions on community development by combining functional samples of Chongqing urban community development planning, based on community assets, since 2010. With regard to the mutual relationship between them, theoretical research, and the obtained deductions, are used as basis and guidance for empirical research. Empirical analysis is used in support of and to expand speculative deductions.

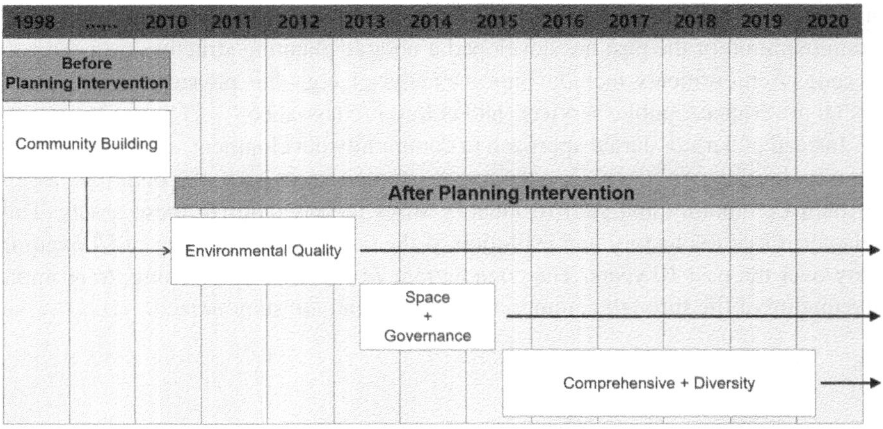

Fig. 7.1 The process of community development and planning intervention in Chongqing City

Moreover, a comparative research method was used to compare the urban community development practices of Chongqing City with leading Chinese cities and analyze their similarities, differences, and applied development laws.

7.3 Development Process of Urban Community Planning in Chongqing

Chongqing municipality was founded in 1997, and its urban community construction was fully launched the following year. At this time, China's urban planning was at the stage of rapid urbanization. Economic development and environmental beautification have become the cores of urban planning. The community development of Chongqing City was proposed by civil affairs departments, which focused on reforming the grassroots-level community management system, community management, community services, and community organizations, as well as creating harmonious communities. Then, the planning department and the civil affairs department were independent of each other. Planning intervention was not part of community development, and community-related issues were also not part of urban planning (Fig. 7.2). In 2000, with the release of China's *Notice on Forwarding the Opinions of the Ministry of Civil Affairs on Promoting Urban Community Construction in the Country*, by the State Council, community construction developed rapidly. In 2006, the basic framework of the community was expected to be completed. Since 2007, the urban community development of Chongqing City has gradually improved both the community service role and community management network and has created a civilized and harmonious community under the guidance of a series of macro-policies.

After more than 10 years of development, the community construction of Chongqing City gradually became successful. However, similarly, many problems

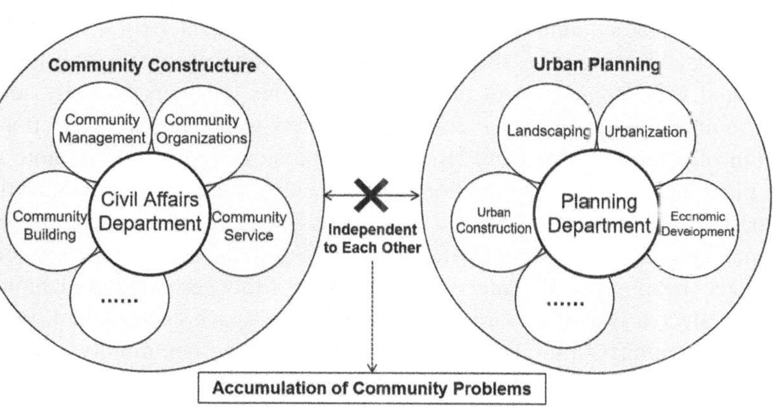

Fig. 7.2 Independent relationship between the civil affairs department and the planning department

still remain, such as the lack of close contact between the community and residents, unclear roles of both the community and the government, and unclear responsibilities between the community and civil powers.

7.3.1 Germination Stage (2010–2013): Focus on the Spatial Environmental Renewal of the Community

Since the beginning of the new century, and in response to the concept of "civil society" and "harmonious community,"in China, many cities with high economic development have established community planning practices to promote the development of their communities. The physical and spatial environment became the focus and the point of dispute of community's interest. The construction of space and the physical environment has always been an essential part of community planning (Zhenghan and Huien 2007). Moreover, influenced by the concept of "space orientation," in urban planning and for both governmental departments or planners, the primary concern of community planning still remained the physical and spatial environment. At around 2010, Chongqing City issued several documents related to community formation, and gradually strengthened the role and status of communities in the city. In 2008, work toward improving the urban space was started with the goal of "Five-Chongqing." In 2009, the "*Notice on the Comprehensive Improvement of Residential Areas in the Main Urban Areas*" was released. By taking Yuzhong district as an example, the government launched work to unify building facades on either side of the main road and started to improve residential areas.

In 2010, the municipal administration bureau of Yuzhong district selected Jiaxicun Community and Dajingxiang Community as pilot projects for community environmental renovation planning. In contrast to previous community environmental renovation plans, these pilot projects changed pure "project" trends and highlighted the development of community planning. This comprehensively improved the quality of the community environment and the characteristics of livable communities. By understanding these community pilot improvement plans as opportunities, and with the support of relevant departments, planners formally addressed the community level. The intervention by planners was not only a reflection of the opinion of relevant departments (i.e., that implementing projects is more critical than planning), but also a reflection of the "consciousness and self-reflection" of planners. Planners formally shifted their focus from the urban macro space to the community's daily micro space. And began to respect the needs of the residents and the "place meaning" itself. Intervention features of this germinating planning were mainly reflected in guidance regarding community space environment, identification of community cultural assets, improvement of the community environment, and the achievement of multiple space management goals. This period formed the beginning of the linkage between urban planning and urban community development in Chongqing City. At that time, planners began to apply the concept of "community assets" to community planning. By mining the "cultural assets" of the community (i.e., mainly the scattered historical and cultural buildings), the overall regeneration of the community space environment was promoted with a strategy of cultural revival.

The idea and practice of "Asset-based Community Development" (ABCD) emerged in foreign countries since John Kretzmann proposed the ABCD model in 1993 (Kretzmann John and McKnight 1993). Since then, the ABCD model has been further developed and widely applied in many countries (Tara O' Leary 2005). Numerous successful experiences and practices corroborated that the ABCD model exerts had remarkable effects on activating community economy, promoting social integration, and developing a sustainable community (Sherraden and Ninacs 1998). Relevant theories about the ABCD have been discussed in connection with Chinese sociology and other fields (Hong 2008). However, only few examples applied an "Asset-based" approach to community planning (Jun and Rui 2008; Ling 2012). Therefore, it remained challenging to learn from the experience and directly apply it to the community development and planning in Chongqing City.

At the germination stage, planners considered what should be retained by the old community, and how to complete its comprehensive management. Faced with the rich but diminishing historical and cultural heritage of communities, planners attempted to introduce the concept of community assets and a specific cultural revival strategy to promote the overall regeneration of the community space environment.

The two communities of Jiaxicun and Dajingxiang only applied the germination-stage practices. Combined with the abundant daily life space of the community, planning was conducted in two ways. First, preliminary research and review of historical data showed that the former residence of the Marshall and the former residence of Xianying in Jiaxicun community were well-preserved and had high historical value. Consequently, these could be applied to promote community regeneration as a "catalytic point." Based on environmental construction, these two residences of former celebrities could become part of public event planning with the aim to stimulate community vitality and promote communication among residents. Thus, a community space system could be formed by constructing community cultural routes and living routes that connect spatial nodes (Xiao Hongwei 2012). This enhanced the accessibility of space nodes, which enabled their effective use by residents. Second, the refined design was further strengthened. For public spaces along living routes, the focus was to meet the daily needs of residents. Therefore, plans mainly targeted the improvement of the quality of sanitation and facilities and the enhancement of landscape, accessibility, and openness. For space nodes along cultural routes, in addition to the above-mentioned methods, ground guidelines and relevant cultural identification systems were also applied (Fig. 7.3).

At about 2010, street vendors in Yuzhong district became more active. This increased activity created numerous job opportunities, provided convenient services for the residents of the community, and increased the vitality of the city. However, this increased activity also induced various inevitably adverse external effects. For example, these street vendors occupied the public space of the community and

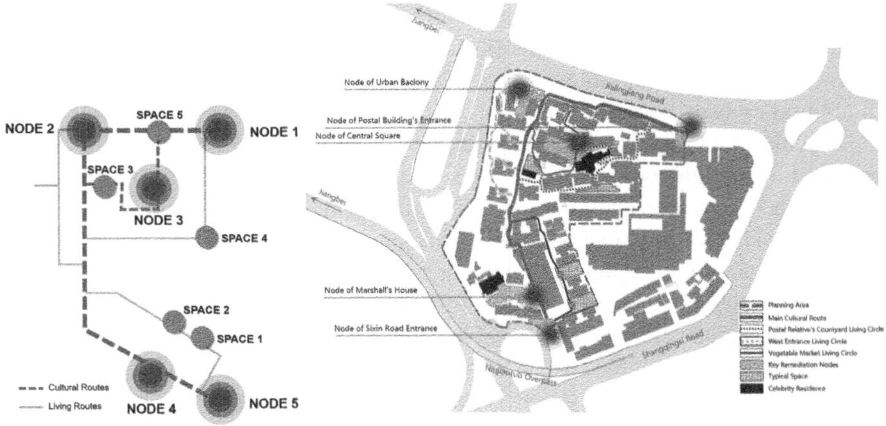

Fig. 7.3 Cultural route and life route planning of Jiaxicun Community

obstructed both transportation and residents' leisure activities. Moreover, the caused sanitary and environmental problems affected the city's appearance in a way that was inconsistent with the city's image. Food safety and problems with product quality also caused social disputes. These disadvantages caused difficulties and challenges for urban management. The survival needs of street vendors, the living needs of community residents, and the management interventions of relevant urban departments have suffered from long-standing but complicated contradictions.

Based on this description, the Yuzhong municipal administration bureau, other relevant departments, and planners took a series of actions to overcome this dilemma. In this process, multi-interest groups are appreciated. Unified booth facilities were designed first, and then, planners implemented space route management as an essential measure to weaken undesirable externalities. These methods realized two visions.

1. Beautifying the public space environment of the community and transforming mobile booths into the movable cultural landscape of the city.
2. Including the booths in the goal of inclusive urban growth to respect residents' daily living habits and the social role of public space.

During planning, the social role of space played an important part in space design and urban development. Moreover, the multiple participating-parties reached the consensus that each side can fairly share public space. Moreover, by integrating the street space and management mode, governmental departments could exert their duties more efficiently and effectively (Fig. 7.4).

In summary, this period aspired to use the community asset theory to promote the overall regeneration of the community. The community's "cultural assets" were used as entry point, and the "cultural route" was introduced to build a community public space system. However, in general, this period focused on the transformation of physical space and the establishment of community entrances and exits,

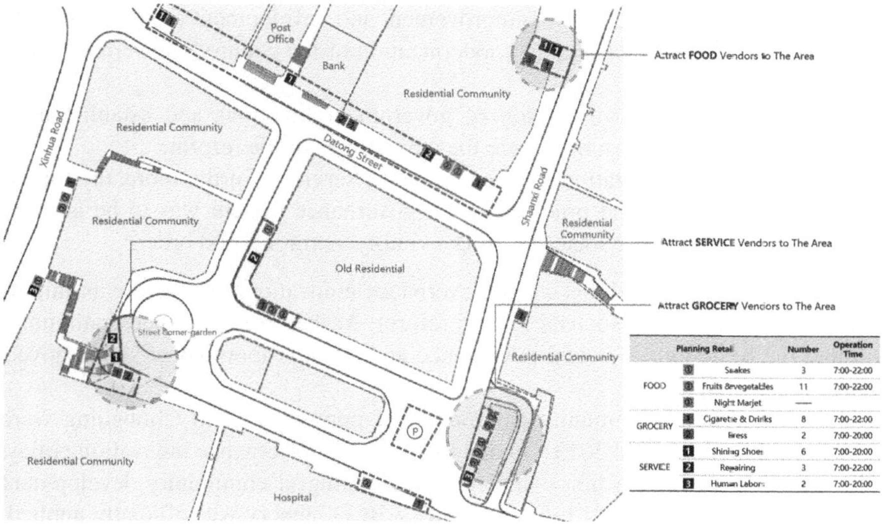

Fig. 7.4 Distribution of Datong street stalls planning in Yuzhong district

community laneways, courtyards, residents' leisure spaces, and urban balconies. The utilized planning method was primarily aimed at community space design. At the same time, the connotation of community assets remained relatively vague, while the content of the assets used for community planning was relatively distinct. Moreover, community "software", e.g., community services and community governance, was not yet involved.

7.3.2 *Rising Stage (2013–2016): Paying Equal Attention to Space Optimization and Governance Promotion*

In 2012, the State Council of China issued a notice that called for the establishment of a relatively complete community service facility, service content, service team, service network, and an operational mechanism by 2015.[1] Chongqing City issued the "*Twelfth Five-Year Plan for the Construction of Community Service System in Chongqing*," which prioritized the promotion of the coverage of governmental public services to communities and the development of community convenience services. In November 2013, during the Third Plenary Session of the 18th Communist Party of China Central Committee, the party proposed a number of action goals.

[1] This information originates from the *Notice of the General Office of the State Council on Printing and Distributing the Community Service System Construction Plan (2011–2015)*,2011-12-20

1. The general objective was the improvement and development of socialism with Chinese characteristics and the modernization of the country's governance system and capabilities.
2. Accelerating the transformation of government functions and establishing a service-oriented government were the main points of the reform.
3. Promoting the equalization of essential public services. Furthermore, the formation of a scientific and practical social governance system was to be accelerated to ensure that the society is full of vitality, harmony, and order.

The proposal signified that social governance innovation has become an important direction for future societal system reform. At the same time, the community formed the basic unit of social governance and the first foothold of social governance innovation.

The projects for community environmental improvement in Chongqing were mostly completed around 2013. Guided by the social governance innovation policy, Chongqing launched the first complete and meaningful community development plan, which lasted from 2013 to 2016. The "ABCD" theory was officially applied. In addition, the city-wide planning of community public service facilities has been launched. During this period, community governance innovation applied equal weights to the optimization of the community space environment and the improvement of public service facilities. The characteristics of planning intervention were mainly embodied by a change of concept, an extension of connotation, the construction of a relevant platform, and the improvement of facilities.

At the last stage, planners focused on solving problems associated with space and the environment of the community in Jiaxicun, Dajingxiang community environmental improvement planning, and booth layout planning in Yuzhong district. Moreover, better planning results were achieved by introducing the concept of "community assets". The typical urban stock community present in Yuzhong district had many spatial and social problems. Therefore, it was difficult to identify effective ways after years of problem accumulation. Moreover, Yuzhong District offers rare and excellent resources, such as sufficient physical assets, human assets, and social assets.

Under the condition of the stock development of the central city in Chongqing, the traditional physical space environment of the community always applied an aging-regeneration construction cycle. If the problem of space aging is addressed from the traditional "problem" and "demand" perspective, endless contradictions would result. The "asset-oriented" concept was merely an exploration that introduced advanced foreign ideas and attempted to apply these to Chongqing's stock community. In the community development planning of the Shiyoulu subdistrict of Yuzhong district, the ABCD concept was officially applied, which marked a shift in perspective from "problem" to "advantage." The residents' needs were as urgent as the community issues in this plan. The formation of community organizations and public participation in the whole process of community development was emphasized. Consequently, a survey of community assets, an assessment of community capacity, community development planning, and the implementation of the final community action plan were all based on this concept.

At the asset survey stage, the overall value of the community was re-evaluated with regard to three aspects: physical assets, human assets, and social assets (Fig. 7.5). The overall community asset database was finally built, and asset maps were drawn based on on-site surveys and data statistics. At the same time, the problems of various assets and the needs of community residents were not neglected, which formed breakthrough points for the preparation and implementation of the subsequent community development plan. Over the process of establishing a community development plan, the goals and measures were formulated in two ways: community space optimization and community governance upgrading. Considering community assets investigations and residents' needs, this development plan could be implemented more effectively. Finally, after drafting a community action plan, planners cooperated with relevant government departments and community residents to implement a number of short-term goals.

The involvement of planning in the social governance area of community development was a milestone. It overcame the community physical space, and responded positively to the policy and demand for community development in social governance. Space planning moved toward comprehensive social planning, and community planning has become a vital basis for innovation in social governance. Thus, urban planning from physical space to comprehensive social planning, became an essential symbol of the progress of planning development.

The history of planning and development in the USA has also experienced a transformation from urban physical construction planning to social integration. This transformation included physical construction, economic development, community development, and social welfare (Zhang Tingwei 1997). In the process, community governance has gradually developed from the obsolete public spirit of villages and towns toward a relatively perfect mechanism dominated by the civil society and formed by public organizations. Community governance showed characteristics of democracy, marketability, organization, and pluralism (Richard C. Box 1997;

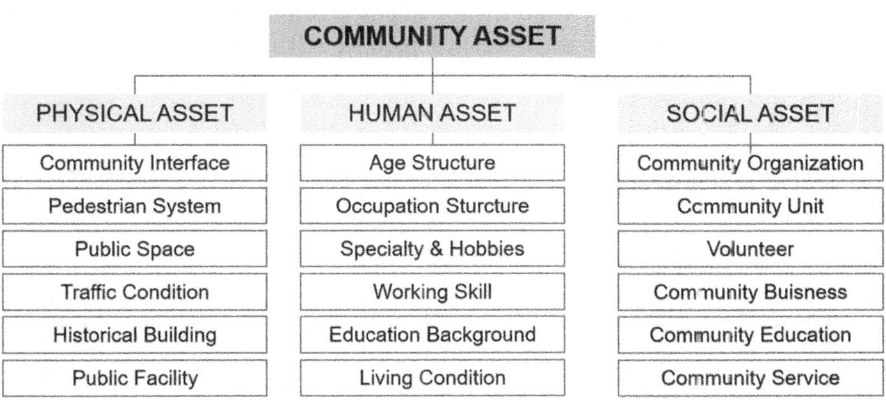

Fig. 7.5 Community asset survey content

Xuehua and Fany 2008). In China, community governance is mostly assigned to the civil affairs department. Furthermore, social development policies are often formulated at an urban scale. However, because of differences between urban communities and the ambiguity of community governance work, fulfilling the development goal was difficult. For urban planning, the Reform and Opening-up policy over the past 30 years has aided the rapid urbanization of China and rarely touched on the field of community governance. Since entering the new century, China has applied a number of planning reforms in the economically developed coastal cities such as Shanghai and Shenzhen (Huxiang 2005; Jiaxi and Lin 2015; Xiaozhuang 2015). On the one hand, via Planning interventions of community governance, governance agencies wanted to advance community development onto the standard track by working with non-governmental organizations. On the other hand, the urban planning discipline began to review the characteristics of this subject based on the philosophy of the developer (Min and Wei 2006). The leading community development plan of Chongqing City was also formulated and implemented based on this.

The community governance planning of the Shiyoulu sub-district was based on the combined effects of government support, residents' needs, and planning reflection (Fig. 7.6). Planners have explored a governance path suitable for Chongqing's urban community. Relying on already existing social assets within the community, the planners analyzed domestic representative community governance models, summarized its advantages and disadvantages, and investigated the community situations on-sites (Ling and Yanhong 2014).

During this stage, the government had sufficient and comprehensive resources. However, the residents had little community consciousness about the available resources. A "mixed-mode" of government guidance, residents' participation, and resident-government cooperation was explored. This approach formed a combination of top-down and bottom-up approaches by fully utilizing governmental public resources and cultivating residents' sense of participation (Fig. 7.7). Therefore, three levels were formed from the planning preparation to the implementation phase. (1) The public level was composed of community residents, community organizations, and social units, which strongly impacted the final result of community planning. (2) The expert level was composed of planners, representatives of scientific research institutes, and relevant departments. This level mainly focused on

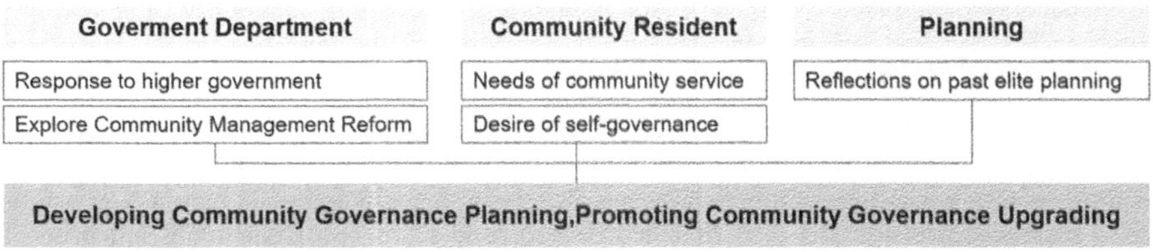

Fig. 7.6 motivation of Shiyoulu Street for community governance planning

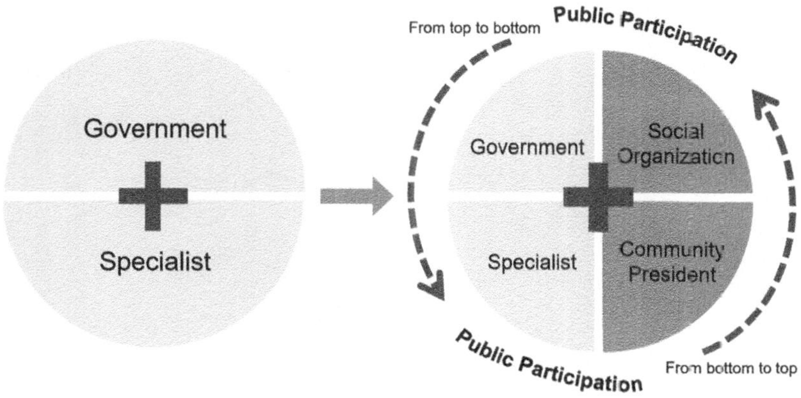

Fig. 7.7 Top-down and Bottom-up Work styles

technical support and project planning. (3) The management level was composed of district governments, street management committees, and committes of community residents that ensured smooth running of the business. This community governance planning was based on the existing "community management" framework, identifiable community human assets, and social assets. Using the Minlecun community as practice project, community governance planning was formulated from the perspectives of community autonomy, community service, and community organization training. The government supplied community governance planning by purchasing services.

Compared with the germination stage, planners have realized the importance of both "human assets" and "social assets" within the community (at this stage). By encouraging community residents to join in and build relevant participation platforms with the relevant departments, public participation in community development has achieved significant progress. It processed from the initial stage of community asset research, the formulation of development planning, to the community action plan (He 2014). The role of planners, who had offered a diversity of services such as coordinating, communicating, technological supporting, and organizing, also changed. They worked with other social workers and encouraged multi-stakeholders to actively engage with community development and promote the realization of inclusive community development goals.

Although public participation had positive effects at this stage, it still faced many bottlenecks. In addition to the limited number of channels of participation, several participant categories remained (such as vulnerable groups that had not been involved in the planning). Because of the immature and undeveloped participation mechanism, the participation depths and effects could not achieve the optimal effect.

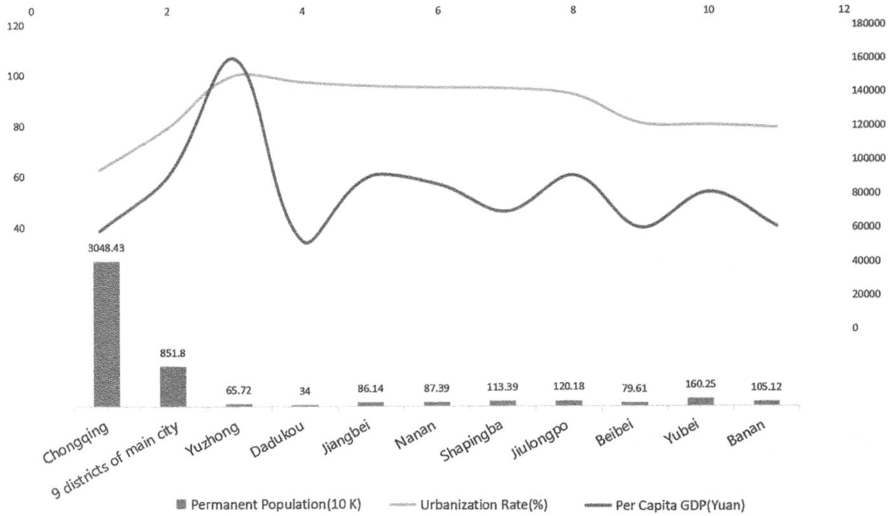

Fig. 7.8 Per capita GDP and urbanization level of Chongqing municipality and central city and nine districts in 2016

7.3.3 Developing Stage (2016–2020): Toward Comprehensive and Pluralistic Collaborative Planning From the Perspective of Governance Innovation

In 2016, the urbanization rate of Chongqing City reached 62.6%, and the per capita GDP reached \$8717. The urbanization rate of nine districts of Chongqing's central city was close to 80%, and the per capita GDP reached \$13,520 (Fig. 7.8). The stock development of nine districts of the central city was further promoted. According to the experience of developed countries, when the per capita GDP of a region exceeds \$10,000, the government begins to focus more on community development and community building in the region. Although significant differences exist between regions caused by the influence of multiple factors, the general law of urban progress still remained in the former condition. This showed that the higher the economic and urbanization levels, the more attention is focused on the community. Moreover, urban planning considers more social and economic factors and pay more attention to the scale of human nature and the needs of residents.

In one specific aspect, the urban community development of Chongqing City has responded to relevant policies of the state and higher-level authorities, thus further promoting social governance innovation. Major conferences, including the 18th and the 19th National Congress of the Communist Party of China, have repeatedly proposed innovative social governance and improved both urban and rural community governance systems. The "13th Five-Year Plan" of Chongqing further clarified that grassroots-level social governance innovation, socialized services, and residents autonomy should be promoted, while significant attention should be directed to grassroots communities.

In the meantime, the contradictions and requirements of community development became more diversified. The community is the principal carrier of the environment of the urban human settlement, urban culture, and urban humanistic ecology. However, in China's unique and rapid urbanization process, the natural progress and evolutionary processes of the community have been interfered with by China's decisive intervention and local development initiatives (Dongquan and Zhiyong 2012). With regard to Chongqing, the main problems of the community have accumulated in various aspects such as space environment, service management, historical context, residents' consciousness, and neighborhood relationships.

Furthermore, the needs of the community in response to the continuous development of both economy and society also became increasingly diversified, especially the comprehensive regeneration of the old city community. Yuzhong district has proposed demands to conduct community planning, especially of the old community regeneration projects in the "13th Five-Year Urban Regeneration Framework." This implies that community development would be associated with overall urban regeneration, and would thus no longer be restricted to the planning and transformation of a single community. Both community development and community planning would become an important starting point of the promotion of urban regeneration and would effectively upgrade urban governance.

The urban planning literature constantly reviewed the drawbacks induced by large-scale spatial production in the past. Moreover, the diversification of community conflicts and community needs, as well as the breadth and depth of planning interventions directed at community development were increasingly strengthened. As a result, more and more planning practices have been involved in urban community development since 2015. The main body involves many departments and governmental levels, and a number of developers have participated in community development projects, such as the planning of community neighborhood centers. The fundamental reason for this development was perhaps the profound transformation of social, economic and urban development.

The diversity of community development requires an increasing level of intervention in planning approaches after the preceding experience. Moreover, the urban planning discipline itself also needs to improve in terms of "daily life," "people orientation," "micro space," and "multiple perspectives." Since 2015, the government has promoted community development and planning practices at multi-perspectives, multi-level, and multi-class. Planners were guided by National policies and focused more on sustainable urban development by respecting the law of urban development and the community situation.

In 2010, the community "cultural assets" concept was initially established in the comprehensive environmental renovation plan of Jiaxicun community. Moreover, the idea of community cultural routes was first initiated to promote the renovation of the community environment, by relying on its historic and cultural resources. However, in general, the connotation and content of the "cultural capital" remained relatively simple, and the planning objectives mainly served the optimization of the space environment at that time. With further research and practice summary during the later period, the connotation of community "cultural assets" was continuously enriched and thus developed a much closer integration with community planning. In addition, after the 18th National Congress of the Communist Party of China, "Cultural Self-confidence" has become one of the national strategies. Planners thus face increased responsibility to know and understand the importance and power of culture and cherish its value. Since 2015, community spatial cultural planning has become an essential feature of planning interventions in the community development of Chongqing City. The cultural characteristic of Chongqing City and its historical and humanistic connotations in traditional old communities were highlighted by the "13th Five-Year Plan" in 2016. This showed that one of the essential community planning goals was to activate the community's cultural assets from a policy point of view. After this, a new type of livable urban community was built with both cultural display and urban tourism functions (Table 7.1).

The cultural assets of the community have significant regional characteristics. The old city of Chongqing has a long history and numerous cultural relics are scattered throughout various communities. Their protection and utilization have become a severe problem that has not been effectively addressed for a long time. Moreover, community planning provided an opportunity to reactivate cultural assets in Chongqing City. The critical point was to help residents to learn about and better understand the culture of their city through public participation platforms and specific opportunities. The inhabitants thus had more chances to actively interact with various cultures to acquire the best. The spatial and cultural planning aspect of urban planning is not merely a simple landscape design but an all-round spatial decision, based on the urban culture, especially when its in-depth understanding has been developed (Huang Ling et al. 2008). Similarly, cultural planning within community planning is not a simple micro-landscape design, but a comprehensive development decision of community culture based on community spirit and

Table 7.1 Planning and empirical comparison of community assets (2015–2017)

Planning name	Urban restoration action plan of Caohuajie community in Hechuan district	Xuetianwan area regeneration concept planning of Yuzhong district
Community cultural assets	Ancient city wall Daiji dye-works Caohuajie Derun square	"A piece of embroider shoes" "Herb shop" "Liuzhen photo studio" Chongqing auditorium Cultural palace Historic block
Planning goals	Enhance community vitality by connecting historic culture with civil culture, and strengthen community-regional links	Connect high-quality resources around the community, activate community cultural assets, and strengthen linkages between community and region
Community spatial cultural planning approach	Build community cultural routes (including community living routes) and urban tourism routes	Create a community vitality catalyst and build a cultural tourism route in the district

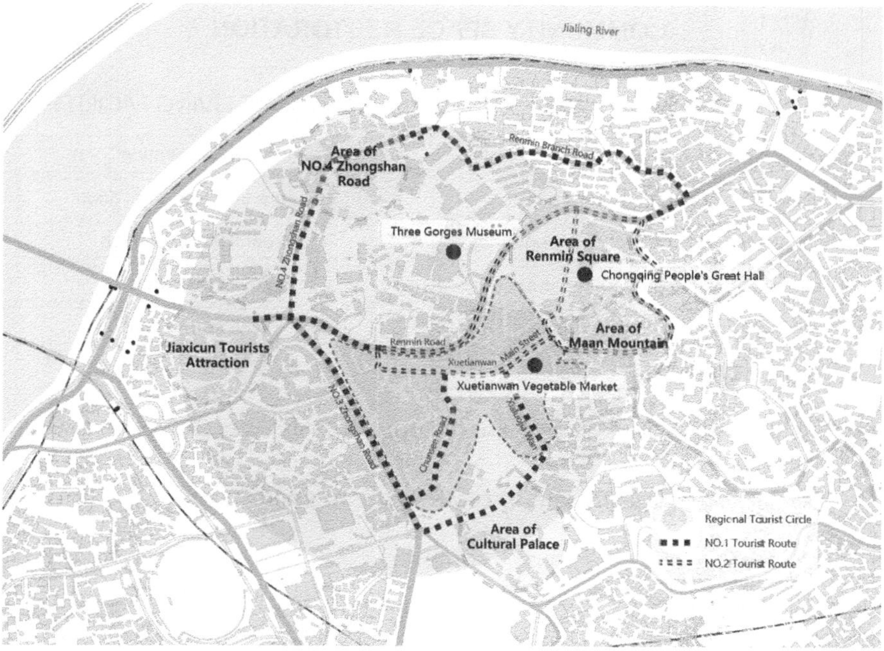

Fig. 7.9 Xuetianwan area cultural tourism route planning

community identity. These are realized through the interaction between people and space.

In the Xuetianwan area regeneration concept planning of Yuzhong district, the concept of "cultural catalyst" was established based on the integration of cultural assets (Huang Ling et al. 2017). Cultural resources within the community were linked up with excellent cultural resources, including the surrounding Chongqing People's Great Hall, the Three Gorges Museum, and the NO.4 Zhongshan Road. These high-quality cultural resources were connected in series to form a systematic cultural tourism route (Fig. 7.9). At the same time, the Xuetianwan farmers' market was identified as a cultural catalyst that could integrate daily life into the cultural tourism route. This shortened the gap between residents and culture and enhanced their sense of identity in the community.

In the urban remediation action plan of Caohuajie Community in Hechuan District, the core of cultural planning combined community cultural routes and constructed community cultural structures. The community's cultural routes include routes of daily life of the community.

Moreover, community cultural structures were assumed to involve displays of the city wall's history and culture, community folk culture, Party-building culture, and celebrities. These cultural contents were all real stories that happened in this community, which were presented by planners through the memories and oral traditions from residents of the local community. Moreover, planners went one step

Fig. 7.10 Illustration of the community space environmental structure of Caohuajie Community

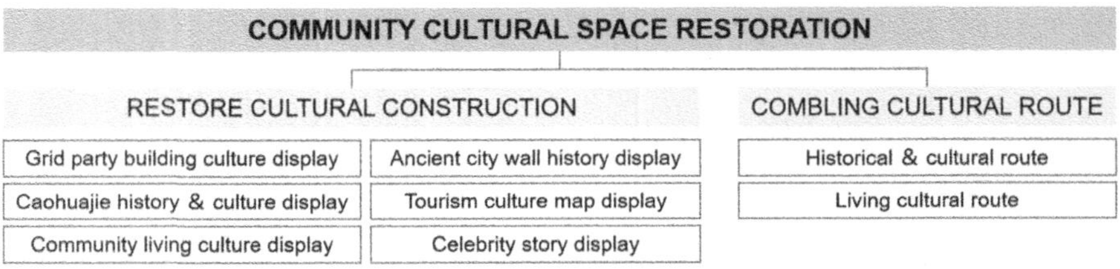

Fig. 7.11 Illustration of the Cultural Restoration Structure of Caohuajie Community

further: they studied the surroundings and urban cultural resources to establish spatial and cultural networks. Furthermore, with the help of the all-for-one tourism perspective, these community cultural routes were linked with the urban cultural tourism network of Hechuan district.

Already at the very beginning, the "space renovation and cultural space restoration" action was combined with the Caohuajie community action plan of Hechuan District in response to the "Urban Renovation and Ecological Restoration" work. In 2015, this work was initially carried out at the Central City Work Conference. It was divided into two parts in the Caohuajie community planning with spatial facilities for repair and cultural restoration (Figs. 7.10 and 7.11). With regard to spatial facilities, the repairs focused on refining and humanizing the spatial nodes where daily activities frequently happened, e.g., street lamps, public furniture, and landscapes. While in community culture, the plan identified community cultural tourism routes and created many cultural exhibition areas with diverse themes along these. The "Urban Renovation and Ecological Restoration" has achieved a great result as a community action plan. It was also the first attempt to merge the new concept of urban transformation with the development of the local community.

Planning intervention during this period has overcome the "individual thinking" of "single community" of the past and has established new "community asset links" guided by planning. In the regeneration concept plan of the Xuetianwan area in Yuzhong district, cultural tourism route planning could play a crucial role in

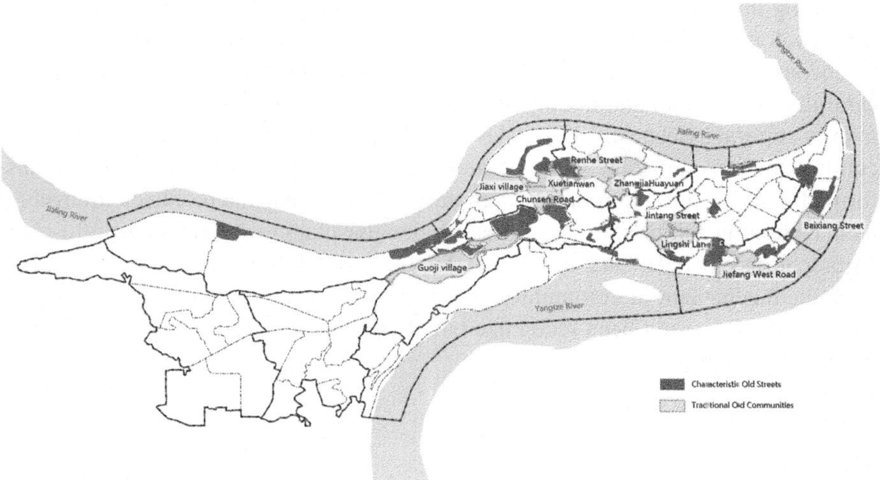

Fig. 7.12 Distribution of characteristic old streets and old communities in Yuzhong district

stimulating the vitality of this area. In the community planning of Caohuajie community in the Hechuan District, a macro pattern extended the cultural and living routes out to the city. Therefore, a complete urban cultural tourism network was gradually shaped. However, in the overall study of community regeneration of Yuzhong district, planners assessed communities from many aspects, mainly based on community assets and community needs. Moreover, the results of the assessment would be further refined when factors such as locations, resource advantages, and development conditions are considered. This could help to formulate sustainable development goals associated with the characteristic old streets and communities as well as other regions of Yuzhong District (Fig. 7.12). The "cultural assets network links" realized the city's vision to connect urban scattered culture from individual points to lines and gradually expand its coverage to establish a cultural city network.

Social demands have promoted the further improvement of the "Asset-based" theory during this period, while the contradictions and demands of community development have become more complex and diverse. The connotation became gradually evident, and the content was continuously enriched. Planners began to realize the importance of space governance, which was involved in the community planning of Caohuajie. Consequently community governors, service providers, and residents were invited to discuss parking problems and then, a parking treaty was created. The spatial resource sharing consensus helped people within the community to increase their quality of life with less difficulties and more public space. Moreover, this planning further explored the social and human assets of the community and attempted to utilize these by matching community needs. Guided by the government, social resources, such as a hospital, a university, and a law firm in the community, have helped residents to build various local community organizations. Long-lasting multi-participation and multi-governance are key to keeping communities sustainable.

During this period, the breadth and depth of planning interventions on community development were further reinforced. Planning has become an effective means to encourage community development. Moreover, community multi-assets were carefully explored for community planning. Spatial culture planning, based on community cultural assets, was further developed and gradually integrated into an urban cultural tourism network. Cultural and spatial planning has considerably transformed from single community regeneration to regional interconnected development by linking community assets. The cooperative planning model with multi-party participation has been established to effectively promote community governance. Still, the mechanism of community planning has not yet been sufficiently formed, its legal status has not been recognized, and a standard system has not been established. There is still a long way to go.

7.4 Conclusion

In the past 10 years, the "community" has exerted a significant role in the urban planning of Chongqing City. With the unique physical and spatial characteristics and different localities, both planners and policymakers have gradually realized the importance of the "community" for comprehensive planning. Moreover, this basic unit of society can also directly determine the happiness and security of community residents as their place of daily life.

This chapter summarized the three stages of community planning in Chongqing City, which significantly influenced and contributed to the urban development and planning of Chongqing City in different periods. At the germination stage, community development began to focus on the quality of the community's spatial environment. The planning introduced the concept of "community assets" for the first time to help planners change their mindset from considering "problems" to "needs." The community environment was renovated through community cultural assets with the aim of cultural revival. This achieved a good effect on promoting communication among resident and strengthening neighborhood relationships. Moreover, space management means the beautification of the community environment and promotes space justice to accommodate economic development. However, the concept of community assets during this period remained relatively vague, and the content of assets was relatively simple.

Subsequently, at the rising stage, community development has shifted from environmental remediation to service facilities and governance innovation needs. The "Asset-based" theory has been officially applied to the first complete community development plan of Chongqing City. By identifying physical assets, human assets, and social assets of the community, the strategy was formulated and implemented on two levels: spatial optimizing and governance upgrading. So far, the ABCD has become a functional theory to guide the development and planning of Chongqing's community. All districts of the city have begun to focus on improving their community public facilities planning based on community stock resources. Consequently, public facilities can be fully and more quickly covered across the city.

At the developing stage, community planning has entered a period of rapid progress. The contradictions and demands of community development have become increasingly complicated and mixed. The "Asset-based" theory has been widely applied to various types of community regeneration plans. The connection between community development and planning has become closer by inspiring the community's endogenous power. Moreover, exploring and cultivating diversified assets may help to realize various goals, e.g., goals associated with spatial planning, cultural planning, and community governance. This increases the overall development of the city as a whole. At this stage, the theory of "community assets" became apparent, and the content of assets was gradually enriched. However, because of the differences between the communities of Chongqing City and the processes underlying their development, the urban community development and planning based on community assets remain in the process of continuous exploration and improvement.

Chongqing's community planning enters the next 10 years, and the arrival of a new era also implies the arrival of new challenges and opportunities. How to properly apply the ABCD theory and the cultural structure of urban space ideas still remain an important part. The critical point is to explore realistic concepts, methods, and public participation mechanisms that are genuinely suitable for the spatial characteristics of the urban communities of Chongqing City.

Furthermore, as the governance modernization idea becomes more important in China, the quality of community governance capacities will play an essential part in this system. Several attempts and reflections on numerous planning practices have become available in recent years. The concept of the "Human-Space-Service" trio within community comprehensive governance achieves a good effect in real practice. Here, "human" forms the principal part of the community, and the happiness of humans is related to the sustainability of their community. Moreover, the primary goal of community planning and community regeneration is to improve residents' living quality. "Space" represents the last foothold of community planning and regeneration. It will be directly present in many forms, such as the arrangement of public facilities and the improvement of traffic systems within the community. At the same time, "service" indicates the management ability of managers and governors in the community. An acceptable "servicer" can meet various needs of residents, and enhance their cohesion and belongingness during the process of governance. The "Human-Space-Service" concept is a process of interactions and influences among residents, physical space, and community governance. Moreover, the next stage enables more explorations and practices to build an improved assessment framework and practical mechanism so that the community can genuinely integrate into the city and serve the people.

While emphasizing the localization characteristics of the community, its role toward realizing the community of shared destiny for humanity should not be ignored. Recognizing the relationship between community planning and global climate change is a further challenge planners and decision-makers must face. How to correctly guide community residents to implement positive changes in their daily life and working styles at the community level is a new research field. It is essential to encourage environmentally friendly behaviors through space construction. This can alleviate global warming and finally achieve global green sustainable development. The community also forms the most basic level to achieve the goal of a healthy city, and to achieve the Healthy China. It is the first step of actual actions and the last hierarchy of national governance independent of the social topics. Chongqing City, a representative of mountainous cities, should not be overlooked when planning to accelerate the establishment of green, low-carbon, livable, and healthy cities. The communities of Chongqing City are both the most significant work areas and practice places to achieve these goals.

Acknowledgments This work was supported by a grant from the National Nature Science Foundation of China (No. 51778078) and China Postdoctoral Science Foundation (ID. 2019M663438).

References

Chen Jiaxi, Lin Dianfeng, 2015. Practical Exploration and Theoretical Reflection on the Negotiation Governance Model of Urban Communities: Observation on the Innovation of Community Management in Nanshan District, Shenzhen, *Social Governance Review*, (01), 84-92. Available at: https://doi.org/10.16775/j.cnki.10-1285/d.2015.01.012

Huang Hong, 2008. Asset-based Development of Community Economic Development: Hong Kong's Experience and Practice, *Jiangsu Social Science*, (02), 227-231.Available at: https://doi.org/10.13858/j.cnki.cn32-1312/c.2005.02.038

Huang Ling, 2012. From Needs-based to Asset-based: Inspiration From Contemporary Community Development in US, *Interior Design*, 27(05), 3-7. Available at: https://kns.cnki.net/KCMS/detail/detail.aspx?dbcode=CJFQ&dbname=CJFD2012&filename=SNSH201205003&v=MjgyMDVyRzRIOVBNcW85Rlo0UjhlWDFMdXhZUzdEaDFUM3FUcldNMUZyQ1VSN3FmWk9kcUZDcmtWYjdQTmlQWVo=

Huang Ling, Luo Yanhong, 2014. Community Planning and the Local Approach in the Perspective of Social Governance, Innovation-A Case Study of Shiyou Road, Yuzhong District of Chongqing, *Journal of Human Settlement in West China*, 29(05), 13-18. Available at: https://doi.org/10.13791/j.cnki.hsfwest.20140503

He Wencui, 2014. *Public Participation Research in Community Development Planning*, Chongqing University, Chongqing. Available at: http://kreader.cnki.net/Kreader/CatalogViewPage.aspx?dbCode=cdmd&filename=1014044626.nh&tablename=CMFD201501&compose=&first=1&uid=

Huang Ling, Zhao Wanmin, Xu Jianfeng, 2008.Cultural Mapping and Cultural Planning, *Planner*, (08), 67-71. Available at: https://kns.cnki.net/KCMS/detail/detail.aspx?dbcode=CJFQ&dbname=CJFD2008&filename=GHSI200808020&v=MzAzOTZyV00xRnJDVVI3cWZaT2RxRkNybFU3ckpJaVhZWjdHNEh0bk1wNDlIWklSOGVYMUx1eFlTN0RoMVQzcVQ=

Huang Ling, Wang Sijia, Lin Sen, 2017.Urban Community Regeneration with the General Thought of Regional Linking and Catalyst Building: An Empirical Study of Xuetianwan Area in Yuzhong District in Chongqing, *Design Community*, (02), 140-147. Available at: https://kns.cnki.net/KCMS/detail/detail.aspx?dbcode=CJFQ&dbname=CJFDLAST2017&filename=ZUQU201702021&v=MTk3NzZyV00xRnJDVVI3cWZaT2RxRkNybFVidkpQemphZTdHNEg5Yk1yWTlIWllSOGVYMUx1eFlTN0RoMVQzcVQ=

Kretzmann John P and John L·McKnight, 1993. *Building Communities from the Inside Out: A Path Toward Finding and Mobilizing a Community's Development Institute*, ACTA Publications, Chicago. Available at: http://www.povertystudies.org/TeachingPages/EDS_PDFs4WEB/ABCD-Bldg-communities-from-inside-out.pdf

Li Xiaozhuang, 2015. Study on the Reform and Innovation of the City Community Governance System: Based on the Survey of Dongshengyuan Community, Zhongguancun Sub-district of Beijing, *Urban Development Studies*, 22(01), 94-101. Available at: https://kns.cnki.net/KCMS/detail/detail.aspx?dbcode=CJFQ&dbname=CJFDLAST2015&filename=CSFY201501017&v=Mjc4MTVUcldNMUZyQ1VSN3FmWk9kcUZDcmxWYnJNSmo3TmQ3RzRIOVRNcm85RVk0UjhlWDFMdXhZUzdEaDFUM3E=

Li Dongquan, Lan Zhiyong, 2012.Challenges of Community Development in the Rapidly Urbanizing China, *Journal of Public Management*, 9(01), 104-110+127-128. Available at: https://kns.cnki.net/KCMS/detail/detail.aspx?dbcode=CJFQ&dbname=CJFD2012&filename=GGGL201201013&v=MDE1MDFpck1Zckc0SDlQTXJvOUVaNFI4ZVgxTHV4WVM3RGgxVDNxVHJXTTFGckNVUjdxZlpPZHFGQ3JtVXJ2S0k=

Qian Zhenghan, Niu Huien, 2007. Community Planning-Theory and Suggestions on its Popularization in China, *Urban Planning Forum*, (04), 74-78. Available at: https://kns.cnki.net/KCMS/detail/detail.aspx?dbcode=CJFQ&dbname=CJFD2007&filename=CXGH200704012&v=MTE5NjR1eFlTN0RoMVQzcVRyV00xRnJDVVI3cWZaT2RxRkNybVVMN1BKalhNWnJHNEh0Yk1xNDlFWm9SOGVYMUw=

Richard C. Box, 1997. *Citizen Governance: Leading American Communities Into the 21st Century*, SAGE Publications, Los Angeles. Available at: https://doi.org/10.4135/9781483328249.n5

Shi Xuehua, Kong Fany, 2008. The Administration of Communities in the U.S.A. and Its Enlightenment, *Journal of Shanxi University(Philosophy & Social Science)*, (04), 90-95. Available at: https://doi.org/10.13451/j.cnki.shanxi.univ(phil.soc.).2008.04.021

Sherraden MS and Ninacs WC, 1998. Community Economic Development and Social Work, *Journal of Community Practice*, 5(1-2), 1-9. Available at: https://www.tandfonline.com/doi/pdf/10.1300/J125v05n01_01?needAccess=true

Tara O' Leary,2005. Asset-Based Approach to Rural Community Development: Literature Review and Resources, *Seminar Report by International Association for Community Development for Carnegie UK Trust*. Available at: https://resources.depaul.edu/abcd-institute/publications/publications-by-topic/Documents/ABCD-IACDGlobal.pdf

Wen Jun, Huang Rui, 2008. On Asset-Based Development Strategy and the Enlightenment to China, *Journal of Social Science of Hunan Normal University*, 37(06), 74-78. Available at: https://kns.cnki.net/KCMS/detail/detail.aspx?dbcode=CJFQ&dbname=CJFD2008&filename=HNSS200806016&v=MjkwNDJQWWZiRzRIdG5NcVk5RVlvUjhlWDFMdXhZUzdEaDFUM3FUcldNMUZyQ1VSN3FmWk9kcEZpbm1WcjNOTFM=

Xiao Hongwei, 2012. *Research on Urban Old Community Regeneration Strategy Based on "Cultural Route" Thought*, Chongqing University, Chongqing. Available at: https://kns.cnki.net/kcms/detail/detail.aspx?filename=1012049430.nh&dbcode=CMFD&dbname=CMFD2013&v=

Zhang Tingwei, 1997. China's Planning to the World--From Physical Construction Planning to Social Development Planning, *Urban Planning Forum*, (01), 5-9.Available at: https://kns.cnki.net/KCMS/detail/detail.aspx?dbcode=CJFQ&dbname=CJFD9697&filename=CXGH199701001&v=MDQwNjlPZHFGQ3JuVkw3UEpqWE1ackt4RjliTXJvOUZaWVI4ZVgxTHV4WVM3RGgxVDNxVHJXTTFGckNVUjdxZlo=

Zhang Huxiang, 2005. *Community Governance and Reconstruction of Social Orders: A Study on Kangjian Block of Shanghai*, Shanghai University, Shanghai. Available at:https://kns.cnki.net/KCMS/detail/detail.aspx?dbcode=CDFD&dbname=CDFD9908&filename=2005120628.nh&v=MTk5NjBGckNVUjdxZlpPZHFGQ3JuVmJyS1YxMjdHN0s2SHRmT3A1RWJQSVI4ZVgxTHV4WVM3RGgxVDNxVHJXTTE=

Zhao Min, Zhao Wei, 2006. Focus on Social Urban Planning-An Introduction to Urban Community Development Planning, *Shanghai Urban Planning Review*, (06), 8-11. Available at: https://kns.cnki.net/KCMS/detail/detail.aspx?dbcode=CJFQ&dbname=CJFD2006&filename=HCSG200606002&v=Mjc4MDFYMUx1eFlTN0RoMVQzcVRyV00xRnJDVVI3cWZaT2RxRkNyblc3L0pMUzdZYWJHNEh0Zk1xWTlGWm9SOGU=

Part III
Green and Eco-City Planning for Sustainability

Chapter 8
Green Low-Carbon High-Density Urban Center Planning Wuhan Wangjiadun Area

Shen Xu

Abstract High-density urban centers have provided a concentrated space for politics, economy, culture and service, yet have also brought many problems. From the perspective of green low carbon planning, this chapter summarizes the experiences and lessons of planning and construction of typical urban centers around the world, and proposes three aspects of high-density urban centers' planning and construction: municipal planning, transportation planning and building energy planning. Taking the high-density urban center—Wuhan Wangjiadun Central Business District (CBD) as an example, the specific strategies regarding green low-carbon planning are proposed.

The chapter aims to analyze the achievements of planning and construction of Wuhan Wangjiadun CBD, and offers suggestions and instructions on the planning, development and construction of high-density urban centers in China, from the perspective of green and low carbon planning.

Keywords Green · High-density · Urban center planning · Wangjiadun

8.1 Introduction

High-density urban centers are areas with high building volume ratios, high building coverage, and low open space rates, whose development provides a central space for politics, economy, culture and general service (Dong 2012). The typical high-density urban centers are Manhattan, Tokyo, Hong Kong, and Shanghai. Manhattan, one of the most famous urban centers, and the icon of New York City,

S. Xu (✉)
School of Architecture and Urban Planning, Huazhong University of Science and Technology, Wuhan, China
e-mail: xushen@hust.edu.cn

L. Bian et al. (eds.), *Chinese Urban Planning and Construction*,
Strategies for Sustainability, https://doi.org/10.1007/978-3-030-65562-4_8

has a regional plot ratio of 7.14 and a population density of 27,000/km^2. The land is concentrated for commercial and residential use, with relatively less public green areas and open space (Lu and Song 2010). Shinjuku, Tokyo, is the main urban center of Japan, and has a regional population density that reaches 19,000/km^2. Hong Kong, China, is an important financial center in the world. Its central area is one of the world's most densely populated urban centers, with a population density over 17,500/km^2. The regional plot ratio is 6.87, and the building density is as high as 31.81%, while the air-ground rate and public activity space account for only 11.85% and 12.73% (Jing et al. 2018). Shanghai Lujiazui area is one of the most important financial centers in China, and its regional population density is 16,000/km^2.

High-density development provides urban centers a centralized political, economic, cultural and service space, which guide regional capital flows and boost the economy. Manhattan attracts headquarters, stocks and trading centers of most of the top 500 corporations, and is a world-class financial center (Lu and Song 2010). Shinjuku, Tokyo has concentrated the main economic, administrative and commercial functions in the Tokyo metropolitan area, including the headquarters of nearly 200 large banks in the world, major Japanese corporate headquarters, and some government agencies (Shinjuku City Government (n.d.)). Hong Kong, China is home to 106 banking institutions, 57 offices in Hong Kong and 190 insurance companies. This area is an important financial center in the world. Shanghai Lujiazui area has more than 380 various types of financial insurance institutions, and is one of the most important financial centers in China (Fig. 8.1).

Fig. 8.1 Manhattan, Tokyo, Shanghai, and Hong Kong

The construction of high-density urban centers has also brought many problems in terms of environment and energy consumption. The intensity of land development and height of buildings in urban centers have been increasing, while public space and green areas have been shrinking (Hua 2019). The gathering of enormous populations means aggravated traffic congestions. And the intense development also causes serious pollution and energy issues. Studies have shown that high-density urban centers are prone to generate heat island effects. According to statistics from more than 20 cities in the world, the annual average temperature in urban centers is 0.3–1.8 °C higher than that in urban suburbs, while humidity is relatively reduced by 4–6%. The development of high-density urban centers has also brought about regional congestion, high energy consumption, and high waste discharge. The use of vehicles in the region has resulted in high carbon emissions and waste emissions, which seriously affect the environment in urban areas and the quality of life of urban residents (Hu 2020). The long-term working and living conditions in the closed space of the urban central area have increased the use of air-conditioning and lighting equipment, which directly led to an increase in building energy consumption (Trepci et al. 2020). The problems associated with the development of urban central areas are universal in the world. Under the current trend of advocating green development, promoting energy conservation, and emission reduction in China, it is necessary to conduct in-depth research on high-density planning in urban central areas from the perspective of green low-carbon planning.

8.2 Method

The development of green low-carbon planning in high-density urban centers focuses on the harmony between people and the environment, rational land use, and low energy consumption in life-cycle construction and operation, which will form a benign and harmonious relationship between economic development, social progress, and environmental protection. The planning strategies need to be carried out in three aspects, which are detailed in the following sections.

8.2.1 Green Low-Carbon Municipal Planning Method

Green low-carbon municipal planning aims at building green, low-carbon, ecological municipal facilities and systems, from the comprehensive utilization of water and solid waste resources, utility tunnel and ecological landscape environment planning.

1. Water resources comprehensive utilization planning based on the principle of open source and throttling, water resources are planned from both water saving and sponge cities. The solid waste planning is based on the principles of reduction, resource utilization, harmlessness and ecologization. It proposes planning guidelines for source control, process control and end control for different types

of solid waste. Planning of the West Coast New District of Qingdao optimizes the water source allocation through comprehensive utilization of water resources, and improves the quality of drinking water, and provide support for the construction of reservoir projects.
2. The utility tunnel planning is an overall plan for integrating the trunk tunnel, branch tunnel and connecting tunnel, and proposes design guidelines for the pipeline and section form. The Beijing Winter Olympics Yanqing District utility tunnel provides access to hardware for snow, water, domestic water, reclaimed water, electricity, telecommunications and cable TV in the Winter Olympics.
3. Ecological landscape environment planning based on the principle of respecting nature and adapting to local conditions, a good ecological landscape system is formed through the combination of urban spatial planning and eco environment concepts. Taking the Sino-Singapore Tianjin Eco-City as an example, the Yingcheng Sewage Reservoir will be treated as a clean lake, the aquatic ecosystem restored, and the vertical system adjusted to ensure drainage safety. Finally, the eco-city is provided with a beautiful and liveable ecological environment (Zhao et al. 2019).

8.2.2 Green Low-Carbon Transport Planning Method

The high-density green transport planning in the central area of the city requires a compact and efficient layout within, and accessibility analysis of public service facilities to increase the proportion of green travel, forming a multi-coordinated composite transportation system.

1. Promote the rail transit facilities, establish a multi-level and high-quality public transport system, improve the punctuation and reliability of public transport, and achieve multiple low-carbon transportation modes, such as rail + bus/rail + public bicycles, etc. Chongqing Yuelai Green Eco-city promotes the "city guided by public transportation" TOD model, and light rails account for 50%. Guangdong Guangming New City will build an external rail transit and rapid transit system (BRT), a ring express bus line, a convenient bicycle trunk line system, all connected to the pedestrian priority zone; will improve the transfer connection, strengthen traffic demand management and intelligent traffic System construction, etc. (Liu 2013).
2. Build a dedicated slow-moving system. With the street, slow path and green lane as the core, supplemented by a perfect bicycle road network and supporting facilities, a continuous and comfortable slow network system is formed. Take Wuxi Taihu New City as an example, the bicycle system was designed in detail, and a bicycle rental system was planned and set up with bus hubs, track stations, residential areas and public service facilities.
3. With integrated development of transportation and underground space, a 3D multi-level circulation is established, forming a "car-free on the ground, corridor into the ground, and seamless docking" area. The Beijing Financial Street Underground Traffic Corridor connects about 8000 underground parking spaces in 11 buildings

in the Financial Street area to the underground tunnel and the West Second Ring Road and Taipingqiao Street to achieve ground–underground traffic diversion. It is expected to reduce ground traffic pressure by 15% (Yuan and Yao 2011).

8.2.3 Green Low-Carbon Building Planning Method

High-density urban centers shall consider green building technology and energy supply comprehensively. In the area planning phase, with the master and control planning, and based on the green diagnosis of the function defining, regional environment, current status of energy resources, and the eco capacity, the green building special plan shall be formulated. The development goals, paths and related measures of green buildings cannot do without local conditions.

1. Land development.
   Starting from the effective use of land resources and advocating the development and utilization of underground spaces, we could set reasonable land use intensity, and define population density, floor area ratio, building density, and employment housing balance index of construction land reasonably. Taking Sino-Singapore Tianjin Eco-city as an example, its planning requires "an employment housing balance index ≥ 50%" to ensure occupational balance and living convenience.
2. Land—use planning.
   Analyze the goals of different plots, and fully consider the green low-carbon evaluation factors such as eco basement, functional location, underground space, building type, transportation and living convenience, and carry out the green building layout, and the planning and construction of prefabricated buildings according to local conditions.
3. Energy planning.
   Optimizes energy supply, consumption, and working efficiency. Enrich the source of energy supply with all available resources. Through the analysis of the potential of traditional energy resources (electricity, gas), renewable energy resources (solar, shallow geothermal energy, wind), the multi-source of regional energy supply will be enhanced.
4. Technical application.
   On the premise of balancing economic, social and ecological benefits, combined with the local geographical, climatic and land characteristics of the planning area, consider the appropriateness and economics of the building technology of the area to formulate specific technical solutions. Consideration of the rationality and economy of building energy-saving technologies, through the implementation of energy-saving standards, passive building energy-saving technologies, building intelligent management and other measures to promote the transformation of energy consumption patterns and reduce building energy consumption. Maximize the efficiency through renewables, efficiency design, and other green energy technologies such as wind and solar hybrid streetlights and intelligent energy microgrids.

5. Engineering implementation.
   Green construction and operation management helps to achieve the requirements of the 100% green building target. The second-star green building area of Meixi Lake New Town in Changsha is about 3.438 million square meters, accounting for 33.1% of the total construction area. There are 7 schools in Meixi Lake New City, all of which are expected to be built as green schools; the utilization rate of non-traditional water resources exceeds 10%, through a series of measures such as rainwater harvesting and reclaimed water reuse, the annual reusable water volume reaches 2.5184 million tons (Wang and Wang 2013).

8.3 Case Study

8.3.1 Background

Wangjiadun is in the geographic center of Hankou, between the inner and second loop (Fig. 8.2). It is adjacent to Changqing Road and Qingnian Road in the east, Jianshe Avenue in the south, Hanxi Road in the west and Development Avenue in the north. 3.3 km east-west, 3 km north-south, 21 km from Tianhe Airport and roughly 1 km from Hankou Railway Station. With convenient transportation, closely tied to the city development, Wangjiadun is the only plot in central Hankou that is available for large-scale and high-standard development.

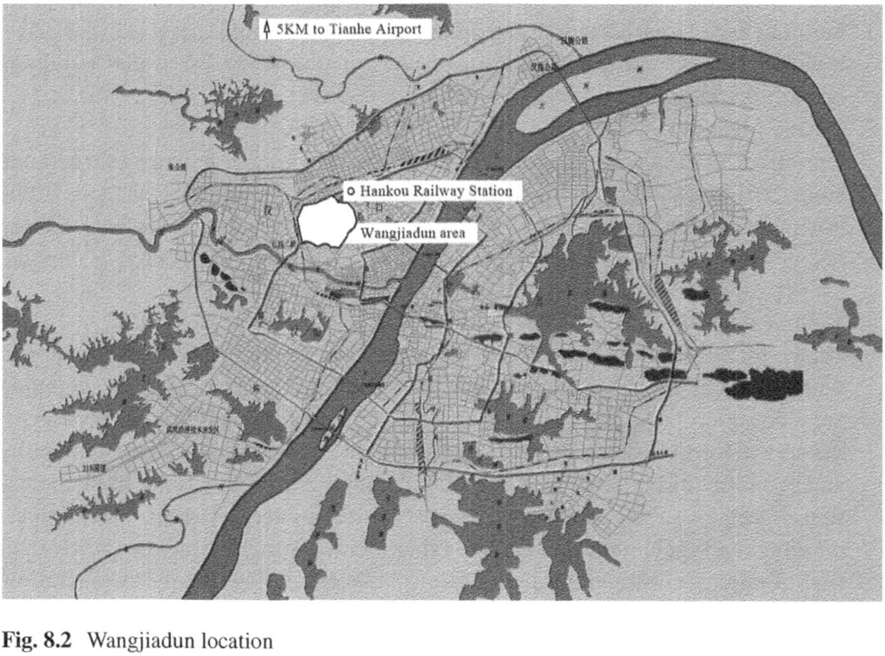

Fig. 8.2 Wangjiadun location

The Wangjiadun area was initially developed as a military airport. With the rapid urban development, Wangjiadun Airport was surrounded by the city, and it became the urban center of Hankou. The use of military airports was to some extent affected. The development of the city was also severely limited since the airport required clearance. In July 2000, the State Council and the Central Military Commission approved Wuhan to migrate Wangjiadun Airport. The Wuhan Government proposed to build the Wuhan Central Business District (Wuhan CBD) in Wangjiadun area.

Wuhan Wangjiadun CBD covers an area of 741 hectares, planned to be a modern service industry center focusing on finance, insurance, trade, information, consulting and other industries. It is formed by investors such as banks, insurance, and the headquarters of the group. It is a comprehensive modern urban center with conference, exhibition, retail, hotel entertainment, residential and leisure functions. By 2020, the size of the working and living population in the CBD is expected to reach 150,000–200,000.

The total planned area is 741 hectares, the average floor area ratio 2.58, which is relatively high as a commercial land in CBD. And some plot ratio is as high as 14, which ensures a high construction and development intensity. The average building density of commercial land there is 40–60%, the average building density of residential land 15–30%, and the underground space development rate is 20%. There are 132 high-rise buildings, 29 super high-rises (of which 18 are over 200 m), among which the Wuhan center, capped in 2015, reaches 438 m.

8.3.2 Green Low-Carbon Municipal Planning Method

8.3.2.1 Unconventional Water Utilization

Unconventional water utilization planning includes rainwater harvesting and utilization planning and recycled water use planning.

The area mainly relies on Wangjiadun Park and Mengze Lake Park for rainwater collection and reuse. There are four water intakes, one of which is located in Wangjiadun Park, two in Mengze Lake Park and one in the Wetland Theme Park. After the rainwater is processed, it will be used as municipal greening, road water and outdoor miscellaneous water. In addition to the two parks, the plots farther from the water intake can be combined with the rainwater collection system, to supply the outdoor miscellaneous water there.

There are two ways to use recycled water, one is reclaimed water from buildings and the other is urban sewage recycling. Reclaimed water in buildings is mainly used in hospitals, schools, hotels and some residence. Recycling of urban sewage will introduce the main pipe of recycled water from the outside, then the branch pipe. The recycled water can be used for building indoor flushing, outdoor greening, road pouring, landscape hydration, and car washing.

8.3.2.2 Sponge City

The construction of Sponge City in Wuhan Wangjiadun CBD aims to create a sponge demonstration zone, and achieve symbiosis between man and nature. Through the three aspects of excessive runoff discharge system, urban pipe system and source sponge system, the integrated runoff coefficient reaches 0.6, and the annual runoff control rate reaches 60% (Table 8.1).

Considering the planning and design conditions, the concave green space, permeable pavement and green roofs are reasonably arranged, to achieve various requirements of annual runoff control rate.

The planned roads can be low-impact development facilities such as permeable paving, recessed green spaces, initial rain sedimentation tanks, and ecological tree pool, and the integrated runoff coefficient of the site should not be more than 0.7.

The green land fully uses the infiltration rainwater, combined with the landscape design of rainwater garden and landscape lake, the annual runoff control rate reaches the target, 85%, and the site comprehensive runoff coefficient control reaches 0.15.

The urban pipe channel system will gradually improve the construction of the rainwater pipeline system. Through the assessment of the extent of water accumulation, once the main pipe is completed, the prevention and control of the rainstorm and floods would improve obviously in 5 years.

The excessive runoff discharge system consists of the aboveground and underground part. The aboveground part is planned to construct flood-resistant squares, ground discharge channels and lake storage, while the underground parts constructs storage tanks and drainage tunnels.

Table 8.1 Sponge City Planning Index System

Category	Index		Target value	Property
Water ecology	Annual runoff control rate		60%	Mandatory
	Water body	Lake area	11.87 ha	Mandatory
		Ecological shoreline rate		
	Urban heat island effect		Heat island is relieved	Guided
Water environment	Water environment quality		Surface water quality meets class V water quality requirements	Mandatory
	Urban pollution control	Pollution control	Sewage collection and treatment rate 100%	Mandatory
		Surface-source pollution control	≥50%	
Water resources	Rainwater resource utilization		≥2%	Guided
Water safety	Urban flood prevention standards		Every 50 years	Mandatory
	Rainwater pipe network design standard		New-built every 3–5 years	Mandatory

8.3.2.3 Solid Waste Utilization

Solid wastes in Wuhan Wangjiadun CBD are mostly domestic waste, construction waste, and medical waste.

Separate disposal of household garbage, kitchen waste and construction waste by means of classification and reduction. For the domestic garbage in the residence, the government or related corporations issue versatile boxes for collection, and establish a developed garbage measurement system and a basic data statistics system, collecting data and apply it to management and decision-making. The kitchen waste requires that the catering hotel and the local urban management department sign a clearing and transportation standard agreement for the kitchen waste, and waste will be collected and transported by intelligent vehicles. Unify the management and recycle the construction waste. At the same time, promote the sale of new commercial houses in fully decorated forms, reduce the sale of rough houses, reduce decorating waste, and strengthen the unified collection and management of construction waste.

Recycle and utilize the solid waste. For all the waste, strengthen the management, reduce the process load from municipal waste transfer, and the rest domestic waste will be transported to the waste transfer station for further treatment. For e-waste, the plan is to establish an old appliance recycling and reservation system in a public place such as a supermarket or plaza by contacting the old appliance recycling company. The residents can register for an appointment, wait for the person to pick up the goods, and strengthen the recycling and utilization of the e-waste. For construction waste, different recycling purposes will be clarified based on the composition, and the target of construction waste recycling utilization rate $\geq$40% will be achieved.

Meanwhile, the solid waste will be treated centrally and harmlessly. The garbage will be transported to the nearby Guodingshan waste incineration plant, through the transfer station for sorting, recycling and incineration, to ensure a complete disposal.

8.3.2.4 Utility Tunnel

The construction of a utility tunnel can effectively eliminate the "pull link" phenomenon, ensure the safety of urban underground pipelines, improve urban functions, embellish urban landscapes, improve the comprehensive land utilization, promote the intensive and efficient development, enhance underground space utilization, attract capital investment, create a new impetus for economy, and generate enormous social and economic benefits.

The utility tunnel of Wuhan Wangjiadun CBD adopts the combination of trunk lines and branch lines, having accomplished the very first utility tunnel in Central China (Fig. 8.3). The total length is 6.2 km, which consists of a trunk and a branch. The power, communication and water supply pipes are arranged for reasonable and efficient use of underground space.

Fig. 8.3 Construction of utility tunnel of road no. 305 in Wangjiadun area

Fig. 8.4 Location and renderings of Mengze lake park

8.3.2.5 The Cases: Mengze Lake Park and Wangjiadun Airport

Mengze Lake Park is in the south of Wuhan Wangjiadun CBD, with a total area of 26.32 hectares, of which water covers 10.70 hectares and land covers 15.62 hectares (Fig. 8.4). Among the land, the green area is 12 hectares, the square and garden road area is 3.31 hectares, and the building (structure) is 0.31 hectare (3080.27 m2).

The park mainly includes a lake body, lawn, autumn water walk, large green area and other major sponges. The theme is to build a systematic, external pollution control system, a stability endogenous source, an integrated, coordinated and well-managed comprehensive system for controlling water ecosystems and controllable forced external circulation systems. Lake water can absorb, store, infiltrate and purify the runoff rainwater, replenish groundwater and regulate water circulation. In the case of drought and water shortage, the stored water is released and used. At the same time, the green road around the lake and the waterfront landscapes such as toons, cherry blossoms and persimmon trees are set up throughout the park, to create a dense forest of greenery, with flowers in spring, fruits in autumn, and evergreen landscapes in winter.

The business district adopts the advanced waste concrete recycling technology to reclaim about 100,000m^2 of concrete abandoned from the runway, after the Wangjiadun Airport was demolished, turning waste into treasure. A waste concrete recycling base has been established, and the recycled aggregates are used in the grassroots and pavements for planning roads and temporary road projects. At present, it has been successfully applied to the roadway foundation and pavement of more than 10 municipal planning roads such as Yunfei Road and Huaihai Road.

8.3.3 *Green Low-Carbon Transport Planning Method*

8.3.3.1 Road System

The Wangjiadun area has four main urban roads, namely Huaihai Road, Yunfei Road, Hanjiang Avenue and Huanghai Road Tunnel. simultaneously, the block structure will be optimized to form a dynamic neighborhood of "Narrow Road, Milu Network" to create a comfortable neighborhood environment.

The block road system strengthens the non-isolated belt and bicycle lane design, designs bicycle shelter and shelter from rain, cultivates tree-lined roads, enhances the landscape effect, and improves the comfort of slow travel. It is important to pay attention to the information guiding facilities, design the driving sign guidance system, and set safe, reasonable and comfortable bicycle guiding facilities at the traffic nodes. On the basis of reducing the length of the block, combined with the current sharing of bicycles, public bicycle parking points should be set up to provide convenient conditions for pedestrians to choose bicycle travel.

8.3.3.2 Public Transportation

The public transportation system construction in the planning area mainly includes the public transportation system and the rail transit system (Fig. 8.5).

The Wuhan Wangjiadun CBD Rail Transit Planning is arranged according to the blueprint of "5 Line 10 Stations," a total of 10 stations, of which Line 3 and Line 10 run vertically through the core area, and Line 7 crosses the core area horizontally, causing transfer in a "cross-shaped" network. Lines 2 and 12 are routed along the edge of the CBD.

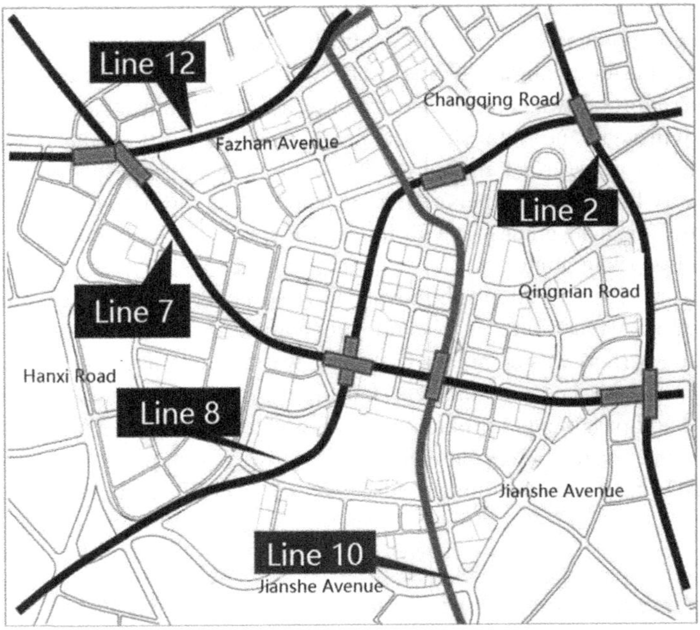

Fig. 8.5 Rail transit route planning

The bus planning combined with the layout of the rail transit line, set up 3 bus hub stations, and more than 30 bus lines, and 23 bus stations, and cooperate with the underground track system to construct a three-dimensional traffic system throughout the Wangjiadun area.

8.3.3.3 Slow Traffic

Implement pedestrian network in the core area to improve the comprehensive performance of transportation and commerce. First, pay attention to the continuity and orientation of the pedestrian system, and realize the "separation of people and vehicles" in areas and nodes with high density of travel in the commercial center. The traffic flow lines of commercial, office blocks and intersections are constrained, and a perfect ground pedestrian transportation system is built with pedestrian inner streets and pedestrian squares as the mainstay, supplemented by corridors and step bridges. The second is to build an underground pedestrian system, establish a pedestrian network of parking system-underground commercial-orbital stations, and form a public transportation system that separates people and vehicles.

The ground walking system is based on the principle of constructing a pedestrian structure system of "three sides, four banks and one shore." The "three sides" are around the large-scale public facilities such as the rail transit station, the scenic park green space and its surrounding areas, and the commercial center. The "four shores" are the four shores of parks and waterfront space and their infiltration areas. "One belt" is the waterfront leisure green belt of Mengmen Lake. The leisure greenway connects the main parks and waterfront space in the central business district, and infiltrates the waterfront space into the inland areas to organize slow corridors. It plans to set up 10 characteristic walkways and 23 avenues to construct two-level characteristic walkways, including 13 km of first-class characteristic trails and 8.5 km of second-class trails.

The underground walking system consists of an underground pedestrian street and an underground pedestrian crossing. The underground pedestrian street is mainly located on the basement level, combined with the construction of large public service facilities. The underground pedestrian crossing is set at the intersection of the underground walking system and the road. The underground pedestrian system mainly focuses on the underground commercial space in the core area and key areas, and combines the layout of important public service facilities, rail transit hub stations, key stations and bus hub sites to form a belt-like underground walking system.

8.3.3.4 Vertical Traffic

The core area of Wuhan Wangjiadun CBD is planned to form a "single-ring and two-axle" overall layout structure of underground space, to build an underground traffic conversion system (single-ring) based on the underground loop, with the subway central station as the commercial core, combined with the subway section. The cut-out section forms two commercial spindles (two-axis) in the shape of a "T." Considering the clearance requirements of underground commercial and underground parking facilities, and adapting to the hierarchical layout of the subway central station, the underground space of the core area is generally divided into four layers.

The basement of the core area is mainly commercial facilities, underground parking lots, equipment rooms, public pedestrian passages, etc. The second basement is mainly underground loop, subway central station hall, underground street, underground parking lot, and equipment room. The second basement has Line 3 platform, underground parking lot and equipment room. Metro Line 7 platform, Huanghai Road Tunnel and underground parking lot are on the fourth floor.

8.3.3.5 Parking System

There are 27 parking lots in Wuhan Wangjiadun CBD, with a total of 5480 parking spaces, including 13 greening and plazas, 9 combined buildings, 5 separate land for control, and parking lots throughout the district.

8.3.3.6 The Case: Underground Traffic Corridor

The underground traffic corridor of Wuhan Wangjiadun CBD is located at the periphery of the central block of the core area. It is the traffic deconstruction ring of the central block of the core area. It consists of Zhujiang Road, Shangdong Road, Fanhai Road, Business West Road and underground ring road. The underground loop has a total length of 1.9 km and a three-lane scale. It uses a one-way counter-clockwise organization. The main service target is the small passenger car, which is mainly based on the parking of the land in the service core area, taking into account the parking of some land outside the core area.

The underground traffic corridor system has a total of 12 lanes of "6 in 6 out." Among them, 4 pairs of ground entrances and exits are distributed at the corner of the corridor. Two pairs of underground orient ways are in the middle of Shangxi Road and East of Business Road, connecting the Huanghai Road Tunnel. The main line adopts the distribution lane + single lane + distribution lane layout, the lane width is 3 m, and the design clearance is 3.2 m.

The ventilation of the ring corridor adopts semi-horizontal mechanical ventilation and natural ventilation, and is integrated with the integrated pipe gallery along the line. At the same time, natural lighting is adopted in some sections, and comprehensive underground development is carried out to solve the problem of dynamic and static traffic. The underground parking and connecting channels account for by 70%, make full use of underground space and saving increasingly tight land resources.

8.3.4 Green Low-Carbon Building Planning Method

8.3.4.1 Guidance for Land Use Development Intensity

The total land area for planning in the Wangjiadun area is 741 hectares, the total population of the planned permanent residence is 172,000, and the population density of construction land is 23,000/km^2.

Based on the planning structure, functional positioning, building properties, and location relationship, the development strength of each block is determined. The core area covers an area of 83.7 hectares, and the gross area ratio of the built-up area is 4.5. The planning area is mainly based on the nature of finance, trade, commerce, hotels, residence, and commercial buildings. The types of land use are mainly the second type of residential land, financial facilities, and commercial facilities. The planned building density is $\leq$40% for residential land, $\leq$50% for financial facilities, and $\leq$55% for commercial facilities. The development intensity of the planning area is controlled according to the seven types of areas (Table 8.2).

Table 8.2 Seven categories of regional development land use intensity

Regional	Classification	Building density	Building height	Floor area ratio
Type one	Finance, trade	≤12%	>350 m	>12
Type two	Trade	8–24%	200–350 m	8.0–12
Type three	Hotel, high-rise apartment	14–16%	100–200 m	4.0–8.0
Type four	High-rise residential area	4–7%	35–100 m	2.5–4.5
Type five	Multi-storey, small high-rise residential area	20–28%	21–35 m	2.0–2.5
Type six	Commercial service facilities, multi-storey residential areas	≤33%	3–21 m	<2.0
Type seven	Road traffic, park greens and waters, etc.	–	–	–

8.3.4.2 Development of Green Buildings

All the civil buildings in the Wangjiadun area that started construction after 2011 are constructed in accordance with China's green building standards. Among them, 29 construction sites are used to build China's one-star green buildings, mainly commercial buildings, office buildings, mixed commercial and residential buildings, etc. Three construction lands for the construction of China's two-star green building, the first phase of the Oceanwide City Plaza has been awarded the LEED standard silver mark certification.

8.3.4.3 Energy Use

All residential buildings and public facilities in Wuhan Wangjiadun CBD use south-facing balcony wall-mounted solar water heaters. The hotels and hospital buildings use roof-mounted solar water heaters, solar energy and construction Integration. The solar energy meets 100% of the user's domestic hot water demand, and its solar thermal utilization is 12,101 tce/a.

Primary and secondary schools, municipal office buildings, commercial buildings, trade consulting buildings, financial and insurance buildings, theaters, book fairs and other public buildings in the Wuhan Wangjiadun CBD use solar photovoltaic power generation, using building roofs, glass curtain walls, skylights, and sun visors, such as the installation of solar photovoltaic and building integration products. According to the power generation capacity of the solar photovoltaic system to meet the electricity demand of 2% of the building, the utilization of solar photovoltaic is 3030 tce/a.

Wuhan Wangjiadun CBD belongs to the ground source heat pump development zone, planning government office land, book exhibition land, and large public buildings above 20,000 m^2 using ground source heat pump utilization technology to meet the building's 65% air conditioning heating needs. The utilization of the ground source heat pump system in the planned area is 9862 tce/a.

It is planned to adopt the wind and solar complementary circuit lamp system on both sides of the secondary road with an annual average wind speed of more than 3 m/s in the Wuhan Wangjiadun CBD to reduce the power consumption required for municipal lighting.

8.3.4.4 The Cases: Wuhan Center and Ocean Wide City Plaza Phase I

The Wuhan Central Project is located at Fanhai Road and West Commercial Road, with a total construction area of 355,800 m^2, including 260,000 m^2 above ground and 80,000 m^2 underground, with a total of 88 floors and a height of 438 m (Fig. 8.6). At present, it has obtained the pre-certification of the US LEED Gold Award and the three-star design logo certification of the Green Building of the Ministry of Housing and Urban-Rural Development.

The project adopts green building technology such as central water treatment, curtain wall system and core energy bar to achieve better green and low carbon energy saving effect.

The Wuhan Central Project will set up a water treatment system on the 31st floor of the tower to collect a large amount of secondary water that can be reused every day in high-rise hotels and serviced apartments. It will be treated with water recycling equipment and then transported downward by gravity, to the office area, bathrooms (6 to 30 floors) in the building for bathroom and urinal flushing. The amount of water reuse is 278.99 m^3/day, saving energy consumption.

The core energy bar system effectively separates the high and low areas of the super high-rise building, not only with the perfect architectural shape, but also effectively and rationally utilizes the empty space of the core tube of the high area, and also relieves the energy supply pressure of the main room.

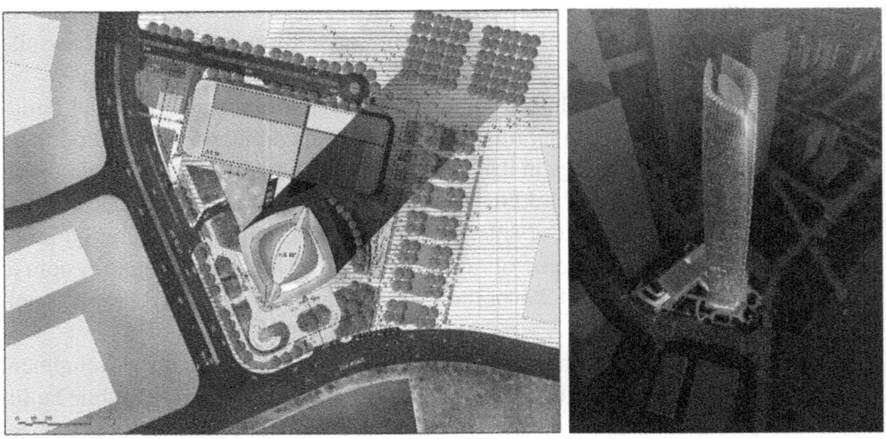

Fig. 8.6 Wuhan center floor plan and renderings

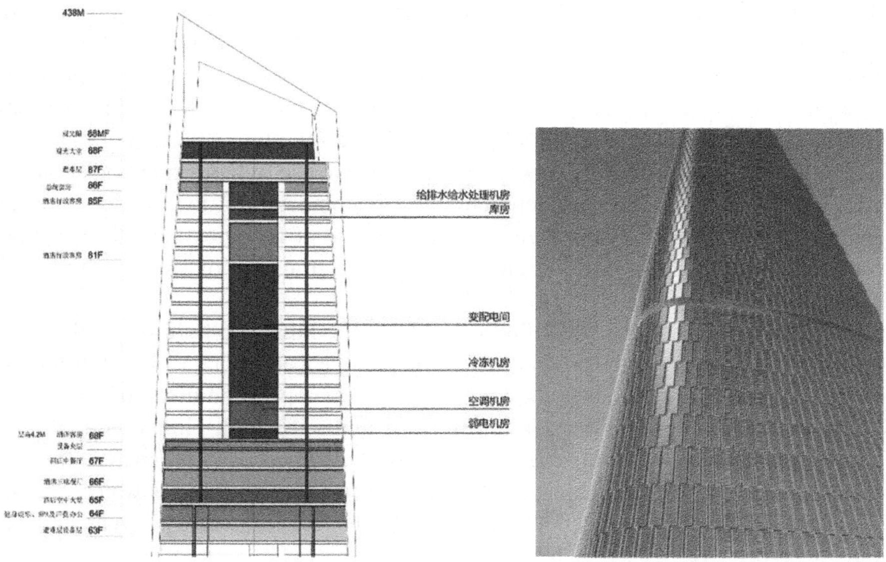

Fig. 8.7 Wuhan center core energy bar and stacked double cell board

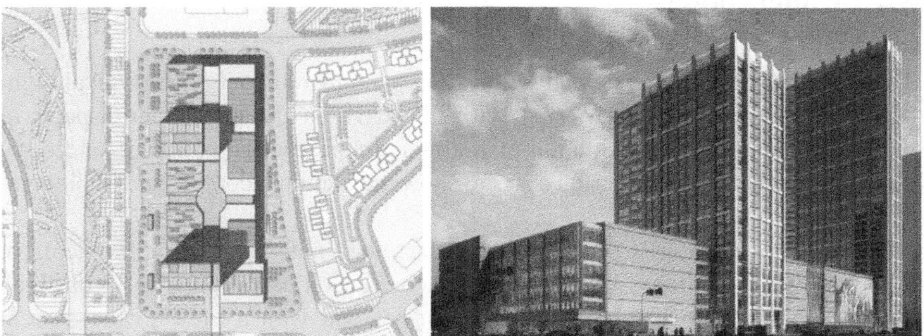

Fig. 8.8 Site plan and renderings of the Ocean Wide City Plaza Phase I

The curtain wall system comprehensively considers the core factors of Wuhan's climatic conditions, façade window-wall ratio, lighting, etc., and adopts a stacked double-unit board to design an energy-saving glass curtain wall system that can be opened (Fig. 8.7).

The Ocean Wide City Plaza Phase I is located at Fanhu Square Road and Baofeng Road and has achieved LEED Silver certification (Fig. 8.8).

The Ocean Wide City Plaza Phase I introduces a public transportation-oriented TOD (Transit-Oriented Development) hybrid development model, using traffic organization analysis technology to set the ground public transportation hub station at the northern end of the first floor of the building. It can stop buses of more than

4 lines at the same time, and stops 38 vehicles, and does not block the surrounding roads. While solving the diversion of people and vehicles including public transportation, it solves the problem of large numbers of people in urban commercial complexes using green transportation and convenient travel, alleviating the impact of urban commercial complex buildings on the surrounding environment and the pressure of urban road traffic flow, shaping intensive use of land and good ecological environment. The multi-functional space is compact and interactive, and the pursuit of multi-functional design and efficient use of facilities, mainly based on walking and bicycle transportation, realizes "zero transfer" of various modes of transportation.

The Ocean Wide City Plaza Phase I utilizes the huge fire-fighting pool of the commercial complex as the cold storage device for the water storage air-conditioning system, and uses intelligent technology to control the operation mode of the water-storage air-conditioning system to ensure that under different climatic conditions and different air-conditioning loads, it can achieve the best comprehensive effect in terms of comfort, energy saving and economy. After the project uses water storage air-conditioning, it can save about 265,000 yuan in electricity costs per year, and the rate of saving electricity is as high as 24%.

8.4 Conclusion

This chapter introduces the green low-carbon planning content of the typical high-density urban center area—Wuhan Wujiadun CBD in the three main aspects of municipal planning, transportation planning and construction planning, and combines some practical engineering cases to show the planning content and implementation effect.

On the one hand, this chapter summarizes the contents and achievements of Wangjiadun's green low-carbon plan, and constructs a basic framework for green low-carbon planning in high-density urban centers, with a view to the current period of rapid urbanization in China, from green and low energy buildings. The carbon perspective provides advice and guidance for the planning, development and construction of high-density urban centers in China. On the other hand, for the planning and development of high-density urban centers, we should actively adopt non-traditional and modern development models that conform to local market conditions. The core is to implement a low-consumption resource production system, a moderate consumption life system, a social system that guarantees social benefits and social equity, and continuously innovates and fully absorbs the application technology system of new technologies, new processes, and new methods. Through the scientific and rational development of green CBD, it could gradually extend to green cities.

References

Chunfang Dong, (2012). Architecture to high density. *China Architecture & Building Press*. (in Chinese)

Jianqiang Hua, (2019). The Composition, Correlation and Distribution of Functional Elements in Urban Central Area—Based on Succession Analysis of Capital Cycle. *City Planning Review*. (in Chinese)

Esra Trepci, Praveen Maghelal, et al. (2020). Effect of densification and compactness on urban building energy consumption: Case of a Transit-Oriented Development in Dallas, TX. *Sustainable Cities and Society*.

Jun Lu, Jitao Song, (2010). Analysis of Urban Land Use Structure and Its Zoning Characteristics in New York City. *Urban Studies*. (in Chinese)

Shinjuku City Government n.d., Japan. *http://www.city.shinjuku.lg.jp*.

Jun Jing, Kang Qi, et al. (2018). Evaluation and Analysis of Compact Metrology in Urban Central Area—A Comparative Study of Guangzhou Zhujiang New City and Hong Kong Central. *City Planning Review*. (in Chinese)

Honggang Zhao, Meng Wang, et al. (2019). Practice and Exploration of Sponge City Construction in Sino-Singapore Tianjin Eco-city. *Construction Science and Technology*. (in Chinese)

Yan Liu. (2013). Exploring the Path of Urban Transformation and Development under the Guidance of Ecological Civilization——Building a Green Ecological City and Promoting Urban Transformation and Development. *China Urban Science Research Association*. (in Chinese)

Tingpeng Yuan, Jian Yao. (2011). Cases and Enlightenment of Construction of Underground Car Lines at Home and Abroad. *Shanghai Construction Science & Technology*. (in Chinese)

Gang Wang, Yong Wang. (2013). Study on Low Carbon Strategy of Ecological City in Meixi Lake New City of Changsha. *Architectural Journal*. (in Chinese)

Chapter 9
Toward a Sustainable City: A Scoping Review of Eco-Cities Development and Practices in China

Zhi Cai and Yan Tang

Abstract With the rapid development of urbanization, China is facing many challenges on environmental problems, such as air pollution, water deterioration and urban heat island effect. To cope with the urban ecological crisis, China has paid more attention to eco-city construction. This chapter traces the origin of eco-city concept at the first part, and then reviews the development of eco-cities in China, in which several typical eco-cities have been taken as case studies to illustrate the practices of eco-city construction. The results showed that there are three periods of eco-city practice development in China: eco-cities (1990s), low-carbon cities (2000s) and smart-eco cities (2010s). In the period of eco-cities, the focus was paid on ecological and environmental issues, and the concept of "harmony between humans and nature" was widely applied to guide eco-city construction. In the period of low-carbon cities, attention was paid to cutting greenhouse gas emissions amid the severe domestic environmental situation and international societal pressure. In the new trend of smart-eco city construction, great importance has been attached to the city with green and ecological concepts, Chinese trying to combine the eco-cities with smart technologies. It is worth noting that the eco-city construction in China has always led by the central government and acted as a top-down approach, which also caused some problems among the practices.

Keywords Sustainable cities · Eco-cities · Low-carbon cities · Eco-smart cities · China

Z. Cai
Department of Urban Planning, School of Architecture, Tsinghua University, Beijing, China

Y. Tang (✉)
School of Architecture, Tsinghua University, Beijing, China
e-mail: yantang@mail.tsinghua.edu.cn

L. Bian et al. (eds.), *Chinese Urban Planning and Construction*,
Strategies for Sustainability, https://doi.org/10.1007/978-3-030-65562-4_9

9.1 Introduction

With the rapid development of urbanization, China has about 60% of the population (800 million) living in urban areas, and the number will rise to 70% in the next few decades, with more than 30% of the population will live in metropolitan areas. In 2019, 16 Chinese cities had a population exceeding ten million, and more cities will reach this level in the following years. A study showed that the 35 largest Chinese cities contribute 40% of China's energy uses and carbon emissions (Dhakal, 2009). As a vast amount of population and energy consumption concentrated in urban areas, it has a high impact on the environment and caused many environmental problems, such as extremely heavy air pollution, water deterioration and urban heat island effect (Manoli et al., 2019; Cai et al., 2019). Thus, to cope with the urban ecological crisis, Chinese government has been concerned more about environmental development and put forward a series of strategies like "sustainable development" and "ecological civilization" since the 1990s (Fu and Zhang, 2017). Later, eco-city construction was boomed in China mainland around the 2000s (Wang & Ye, 2004). Since 2008, with the rising consensus of global climate change, Chinese government has adopted the low-carbon emissions concept and combined it with eco-cities (Qiu, 2009b).

To offer an overview of the Eco-city development progress in China, this chapter firstly introduces the development of eco-city in China, and selected Shanghai city as a case study to illustrates the practices of Chinese eco-city construction. Then it reviews the development of the low-carbon cities in China and takes several demonstrating low-carbon eco-city as case studies, including Sino-Singapore Tianjin Eco-city, Chongqing Yuelai Eco-city and Tangshan Bay Eco-city. In the last part, it presents the new trends of Chinese eco-city practices: Smart-eco cities.

9.1.1 The Concept of Eco-City

In the 1960s, environmental problems outbreaks in western developed countries, and attracted global attention. Since then, lots of scholars have focused on environmental and ecological issues, and they are trying to find a sustainable development way. In 1971, the 16th General Conference of UNESCO proposed a long-term research plan for international scientific cooperation between human beings and the biosphere. In the next year, the United Nations Conference on Human Environment officially launched the "Man and the Biosphere" (MAB) plan, which shows that human-environmental issues had been paid attention to by the international community. At the same time, the international scientific and technological community reached a consensus on long-term and extensive cooperation on human-environmental issues, and put forward a research program that still has a profound and essential impact on the twenty-first century. In the program, they especially proposed a research project toward urban ecological problems. Moreover, MAB plan also formed an expert working group of leading scholars in the fields of environment and urban planning. It gradually promoted the interdisciplinary cooperation of natural science and social science in the global scope with urban ecological problems (Castries Francesco and Malcolm, 1984).

As a core member of the expert group in MAB, Oleg Yanitsky took the lead in systematically expounding the basic concept of "ecological city" based on the research results of urban ecology in various countries in the 1970s (Oleg, 1984). After reviewing the research results of many scholars in the same era, Oleg pointed out that eco-city will be the main residential area for human beings, and coordinated with the social and ecological process. The process of eco-city development includes five stages: fundamental research, applied research, planning and design, facilities construction and social structure transformation. In particular, he stressed that the process of eco-city development must be a process of interdisciplinary cooperation and knowledge integration in social sciences, natural sciences, engineering technology and other fields. At the same time, Register Richard, who later became president of the American Eco-City Association, emphasized that eco-city is an "ecologically healthy city" from the perspective of urban planning and construction, and the compact form of urban space is convenient for human activities. Moreover, energy and resource were efficiently utilized and had little impact on the natural environment (Richard, 1987).

Eco-city has rapidly become a hotspot in international research areas with the rise of environmental problems. Since 1980, the combination of the active promotion of the "Man and Biosphere Project" and the advocacy of interdisciplinary cooperation making the eco-city concept gained worldwide recognition, and the practice of eco-city planning and construction has been carried out in various countries.

In 1984, MAB put forward five principles of eco-city planning in a research report. Including eco-protection strategy (e.g., natural protection, flora and resources protection, pollution prevention), ecological infrastructure (e.g., natural landscape and hinterland to support sustainable city development), living standards of residents, preservation of culture and history, integrating nature landscapes into urban areas.

In 1996, based on the five principles of ecological city construction, Register put forward ten principles of the ecological city, such as land development, transportation, natural environment, resource utilization technology, production and consumption patterns, ecological awareness and social equity, government management.

In 1997, the Australian Urban Ecology Association (UEA) put forward the idea of eco-city development (Roseland, 1997). The content refers to several aspects, such as end urban sprawl, optimize energy structure and provide healthy and safe communities.

9.2 Methods

In this study, we were using a multi-case analysis method to capture the development of eco-city construction in China. The multi-case analysis is an effective method to study the common phenomena, and help researchers have a deep understanding about the mechanism of phenomena or system. Which aims to find,

1. What is the development of eco-city in China,
2. What methods that Chinese used to construct an eco-city,
3. What difficulties in developing an eco-city in China?

The materials are collected and processed as following steps. Step 1: Material collection. We searched the articles theme about "eco-city" and "ecological city" in the website of CNKI (China National Knowledge Infrastructure), Web of science, Google scholar and Baidu scholar; Step 2: Data analysis. Based on the material, we were using Citespace software to analyze the core interests and development of eco-city in China. The results show that there are several research clusters, such as "eco-city," "low-carbon city" and "smart eco-city." Since we know the core cluster of eco-city construction in China, we expand the search scopes to other sources like newspapers, government websites, social communication platforms (Weibo, Wechat) and social media. For each main cluster, we list the main content of it, and given several examples to illustrate their details.

9.3 Eco-City in China

9.3.1 The Development of Eco-City in China

China joined MAB and was elected as a member state in 1972. In 1978, Chinese government established the National Committee of MAB. In the next year, Chinese established the Ecological Society of China. Since then, scholars in China have begun to study urban ecological issues systematically.

In 1982, Chinese put forward the idea of "attaching importance to urban issues and developing urban science," which is the first symposium on strategic thoughts of urban development. In the same year, Chinese government drafted the sixth Five-Year Plan[1] (1983–1985). It set the study of urban ecosystems in Beijing and Tianjin as one of the critical scientific and technological projects.

[1] The Five-year Plan, the outline of the five-year plan for national economic and social development of the People's Republic of China, which is an important part of Chinese national economic plan and a long-term plan, the first Five-year Plan began in 1953. In a five-year cycle, major national construction projects, productivity distribution and the important proportion of the national economy are planned to set goals and directions for the development of the national economy.

In 1984, Shanghai held the first urban ecology symposium of China, and established the Urban Ecology Committee of the Chinese Ecology Society. In the same year, UNESCO's comprehensive publication International Social Science (Chinese edition) was first published in China. The fourth special issue systematically introduced the research progress in the field of International Ecology and environment since the initiation of the MAB in 1972.

Since 1990, with the rapid development of China's economy and urbanization, lots of urban ecological and environmental problems have emerged. Urban ecological and environmental issues have attracted the attention of scholars in China's geography, ecology, urban planning and other fields. With the advancement and deepening of the research, Chinese researchers put forward the concepts of "landscape city"[2] (Shang Shui city) and "garden city" based on the Chinese context, it expanded the concept and connotation of the ecological city.

9.3.2 Eco-City Practice and Case Studies in China

As Chinese pay more attention to urban ecological issues, the construction of the ecological cities is emerging nationwide (Wang, 2000). In 1986, the province of Jiangxi, Yichun City, first proposed a goal of building an ecological city. Subsequently, lots of the major cities in the Yangtze River Basin put forward the development strategy of building ecological cities. They began to construct eco-cities from different levels, like ecological cities, ecological counties, ecological demonstration areas, ecological villages and ecological districts. In 1994, Shanghai also explicitly propose the construction of a clean, beautiful, highly harmonious ecological city between human beings and nature. In 1996, the province of Shandong, Weihai City, which put forward the overall construction policy of "ecological environment, civilized quality of life." Chongqing city has also positioned its future development as an "eco-urban and rural metropolis in the twenty-first century." Other provincial capital cities, such as Beijing, Hangzhou and Changchun, have also put forward the strategic objectives of eco-cities.

While, it should be noted that the eco-city construction was guided by an indicator system (Table 9.1), which is proposed by the Chinese central government. There are two kinds of indicator system of eco-city in China (Liu and Ma, 2013). One is based on the analysis of three subsystems like society, economy and nature. The other is based on the consideration of the structure, function and coordination degree of the urban ecosystem. Both of them are using the expert scoring method or weight average method to evaluate all kinds of indicators.

[2] Landscape city, also known as "Shang Shui City," which is based on the Chinese traditional culture and philosophy of harmony between human being and nature. Its purpose is to build a harmonious human settlement of man-made and natural environment.

Table 9.1 The indicator system of Eco-city

Indicator	2000	2005	2010	2020
Economic level				
Per capita gross domestic product (10,000 yuan)	1.05	1.8	2.8	5.8
Land output rate (10,000 yuan/km^2)	711.2	1100	1850	4000
Life quality				
Average life expectancy (a)	72	73	75	78
Urban house index (urban per residential area/rural per residential area)	0.67	0.75	0.85	1
Environmental quality				
The proportion of water quality above than class III	41.40%	60%	80%	95%
Air quality index	83.90%	90%	95%	95%
Forest cover ratio	15%	20%	25%	
Public satisfaction with the environment	69.50%	80%	90%	95%
Economic				
Annual growth rate of GDP	10.50%	11%	8%	7%
Energy output rate (industrial added value/standard coal)	0.85	1	1.6	2.8
Proportion of revenue to GDP (%)	7	10	15	20
Social dynamic				
Gini index	0.31	0.34	0.38	0.34
Environmental protection dynamic				
Degraded land restoration rate	80%	94%	96%	100%
Attainment rate of industrial wastewater	93.80%	95%	99%	99%
Harmless reuse rate of urban domestic waste	40%	60%	80%	100%
Reuse rate of livestock and poultry manure	35%	55%	70%	90%
Economic development ability				
The certification rate of enterprise by ISO14000	10%	30%	50%	90%
The proportion of fixed assets in GDP	27%	30%	33%	38%
Ratio of scientific research	2.90%	8.00%	14%	18%
Social development ability				
Average education year (a)	8	10	12	14
Average year for professional education (civil servants)	2	4	5.5	6.5
The ratio of policy and regulation fit with eco-city planning	70%	90%	100%	100%
Ecological construction ability				
The proportion of environmental protection in GDP	1.69%	2.10%	2.50%	4%
Protected area ratio	5%	12%	15%	20%
Popularization rate of environmental knowledge among citizens	35%	50%	75%	90%

Source: Wu et al., 2005

9.3.2.1 Case Studies in Shanghai

Shanghai is one of the first cities in China to enter the construction of eco-cities. In 1999, the Shanghai government set a strategic goal to build a first-class international eco-city, the primary work is to optimize the urban layout, improve urban infrastructure and environmental quality, promote the sustainable development of economy and society. Its eco-city construction approach has many aspects, such as optimize the urban structure, enhance the function of urban infrastructure and the coordination of urban and rural areas (Wang, 2001).

9.4 Optimize urban structur

Optimize urban structure: Shanghai government set four strategies to optimizing the urban structure. Firstly, Shanghai city needs to optimize the structure and balance the proportion of urban land use, including increasing the proportion of transportation land and green spaces, appropriately reduce the proportion of industrial land, strictly control the number of high-rise buildings and reduce the building density in central urban areas. Secondly, improve the living quality of urban life and control the population density, such as increasing the investment in primary education (above than GDP 3%), adopt the corresponding zoning management and temporary household registration management system for the migrant residents. Thirdly, promote clean energy and green consumption by accelerating industrial restructuring and technological innovation. The measures include strictly control coal consumption, change the original energy structure, utilize local clean energy and natural energy, draft and promote circular economy law, popularize green consumption awareness. Fourth, strengthen urban greening and protect bio-diversity. The measures include appropriately introduce the landscape plants with better adaptability, enrichment of urban plant species, expansion and strengthen the construction of natural and ecological protection areas.

9.5 Enhance the function of urban infrastructure

Enhance the function of urban infrastructure: Enhance the urban function is the main part of promoting eco-city construction in Shanghai city, such as strengthen the protection of the ecological environment, develop the ecological industry and promote cleaner production, build a fast information circulation platform. While the ecological environment aspects mainly focus on the management and treatment of urban sewage, automobile exhaust, solid waste and urban noise, the information platform aspects mainly focus on the construction of the internet infrastructure and develop a GPS communication system.

9.6 Coordination of urban and rural areas[3]

The dualistic structure system of the urban and rural area has a significant impact on social-economic development, and Shanghai city act the coordination of urban and rural areas as an essential issue to promote eco-city construction. The measures include implementing the integration of urban and rural social-economic planning, optimizing urban/rural spaces and form a rational distribution of productive forces and urban networks, promoting the rational distribution of urban industry and population, and integrate urban artificial environment with the rural natural environment, strengthening the construction of urban public service facilities to improve the quality of urban life, centralized development residential areas and service facilities, forming a multi-level, multi-type community service network, improve the building design and the quality of the outdoor environment. Moreover, social governance was also considered, such as enhance the residents' awareness of the ecological environment, strengthen the construction of government organizations, and ensure the orderly decision-making, supervision and management of eco-city construction.

9.7 Low Carbon Eco-City in China

After entering the twenty-first century, the large number of greenhouse gases (mainly carbon dioxide) emissions have led to a global consensus on climate change. Intergovernmental Panel on Climate Change (IPCC) has published several assessment reports, and they pointed out that according to current development speed and world national policies, greenhouse gas emissions will continue to rise in the following decades. Moreover, green gas will lead to global warming, accelerating the melting of glaciers in the north and south poles. With the rise of sea level, lots of coastal cities will under the risk of inundation (Fairbanks, 1989). Global society faces tremendous challenges in mitigating climate change, and it is inevitable to reduce greenhouse gas emissions and find a sustainable development way. To deal with the global climate crisis, IPCC suggests cutting the greenhouse gas emissions, which have got concerns all over the world and involved more countries since the 1970s.

Worth to note is the carbon emission. With the rapid urbanization, China's carbon emissions have reached 6.071 billion tons by 2007, which accounting for 20.96% of the global total carbon emissions. It has surpassed the United States

[3] China has a dualistic structure system of urban and rural area, which mainly reflects in land institution and household registration system. Where the urban land is owed by the state and the rural land is owed by the farmers collectively. The household registration system also divides the population into urban residents and rural residents.

(5.769 billion tons, 19.92%) and becoming the largest carbon emitter (The Global Carbon Budget 2007, 2008).

9.7.1 The Development of Low Carbon Eco-City in China

As the most abundant carbon emission country, China faces great pressure from international society to cut its carbon emissions (Yang and Li, 2013). Moreover, with the grim domestic situation of the Chinese mainland, urban industry and economy in a low efficient energy consumption development way, which could not cope with sustainable development, the low-carbon transformation is inevitable in urban China.

In 2007, Chinese President Hu Jintao advocated the development of a low-carbon economy at the 15th APEC Leadership Meeting. Since then, low-carbon city construction has begun in China. In 2009, the Chinese Academy of Social Sciences issued the Blue Book of Cities: China's Urban Development Report (No.2) in Beijing, and it pointed out that low-carbon development is an inevitable choice for China to control greenhouse gas emissions in the process of urbanization. In the same year, Qiu Baoxing, who is the Minister of Housing and Construction of China, put forward the concept of low-carbon eco-city in the International Forum on Urban Planning and Development (Qiu, 2009a). Qiu pointed out that low-carbon eco-city is the concretization of the idea of sustainable development in urban development and the mode of low-carbon economic development (Li et al., 2010) (Fig. 9.1).

With the vigorous support and promotion of the Chinese Ministry of Housing and Construction (CMHC), CMHC signed a cooperation framework agreement on building a "national low-carbon ecological demonstrating city" with Shenzhen Municipal Government in January 2010. Shenzhen became the first "national

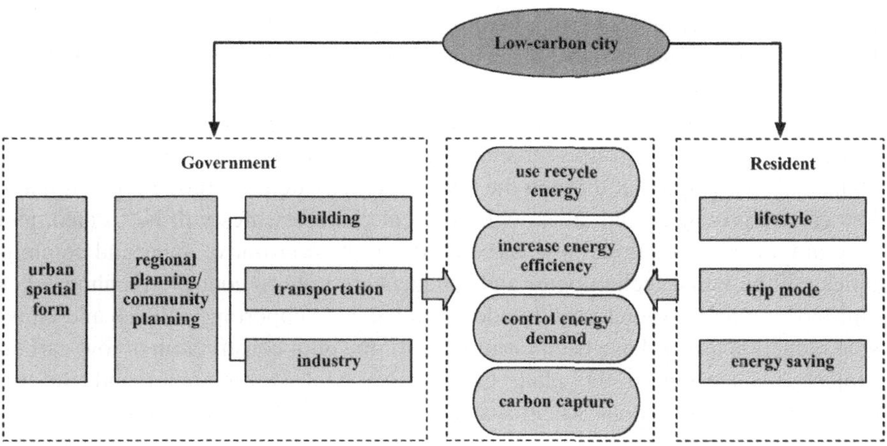

Fig. 9.1 The frame contents of the low-carbon city in China

low-carbon ecological demonstrating city" in China. In July 2010, the National Development and Reform Commission of China issued a notice on the demonstrative construction work of low-Carbon provinces and cities. It decided to start the demonstrative construction work of low-carbon provinces and cities nationwide, which covered eight cities (Tianjin, Chongqing, Shenzhen, Xiamen, Hangzhou, Nanchang, Guiyang and Baoding) and five provinces (Guangdong, Liaoning, Hubei, Shanxi and Yunnan). In the same month, Chinese Ministry of Housing and Construction signed the "co-construction of national low carbon eco-city" with the government of Wuxi City in Jiangsu Province. In October 2010, Chinese Ministry of Housing and Construction signed the Memorandum of Cooperation with Hebei Province to Promoting the Construction of Ecological Demonstration Cities and healthy urbanization. Those cooperations eventually making the eight awarded eco-cities came into being: Sino-Singapore Tianjin Eco-city, Tangshan Bay Eco-city, Shenzhen Guangming New District, Wuxi Taihu New Town, Meixihu New Town, Chongqing Yuelai Eco-city, Kunming Chenggong New District and Guiyang Future Ark Eco-city. The main tasks of the demonstrative cities aimed at the following five aspects:

1. compile a low-carbon development plan;
2. establish the supporting policies for low-carbon green development;
3. accelerate the establishment of an industrial system characterized by low carbon emissions;
4. establish the statistical and management system of greenhouse gas emission;
5. advocate low-carbon green lifestyle and consumption behaviors.

Then, in January 2011, CMHC set up a research group to guide the healthy development of low-carbon eco-cities in China. The group mainly focuses on the issues of low-carbon eco-cities, such as the urban planning, policy suggestions, demonstration technology of low-carbon eco-cities and indicator system. Lately, the indicator system has been widely used to define, guide and assess low-carbon development eco-city in China. The indicator system consists of 30 key performance indicators and divides into four major categories of efficient use of resources, friendly environment, sustainable economy and harmonious society (Table 9.2).

In 2011, China set the goal of reducing carbon emissions per unit GDP by 17% compared with 2010 by 2015. For this goal, Chinese State Council promulgated the work program for controlling greenhouse gas emissions in the twelfth five-year plan. They proposed that through the low-carbon city demonstration project, many low-carbon provinces and cities with different characteristics will be formed, and several typical international low-carbon industry demonstrative areas and communities will be built. Technologies and products are used to control greenhouse gas emissions in an all-round way. In addition, detailed supporting policies and emission reduction targets have been formulated for the pilot construction of low-carbon projects. According to the plan, Chinese National Development and Reform Commission selects about 30 cities as the second round of low-carbon ecological demonstration projects, which covered all provinces except Tibet in China. The second demonstration project emphasizes its feasibility, extensibility and

Table 9.2 Chinese low carbon eco-city indicator system

Indicator	2015	2020
Efficiently use of resources		
1 Recycling of water usage (%)	Severe water shortage area ≥ 25	Severe water shortage area ≥ 30
	Water shortage area ≥ 15	Water shortage area ≥ 20
2 Recycling of industrial water (%)	≥ 90	≥ 95
3 Rate of no-fossil energy usage (%)	≥ 15	≥ 20
4 Carbon emission per GDP unit (ton/10 k yuan)	≤ 2.13	≤ 1.67
5 Energy consumption per GDP unit (ton standard coal/10 k yuan)	≤ 0.87	≤ 0.77
6 Construction land use per capita (m^2/person)	≤ 85	≤ 80
7 Percentage of green building (%)	Existing building ≥15	Existing building ≥15
	New building 100	New building 100
Friendly environment		
8 Days of compliance with air quality (SO NO PM10)	≥ 320	≥ 347
9 Days of compliance with PM2.5 standard per year	≥ 292	≥ 310
10 Water quality compliance rate of the centralized drinking water source (%)	100	100
11 Surface water quality compliant rate for urban water environmental function zones (%)	100	100
12 Reuse rate of domestic wastes (%)	≥ 50	≥ 50
13 Recycling of industrial solid waste (%)	≥ 90	≥ 95
14 Noise level compliance area coverage (%)	≥ 95	100
15 500 meters radius service coverage from parks and green open space (%)	≥ 80	≥ 90
16 Bio-diversity	Comprehensive index ≥ 0.5	Comprehensive index ≥ 0.5
	Local species index ≥ 0.7	Local species index ≥ 0.85
Sustainable economy		
17 The percentage of value-added of the tertiary industry in overall GDP (%)	≥ 47	≥ 51
18 Urban unemployment rate (%)	≤ 4.2	≤ 3.2
19 Percentage of R&D input to GDP (%)	≥ 2.2	≥ 2.6
20 Engel's coefficient (%)	≤ 33	≤ 30
Harmonious society		

(continued)

Table 9.2 (continued)

Indicator	2015	2020
21 Percentage of affordable housing (%)	≥ 20	≥ 30
22 The ratio of housing price/income	≤ 10	≤ 6
23 Gini coefficient	0.33 ≤ G	0.33 ≤ G
24 Income ratio between urban citizen/rural residence	≤ 2.54	≤ 2.41
25 Percentage of green transport (%)	≥ 65	≥ 80
26 The coverage rate of social security (%)	≥ 90	100
27 Land use floor space of public services and utilities per capita (m^2/person)	5.5	6.0
28 Average commuting time (min)	≤ 35	≤ 30
29 Urban disaster prevention and reduction	Space for disaster escape per capita ≥ 3 m^2	Space for disaster escape per capita ≥ 3 m^2
30 Satisfaction rate for public security (%)	≥ 85	≥ 90

Source: Yu, L. (2014)

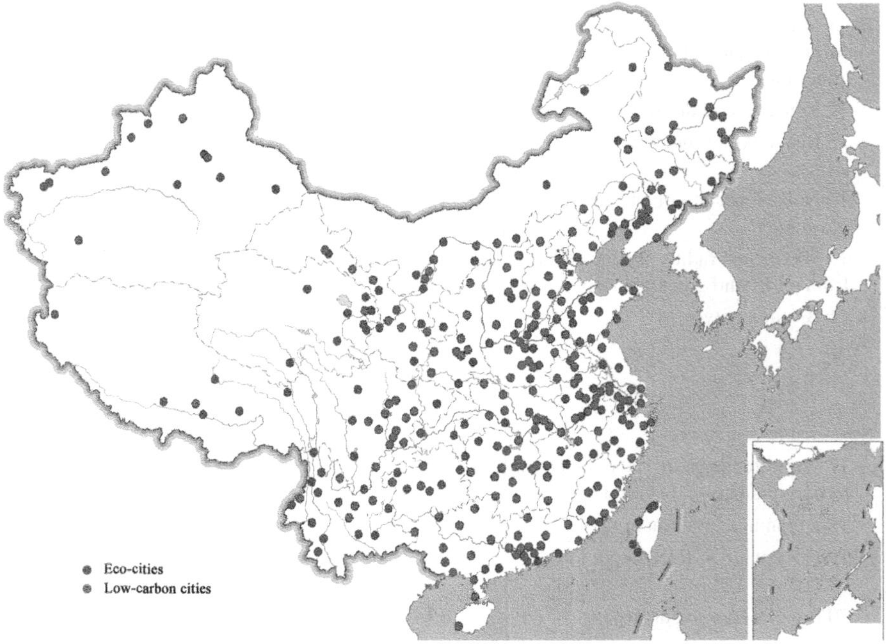

Fig. 9.2 Locator map of China's Eco-cities and low-carbon cities

replicability. The Ministry of Finance of the central government with other ministries pulled 3.7 billion dollars to encourage low carbon development. Under the dual influence of central government financial support and local government promotion, low-carbon eco-city booming in China again. A statistic made by the Academic Exchange Department of China Urban Science Research Association showed that 230 of 280 prefecture-level cities in China had proposed the goal of "eco-city" construction, and 133 of those cities (46.3%) have proposed the goal of "low-carbon city" construction (Fig. 9.2).

9.7.2 Low Carbon Eco-City Practice and Case Studies in China

9.7.2.1 Sino-Singapore Tianjin Eco-City

In November 2007, China and Singapore signed a cooperation agreement to build a new and medium-sized eco-city in the Tianjin Binhai New Area with a planned area of 34.2 km^2. Sino-Singapore Tianjin Eco-city started its construction in September 2008. Binhai New Area has initially been a saline-alkali wasteland with heavily polluted water. In order to protect and improve the fragile ecological environment of

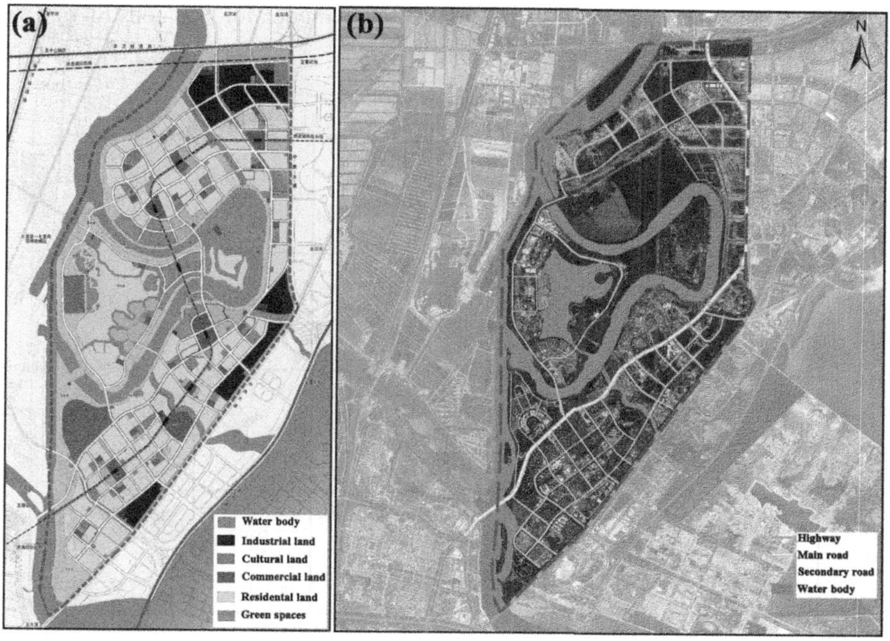

Fig. 9.3 (**a**) The master plan of Tianjin Eco-city, and (**b**) the current construction of Tianjin Eco-city

the area, the plan adopts the method of "bottom-up mapping." Firstly, they evaluated the suitability of land and determined the protection of the ecological area. Then, they calculated the carrying capacity of the natural environment, and determined the area for urban development and construction (21km^2) and the carrying capacity of the population (350,000 residents). By redeveloping and protecting the original water wetlands, Sino-Singapore Tianjin Eco-city finally has formed a radial network pattern with water bodies as the core area (Fig. 9.3).

In order to guide the construction of low-carbon eco-city, an indicator control system was set at the beginning of the construction, which refers to four aspects like ecological environment, society, economy and regional coordination between urban and rural areas. The system covers 22 control indicators and four guiding indicators. The methods including implementing green building standards, strengthening building energy efficiency, building a green transportation system, optimizing the layout of urban land, optimizing energy structure, developing and building a low-carbon energy system, graded allocation of the ecological community model (Fig. 9.4).

By 2019, the permanent population of Sino-Singapore Tianjin Eco-city has reached 100,000 people, more than 8000 registered enterprises and institutions, the proportion of modern service industry has reached more than 90%, and 30,000 talents introduced. The development of society, environment and economy have been harmonious coexistence. Moreover, Sino-Singapore Tianjin Eco-city was elected as "China's most happy ecological city in 2018." The primary significance of Sino-Singapore Tianjin Eco-City is to adapt to the needs of low-carbon development in China at that time, and provide a demonstrating model for the construction of

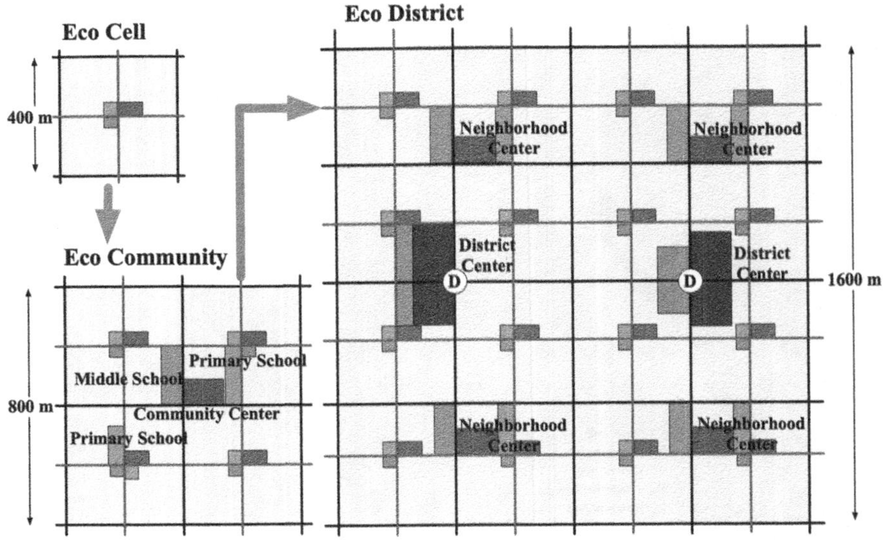

Fig. 9.4 Eco community model in Sino-Singapore Tianjin Eco-city

low-carbon eco-city in China. It is also an inevitable choice under the situation of tight resource constraints.

However, Sino-Singapore Tianjin Eco-city is still far from its goal (Fig. 9.3b). The large scale of residential blocks makes the city have an uncomfortable walking environment, and the lots of tall buildings with few residents (below than 30% of the planning population), making the city looks like an empty eco-city (Caprotti, 2014). The local government has less consideration about the migrant workers, and most of the empty house was owned by multiple home speculators (Chang et al., 2016). What is more, some of the detail and comprehensive sustainability indicators are difficult to reach in urban scale, but applicable to a region or country scales, such as ambient air quality and surface water quality (Li et al., 2019).

9.7.2.2 Chongqing Yuelai Eco-City

As one of the first green Eco-Demonstration cities in China, Chongqing Yuelai Eco-City is located in the core area of the development of new urban areas. It began planning and construction in 2011, the plan area is about 3.44 km^2, and its capacity of the population is 50,000. Chongqing government invites Peter Calthorpe to serve as the chief planner, who is the representative of American New Urbanism. Different from other plain low-carbon eco-cities, the site in Chongqing Yuelai eco-city has a significant topographic fluctuation. Nearly half of the land has a slope of 15 degrees or more, the internal traffic the city is difficult to organize, the land suitable for construction is small, and the land use function is single.

Therefore, Yuelai New Town adopts the strategy of efficient and intensive use of land and construction of a composite ecological city, making full use of mountain terrain, conforming to the mountain situation, and rationally allocating urban industrial land in combination with topographic changes. Firstly, in the land mixed-use, all plots require 10–20% of the land mixed-use ratio. It was emphasizing the mixed layout of residential, commercial, industrial and other types of land to meet residents' daily leisure, recreation, living, work and other needs within the block. Secondly, different from the common urban planning mode of streets and wide roads in China at that time, Yuelai New Town is the first city to use the small blocks and narrow roads to connect urban space and activate urban street space. The public service facilities and green space will be arranged in the residents' 5-minute walking circle. Thirdly, Yuelai New Town, aiming at mountainous terrain, has adopted a mixed-mode of transportation and built a perfect pedestrian system. Based on the TOD model, with rail transit stations as the center, bus, bicycle lane and pedestrian road extend to all urban districts. Fourthly, in the municipal aspect, the introduction of low impact development technology, green space, municipal, rainwater collection and other unified consideration. Finally, using a retreat building layout to increase its visibility to the river within the city.

Chongqing Yuelai New Town provides a safe and efficient way of travel for the city through the mixed-function and TOD development mode. The perfect walking system and the creation of a large number of suitable walking areas reduced the probability of car travel, which has gradually become a primary low-carbon development mode in China.

9.7.2.3 Tangshan Bay Eco-City

Under the circumstances of China's national low-carbon ecological policy and the overall industrial layout adjustment of Tangshan City. Tangshan City started the planning work of Tangshan Bay Eco-city in 2007, and formally started its construction in 2009 (Lv et al., 2014). The plan area of Tangshan Bay is 74.3 km^2, with a resident of 800,000 by 2020. It has 18 kilometers of highly quality coastline and an excellent ecological environment.

At the beginning of its construction, the local government invited five international cooperation to participate in its urban planning and design. They are Swedish SWECO, Dutch DHV, England ARUP, Italian Archa and Beijing Tsinghua Institute of Urban Planning and Design. SWECO won the competition and merged excellent ideas from the rest urban plan. SWECO expects to build a "dark green" eco-city, which will play an exemplary role in energy saving and emission reduction, intensive land use development, circular economy and knowledge of innovation, and create a harmonious city through the guidance of ecological indicators. Its ecological indicators draw lessons from the latest planning theory and practice in Sweden. Based on the concept of "symbiotic city," it covers seven aspects, including urban function, architecture, transportation, energy, water, waste, landscape and public space, with a total of 141 specific indicators. Besides, Tangshan Bay Eco-city also proposed a systematic green building evaluation index, vigorously develop and adopt passive energy-saving green building technology; the proportion of green buildings reached 100%.

Although the Tangshan Bay Eco-city with stronger support of the local government and has carried out scientific and rational urban planning and design, by the end of 2017, the city has invested 32.5 billion yuan in the capital and built an area of 33 km^2, but it is still far away from the goal of high prosperity. Because of the long distance to the old urban center, which making an inconvenience of transportation and the lack of attractiveness of the investment. There is a single function in land use, and mainly for education and residential. Apart from universities and residences, the number of companies and enterprises staying is relatively small, while most of the residential houses are used for speculation. College teachers and students mainly compose the current 48,000 residents. On the other hand, Tangshan Bay Eco-city relies excessively on government investment and bank loans, and local governments are heavily indebted, which has severely affected local sustainable development.

9.8 Smart-Eco Cities in China

In 2008, IBM put forward the concept of "Smart Earth." Later, it developed with the concept of "smart city." The smart city is the deep integration of the new generation of information technology and the transformation of the city, such as the internet of things, cloud computing, big data, spatial information (Hollands, 2008). Smart city construction is also the best way to solve the urban problems and realize the sustainable development of the city, reflecting the essential demand of the city's green, low carbon and sustainable development. Hence, some countries try to combine the concept of "smart city" with "eco-city," which is "smart-eco city." Smart-eco city could be defined as an experimental city that functions as a potential niche where both environmental and economic reforms can be tested and introduced in areas that are both spatially proximate and in an international context (Tan et al., 2017).

9.8.1 Smart-Eco Cities Practice and Case Studies in China

Since several decades of fast economic development and rapid urbanization, the speed of Chinese social-economic has slow down in 2013. Thus, Chinese government tries to find a more sustainable way to develop social-economic. Meanwhile, Chinese central government attaches great importance to the city in a green and ecological developing way. With the development of information technology, Chinese cities pay more attention to smart-eco city construction. In 2013, the Ministry of the housing of China announced the first round of 90 demonstrating cities for smart city construction. At the end of 2017, the smart goal and construction have covered 86% of cities in China (Table 9.3). The central government also drafted an indicator system to guide smart-eco city construction in recent years, including ubiquitous service for the residents, intensive infrastructure, livable environment with low carbon and green technology. The livable environment with low carbon and green technology mainly covers the following aspects: strengthen the synchronous planning of urban functions and industrial economy, promote the coordinated development of urban social economy and ecological environment.

Even though Chinese smart-eco city construction is showing a trend of vigorous development, it still needs to be noted that there are still many problems in its construction. Chinese smart-eco city construction is still in its initial and conceptual stage. The imitation and duplication among governments, the assessment requirements of higher authorities, the political pursuit of government leaders, and the need for urban publicity become the main driving force for smart-eco city decision-making. Besides, Chinese smart-eco city construction is dominated by government investment, while few investments come from private.

Table 9.3 Several smart or smart-eco cities practices in China

Cities	Theme	Practices
Wuhan	Smart transportation	Explore the "internet +" smart traffic mode and cooperate with the internet corporations to build a smart transportation management system. Build a mobile internet government platform to provide convenient services like illegal traffic inquiries and payment of the fines online.
Nanning	The big transit system	Establish a smart transportation APP service platform and build an intelligent bus station board. Sharing urban information data to achieve the integration of resident card and traffic card.
Weifang	Spatiotemporal information platform	Integration of the present and historical urban basic geographic information, population, economy, industry, environment and other data, and constructed a space-time information cloud platform. Based on this, built demonstrating application systems like smart land, smart policing, smart transportation, smart environmental protection and smart street.
Jilin	Big data in agriculture	Geographic information technology was used to synthesize land ownership, soil information and fertilization. Real-time monitoring of crop growth by satellite imagery and remote sensing technology. Based on big data technology, a third-party trading platform has been built to promote the perfect development of the agriculture industry.
Harbin	Smart pipe network	Based on the survey of underground pipelines and establish the platform of a spatial information system, designed six subsystems like basic application, operation and maintenance configuration, pipeline network data management, integrated business, secondary development, monitoring and management. By the platform, 43 functions for the underground pipeline network can be realized like statistical query, data management, operation and maintenance, and assistant decision-making.
Chongqing	Comprehensive urban information system	In order to support data sharing and integration, a standard information resource specification is formulated. Local governments also build a data-sharing platform to promote data exchange and sharing among different departments. The integration of shared resources and the development of various decision applications have realized basic functions such as querying, statistics and analysis of comprehensive urban information.

9.9 Discussions and Conclusions

This chapter introduces Chinese ecological sustainable urban development in recent years. From the initial awareness of environmental protection, China has gone through three stages: eco-cities, low carbon eco-cities, and smart-eco cities development. With the rapid development of urbanization, the transformation of urban development and the development of information network technology, Chinese sustainable urban construction is continuously changing. It can be found that during the initial period of eco-city construction, China was mainly imitating western countries. While, during the period of low carbon eco-city construction, China began to carry out extensive international cooperation. After stepping into the stage of smart-eco city construction, since there were fewer international cases for reference, China began to find out its own way to develop sustainable cities. Judging from the driving factors, unlike the western countries, which took bottom-up approaches, Chinese sustainable urban construction adopt top-down approaches (government-driven). Urban construction mainly depends on the allocation of funds from the higher authorities and the local government's loan to the bank through the way of land mortgage. However, in the process of construction and implementation, this top-down government-led construction mode has gradually exposed several problems. The first is that it just applied a simple indictor system to guide urban construction. China has a vast territory, and there are significant differences in social customs, economic development level, climate and environmental conditions all over the country. This difference leads to some indicators almost impossible to achieve, and all kinds of urban construction goals have finally become empty words. The second is that ecological sustainable urban construction has become a slogan for local governments to develop new districts. From the perspective of eco-city construction, most of the pilot cities choose to build new towns outside the old urban areas. Even though, the government expects to attract population and promote the development of new areas with slogans such as low-carbon, sustainable cities and beautiful landscapes. However, the facts have proved that without stronger support like a pleasant investment environment and convenient transportation system, some demonstrating areas have little attraction to residents, and it eventually turned into ghost cities. The above problems have made those areas failed to achieve the goal of an ecologically sustainable city, and give a heavy debt burden to local governments.

Even though there are still many problems in Chinese sustainable urban construction, it is optimistic that China is still steadfastly taking the road of sustainable development. Ecology, green and sustainable development have also become a national policy. China has accumulated a great deal of experience in sustainable urban construction and has responded quickly to urban construction. We believe that in the future, Chinese sustainable urban construction will not only provide a model for the local city, but also provide a reference model for the world.

Acknowledgments This work was supported by the National Natural Science Foundation of China (grant no. 51978363), Beijing Social Science Foundation (grant no. 19GLB034) and National Key Research and Development Program of China (grant no. 2018YFC0704600).

References

Cai, Z., Tang, Y., Chen, K., & Han, G. (2019). Assessing the heat vulnerability of different local climate zones in the old areas of a Chinese megacity. Sustainability, 11(7), 2032.

Caprotti, F. (2014). Critical research on eco-cities? A walk through the Sino-Singapore Tianjin Eco-City, China. Cities, 36, 10-17.

Castries, F. D., Hadley M. (1984). People in the ecosystem. International Social Science Journal: Chinese Version, 1(4), 3-6.

Chang, I. C. C., Leitner, H., & Sheppard, E. (2016). A green leap forward? Eco-state restructuring and the Tianjin–Binhai eco-city model. Regional Studies, 50(6), 929-943.

Dhakal, S. (2009). Urban energy use and carbon emissions from cities in China and policy implications. Energy policy, 37(11), 4208-4219.

Fairbanks, R. G. (1989). A 17,000-year glacio-eustatic sea level record: influence of glacial melting rates on the Younger Dryas event and deep-ocean circulation. Nature, 342(6250), 637.

Fu, Y., & Zhang, X. (2017). Planning for sustainable cities? A comparative content analysis of the master plans of eco, low-carbon and conventional new towns in China. Habitat International, 63, 55-66.

Hollands, R. G. (2008). Will the real smart city please stand up? Intelligent, progressive or entrepreneurial?. City, 12(3), 303-320.

Li, X., Cao, G., Xu, W., Yang, C., Song, F., Zhao, P. (2010). China's low-carbon eco-city development strategy. Urban studies, 17(01), 32-39. In Chinese.

Li, Y., Commenges, H., Bordignon, F., Bonhomme, C., & Deroubaix, J. F. (2019). The Tianjin Eco-City model in the academic literature on urban sustainability. Journal of Cleaner Production, 213, 59-74.

Liu, Z., Ma, J. (2013). Review on Theory and Practice of Eco-city Construction. Environmental Science and Management, 38(02), 159-164.

Lv, X., Ye, Q., Zhao, Q., & Sun, K. (2014). The implementation strategy and development mechanism of green architecture: To draw lessons from Tangshang Bay. Eco-city Architectural Journal, 11, 130-133. In Chinese.

Manoli, G., Fatichi, S., Schläpfer, M., Yu, K., Crowther, T. W., Meili, N., Bou-Zeid, E. (2019). Magnitude of urban heat islands largely explained by climate and population. Nature, 573(7772), 55-60.

Oleg, Y. (1984). Towards eco-city: A combination of knowledge and practice problems. International Social Science Journal: Chinese Version, 1(4), 103-114.

Qiu, B. (2009a). From Green Building to Low Carbon Eco-City. Urban Studies, 7, 1-11. In Chinese.

Qiu, B., (2009b). The Transformation Trends of Urban Development Model in China——Low Carbon Eco-city. Urban Studies, 8, 1-6. In Chinese.

Richard, R. (1987). Eco-city Berkeley: Building Cities for a Healthy Future. Berkeley: North Atlantic Books.

Roseland, M. (1997). Dimensions of the Future: An Eco-city Overview. Eco-city Dimensions. New Society Publishers, 1-12.

Tan, M., Cheshmehzangi, A., Chien, S., & Xie, L. (2017). Smart-eco cities in China: Trends and city profiles 2016. Exeter: University of Exeter (SMART-ECO Project).

The Global Carbon Budget 2007. (2008). Global Carbon Project, Press release from Global Carbon Project.

The International Ecological Re-building Program. (1996). Village Wisdom/Future Cities[C]// Register R, Eds. The Third International Ecocity and Ecovillage Conference. Oakland: Ecocity Builders, 204-205.

Wang, R., & Ye, Y. (2004). Eco-city development in China. Ambio: A journal of the human environment, 33(6), 341-343.

Wang, X. (2001). On the theories, ways and countermeasures for the construction of eco-city – a case study of Shanghai, China. Journal of Fudan University, 04, 349-354. In Chinese.

Wang, X. (2000). Ecological and environment: The new concept of urban sustainable development and ecological environment. Nanjing: Southeast University Press. In Chinese.
Wu, Q., Wang, R., Li, H., Xu, X. (2005). The indices and the evaluation method of eco-city. Acta Ecologica Sinica, 08, 2090-2095. In Chinese.
Yang, L., & Li, Y. (2013). Low-carbon city in China. Sustainable Cities and Society, 9, 62-66.
Yanitsky, O., (1984). Integration of social and natural science for urban planning. In: Di Castri, F., Baker, F.W.G., Hadley, M. (Eds.), Ecology in Practice, eighth ed. Tycooly International Publishing Ltd. and UNESCO, Paris, 30-46.
Yu, L. (2014). Low carbon eco-city: New approach for Chinese urbanisation. Habitat International, 44, 102-110.

Chapter 10
Healthy City Planning: Insights from China

Lan Wang, Xiji Jiang, and Jie Cai

Abstract City planning, a major means for the allocation of resources and shaping of urban space, is an authoritative guide for the promotion of public health. Healthy city planning is considered one of the most important instruments for building a 'Healthy City', which is an important concept proposed by the World Health Organisation (WHO). This chapter elaborates on the development progress of healthy city, the research progress of healthy city planning, and the theoretical exploration and practical experience in China. We propose both theoretical and practical frameworks for healthy city planning and investigate the potential capacity of health impact assessment (HIA) as a policy tool to integrate health into city planning. It is concluded with the challenges and experiences of research, and the practice of healthy city planning in China, aiming to shed light on the important principles in developing a healthy city in different contexts across the world.

Keywords Healthy city · Healthy city planning · Health impact assessment

10.1 Introduction

As the issue of health becomes increasingly prominent around the world, the development of healthy cities receives increasing attention. With continuous global urbanisation, city populations have surpassed those in rural areas. The WHO has come to realise that the physical environments of cities, important habitats for people, have a significant impact on public health. It has, therefore, proposed the concept of the 'Healthy City', actively promoting the building of healthy cities globally, and extending the scope of support and intervention from the health of individuals to that of an entire city. It is emphasised that a healthy city is a process rather than an outcome. A healthy city is 'a city that is continuously striving to improve the

L. Wang (✉) · X. Jiang · J. Cai
Healthy City WLAN Lab, College of Architecture and Urban Planning, Tongji University, Shanghai, China
e-mail: wanglan@tongji.edu.cn

L. Bian et al. (eds.), *Chinese Urban Planning and Construction*,
Strategies for Sustainability, https://doi.org/10.1007/978-3-030-65562-4_10

health and healthy determinants of residents, not one that has achieved a particular health status' (WHO 1998). Healthy cities focus on measures and policies to improve urban physical and social environments and promote healthy lifestyles.[1]

So far, the development of healthy cities has witnessed great achievements around the world. The WHO has contributed significantly to the development of the European healthy city network and the establishment of a multi-regional and multi-agency alliance. With a history of over two decades, China's healthy city programme can be traced back to the 'Hygienic City' programme of 1989. Despite China's achievements, such as developing a national healthy city network, building healthy cities and enacting supporting policies, problems remain to be solved, such as planning absence, low feasibility, insufficient inter-departmental collaboration and poor understanding of a healthy status. While modern city planning stems from public health issues in industrial cities, public health is no longer a priority in city planning as solutions have been found for issues relating to traditional infectious diseases and sanitation issues. The increasing numbers of patients with chronic non-communicable diseases (NCDs) has placed the role of city planning in addressing public health issues and promoting development of healthy cities under the spotlight again in recent years (Giles-Corti et al. 2016; Neira 2018).

Against such a background, this chapter: (1) analyses the characteristics of China's policies, systems and organisations in relation to its healthy city development; (2) proposes the theoretical framework of healthy city planning and analyses the progress of relevant researches in China; (3) develops a theoretic framework and a working progress for healthy city planning, and summarises the experiences gained so far. It is hoped that the world can benefit from China's experiences.

10.2 Research Methods and Research Design

10.2.1 Research Method and Materials

This chapter adopts event flow analysis to identify the milestones in the development history of global healthy cities. This method is an analytical tool used to describe the incidents, behaviours and people involved in an event. It is designed to enrich understanding of the cause, process and mechanism of the event and thus help to predict and guide future events or behaviours.[2] Meanwhile, based on the CiteSpace software,[3] this chapter uses the bibliometrics method to reveal the research progress and change of hot topics in this field, providing inspiration and references for follow-up research and practice.

---

[1] http://www.wpro.who.int/china/mediacentre/factsheets/healthy_cities/zh/

[2] https://itlaw.wikia.org/wiki/Event_flow_analysis

[3] CiteSpace can convert a large amount of document data into a visual map, which is beneficial for revealing the laws behind the data (Chen Yue et al. 2014).

The research materials for analysing the development of China's healthy cities can be divided into official documents and academic papers. The official documents are mainly obtained from the official websites of the National Health Commission of China and its important subordinates, including the Department of Publicity, the Department of Planning and Information, the Centre for Disease Prevention and Control, and the Chinese Centre for Health Education.

For analysis of existing studies, we have perused the database of CNKI (China National Knowledge Infrastructure), including Chinese journals from SCI or EI sources, Chinese Core Journals, CSSCI and CSCD, setting the time range at 'as of November 2019.' The selected articles in Chinese need to cover both 'health' and 'planning and design,' so 'health,' 'planning,' 'design' and other relevant words have been used as keywords for literature screening. The 'search by theme' method has been adopted and 4 kinds of retrieval modes have been used (Table 10.1). Literature retrieved during the preliminary search amounts to 350, and of these, 202 documents have been considered relevant and selected as base data for this analysis.

Table 10.1 List of keyword combinations for literature screening

Search method	Keyword combination	Search result
Mode 1: Direct keywords	SU% 'healthy city panning' + SU% 'healthy city design'	51 articles
Mode 2: Feature words of 'health' and 'planning'	SU = 'health' * ('city planning' + 'city and rural planning' + 'town planning' + 'urban planning' + 'national land planning' + 'spatial planning' + 'land use planning' + 'land utilisation planning' + 'layout of land use' + 'master plan' + 'detailed plan' + 'city design' + 'spatial design' + 'urban design' + 'city construction' + 'implementation of plan' + 'planning for transportation' + 'planning for public service facilities' – 'municipal planning' + 'ecological planning' + 'spatial layout' + 'community planning' + 'neighbourhood planning' + 'street design' + 'street transformation' + 'city renewal '+ 'renewal planning' + 'renewal design' + 'venue design' + 'venue planning' + 'planning strategy' + 'planning optimisation' + 'design strategy' + 'design optimisation' + 'planning governance' + 'planning control' + 'planning evaluation' + 'planning assessment' + 'planning and design guidelines' + 'city design guidelines' + 'design guidelines' + 'planning factors' + 'design factors')	256 articles
Mode 3: 'Health' and 'planning or layout or design' and the factors of planning and design or the factors of environment	SU = 'health' * ('planning' + 'layout' + 'design') * ('built environment' + 'land use' + 'green land' + 'green space' + 'park' + 'greening' + 'open space' + 'outdoor space '+ 'public space' + 'communication space' + 'community Centre' + 'hospital' + 'medical facility' + 'old-age care' + 'sports facility' + 'sport facility' + 'physical activity' + 'slow travel' + 'walking' + 'riding' + 'industry' + 'city street' + 'highway' + 'expressway' + 'waste water' + 'water pollution' + 'garbage' + 'waste gas' + 'air pollution' + 'air contamination' + 'noise' + 'soil pollution')	130 articles

(continued)

Table 10.1 (continued)

Search method	Keyword combination	Search result
Mode 4: 'Health impact assessment'	SU = ('health impact assessment' + 'health impact evaluation' + 'HIA') * ('city planning' + 'national land planning' + 'spatial planning' + 'urban design' + 'spatial design' + 'implementations of plan')	8 articles
Comprehensive search	Mode 1 + mode 2 + mode 3 + mode 4	350 articles

Source: Drawn by author
Note: 'SU' represents subject, '+' represents 'or', '*' represents 'and', = 'str' represents including the search word: str, and % 'str' represents including str or a segment of str

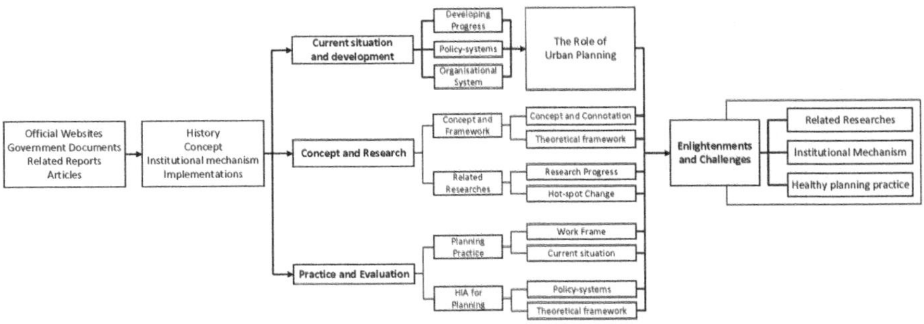

Fig. 10.1 Source: Drawn by author

10.2.2 Overall Research Design

This chapter has three sections, including healthy city development, healthy city planning and its research progress, as well as healthy city planning practice and HIA in China (Fig. 10.1). The first section, through a comprehensive analysis of the concept, institutional architecture and practice of China's healthy city development, gives an overall picture of China's healthy city development progress, policy systems and organisational system. The role of city planning is also discussed. The second section contains two parts. Healthy city planning is explored as a discipline and a theoretical framework for healthy city planning is proposed in the first part. In the second part, the progress and shift of focus of China's healthy city research is unveiled by investigating existing literature on China's healthy city planning. Centering on health promotion, the last section proposes a basic framework for planning practice and HIA, with a summary of China's achievements. This chapter ends with a review of experiences and future challenges of healthy city planning in terms of research, institutional mechanisms and the practice of 'integrating health in planning'.

10.3 Overview of China's Healthy City Development

10.3.1 Development History

In 1989, China launched the national 'Hygienic City' programme, which later evolved into the healthy city programme, covering additional factors that could affect health.[4] According to the *Annual Report on Healthy City Construction in China (2016)*, the past two decades have witnessed three phases of China's healthy city development: the start-up pilot phase, the exploration and development phase, and the comprehensive promotion phase (Fig. 10.2).

Fig. 10.2 History of China's healthy city development. Source: Drawn by author

1. **Start-up Pilot Phase (1992–2000).** In this phase, China's healthy city programme was carried out mainly under the guidance of the WHO, which launched a global healthy city movement and advised China's Ministry of Health to select pilot healthy cities. Beijing's Dongcheng District and Shanghai's Jiading District were chosen as the first batch of pilot cities (districts) for healthy city construction in China. Despite the slow pace of this phase, more cities joined the healthy city programme successively. The Patriotic Health Campaign, lasting for many years, and the Hygienic City Construction Campaign, initiated in the early 1990s, have laid the groundwork for China's healthy city development, which has become a unique advantage.
2. **Exploration and Development Phase (2000–2010).** At the national level, the Office of National Patriotic Health Campaign Committee held a meeting in Shanghai in 2007, announcing promotion of healthy city construction across the country, based on the experiences of the pilot healthy cities. The 'National Construction of Healthy City (District/Town) programme' was officially launched in 2007. Ten cities in the first batch (districts/towns), including Shanghai, Hangzhou, Dalian and Suzhou, released their own guidelines after being appointed pilot healthy cities. For example, Hangzhou issued the *Decision on Building a Healthy City* in 2008, and Suzhou releases the *Essentials of Suzhou's Healthy City Construction* every year. The pilot construction of healthy cities is fully underway and achieving excellent results.
3. **Comprehensive Promotion Phase (2011 to the present).** With the support of national policies and the participation of more cities, the construction of national healthy city network has continued to progress. 'Healthy city' has become an important national strategy, and with the promulgation of a series of policies and plans, the healthy city development has been advancing vigorously. Liu Yandong, Deputy Prime Minister of the State Council and Director of the Office of National Patriotic Health Campaign Committee, proposed the promotion of healthy city development across the country. In 2016, the Central Committee and State Council of China issued the *Outline for 'Healthy China 2030' Plan*, requiring the cities to 'take the construction of healthy cities, villages and towns as an important means to promote the construction of healthy China', putting forward an important goal to be realised by 2030: 'building a batch of model cities, villages and towns for the construction of healthy cities, villages and towns.' In 2019, the Healthy China Promotion Committee issued the *Healthy China Initiative (2019–2030)*. Local governments, including those of the municipalities directly under the Central Government, provincial capitals, prefecture-level cities, counties and towns, have joined in the healthy city development. They have, respectively, formulated relevant development programmes and action plans, aiming to ensure successful implementation of the plans by formulating regulations and supervisory systems, such as the *Twelfth Five-Year Plan for Healthy Beijing*, the *Guangzhou Healthy City Construction Plan (2011–2020)*, the *Liujiang District Implementation Plan for Healthy City Construction (2017–2020)*, and the *Baoting County Healthy County Construction Plan (2018–2025)*.

10.3.2 Features of China's Healthy City Development

China's healthy city development has been gradually expanding over the past two decades, now covering large areas, developing from healthcare policy to health in all policies, and transforming from exclusive leadership to diverse participants.

1. **Expanding to large areas.** With the introduction of the healthy city concept, proposed by the WHO, to China, healthy city development has been in full swing across the country. First, significant experience has been gained during the construction of multiple batches of pilot healthy cities (districts/towns). In 2016, the Office of National Patriotic Health Campaign Committee announced the first batch of 38 pilot healthy cities in China. Second, policies and systems have been put in place to encourage more cities to actively formulate development plans suitable for their own conditions. So far, several cities, including Beijing and Hangzhou, have made relevant plans. China's healthy city development covers a large area, involves multiple types of cities (towns), and boasts appropriate plans for local conditions.
2. **Developing from healthcare policy to 'health in all policies'.** In China, the issue of 'health' used to involve only hygiene and primary medical care, so policies in the past were mostly limited to a narrow field. In 2013, the WHO proposed the concept of 'Health in All Policies (HiAP),' which has been introduced to China and become a key part of China's hygiene and health policies in the new era, as well as an important guideline for healthy city construction in China (Qi 2019). Since then, China has viewed 'health' as a basic issue for development, and factored it in when formulating policies at all levels. With the 'Healthy China' initiative becoming a national strategy and the release of planning outlines and action plans, the integration of 'health' in all policies has become increasingly necessary and evident.
3. **Transforming from exclusive leadership to diverse participants.** The organisation and management of healthy city development in China, similar to that of policy, have also undergone transformation from the domination of the health department to the participation of multiple parties. Having gained a deeper understanding of the healthy city concept, China is gradually establishing a coordinated operating mechanism, led by the government and supported by various departments and social forces. Party Committees tend to take the leading role, and the Patriotic Health Campaign Committee acts as the organiser and coordinator of multiple departments. Although this system has not been established for long and varies between different areas, it has proved to be effective in some cities and regions. For example, Hangzhou has set up a leading group for health promotion as well as 7 special working groups involving 44 government departments, including the Health Commission, the Development and Reform Commission, the Human Resources and Social Security Bureau, and the Planning and National Land Bureau. While performing individual duties, all departments cooperate in working to achieve the common goal of building a healthy city. So

far, Hangzhou has achieved positive results, improving public bathrooms, medical care and IT applications, drinking water renovation in rural areas, and so on. The improvement of healthy governance has promoted the health of local residents.

10.3.3 Role of City Planning

Although healthy city development in China is shifting towards 'multiple participants and collaboration', and the city planning department is on board, the level of participation remains low and limited achievements have been undervalued.

Firstly, some cities have already made plans for 'integrating the healthy city concept into city planning'. Hangzhou is a good example. This city launched a 'spatial planning study for healthy city development' before formulating its 2010 master plan. Later, it undertook further research on 'policy-making suggestions on integrating the healthy city concept into Hangzhou's master plan' before getting started on its new version in 2019. However, the city planning department must engage wholeheartedly with the whole process of healthy city development, with thorough consideration of ways of ensuring the health of residents and how to build a healthy city operating system.

Secondly, the cities 'each work in their own way' in relation to healthy city development, lacking macroscopic guidance for implementation tailored to local conditions. As for inter-departmental collaboration mechanisms, the parallel relationship among participating departments and the differences in terms of terminologies make it hard to integrate health with policies in non-health sectors (Lan 2019). Additionally, some cities develop healthy city plans only for specific projects. In short- or long-term and coordinated health development plans, healthy city development is confined to specific projects, rather than fully integrated with the concept of people-oriented health in city planning and the development of the new planning system, titled 'National Territory Spatial Planning'. Meanwhile, healthy city planning has not gained enough ground across China, and the cities that have embarked on healthy city development vary greatly in planning methods, construction progress, specific practices and expected goals.

10.4 Healthy City Planning and its Research Progress in China

10.4.1 Connotation and Theoretical Framework of Healthy City Planning

10.4.1.1 Connotation of Healthy City Planning

While urban planning was initiated to provide a solution for basic public health in industrial cities back in the 1900s, the 'healthy city' concept proposed by the WHO aims to mitigate health burden with a focus on non-communicable disease. The 1980s saw the return of research and planning for the healthy city by setting human health at the core of urban planning in developed countries. The 'health turn' in planning explored the relationship between the built environment and physical and mental health (Forsyth 2018). Health consideration and research evidence are required as a basis for optimising design principles and standards.

Healthy cities are the unity of 'Health of City (city's health)' and 'Health in City (residents' health in a city)' and can achieve 'City for Health' through planning and design. Specifically, 'Health of City' provides a guarantee for 'Health in City'; and 'City for Health' is the goal of planning and construction. Therefore, healthy cities could refer to cities with a healthy environment, a healthy economy and a healthy society to support the physical and mental health of local residents (Fig. 10.3),

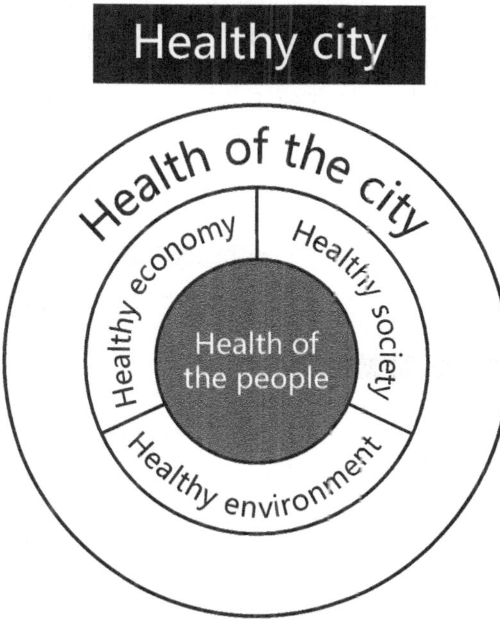

Fig. 10.3 Connotation of city planning for healthy city development. Source: Drawn by author

within which urban planning functions as an important method and policy for providing spatial intervention for better public health.

Healthy city planning is ultimately for people and about people (Giles-Corti et al. 2016). Within the framework of healthy city planning, planners need to be aware that their planning and design may affect health—purposefully or unconsciously (Duhl and Sanchez 1999). Healthy city planning encourages urban planners to support public health considerations and integrate them into planning strategies and projects; emphasising equity, well-being, sustainable development and community safety (WHO EURO 2003). For this, a corresponding system must be established.

10.4.1.2 Theoretical Framework of Healthy City Planning

Based on China's existing spatial planning system, this chapter proposes a theoretical framework for healthy city planning, including 'four factors and three pathways', which refer to the four major factors that influence public health and can be controlled through planning, and the three pathways for improving the physical and mental health of residents through planning and design. The four factors include land use, urban form, transportation system, and green and open space. The three pathways include decreasing human exposure to pollution, providing accessible healthy facilities, and facilitating physical activity and social interaction (Fig. 10.4).

Healthy city planning, therefore, aims to achieve health equity through spatial allocation of urban development elements, reduction of health risks, and provision of health resources. Meanwhile, it also has a broader dimension as a public policy. As related empirical research reveals the complexity and interrelation of health issues in modern times, policy-makers are required to transcend traditional methods of city planning in order to adopt a broader vision (McCarthy 1999; Lawlor et al. 2003). Thus, the basic principles underlying healthy city planning include seeking common understanding and solutions through interdisciplinary, interagency and intersectoral collaboration (Barton et al. 2009).

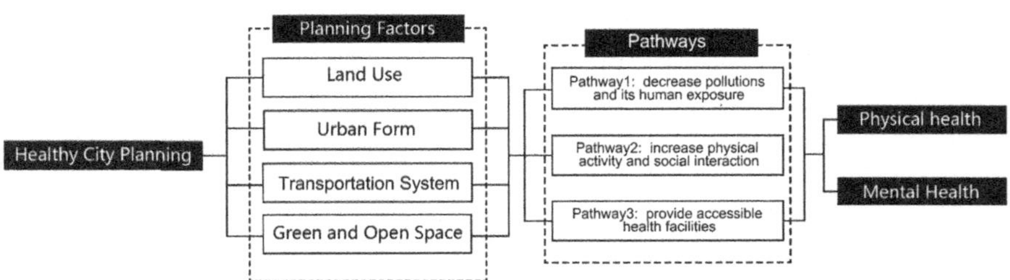

Fig. 10.4 Theoretical framework for healthy city planning. Source: Lan 2018

10.4.2 *Research Focus*

Because studies on healthy city planning have, to a certain extent, provided grounds for, or approaches to, the development of healthy cities in reality, the progress and focal change of China's research on healthy city planning over the past three decades will be traced—based on 202 identified Chinese papers—to provide inspiration and reference for follow-up research and practice. This section then analyses changes and trends in research focus.

10.4.2.1 Basic Statistics

Figure 10.5 shows the annual number of Chinese articles published on healthy urban planning, indicating that overall research interest is continuing to increase. The earliest relevant article appeared in 1994, and there has been a significant increase in the number of articles sent since 2005, with rapid growth in the last decade. In 2018, the number of relevant articles reached 33.

Taking the development history and features of China's healthy city development into account, it can be concluded that: (1) implementation of the pilot healthy city programme and promulgation of relevant policies have facilitated the development of research; (2) the introduction of 'HiAP' and the establishment of an interdisciplinary platform have contributed to its rapid development.

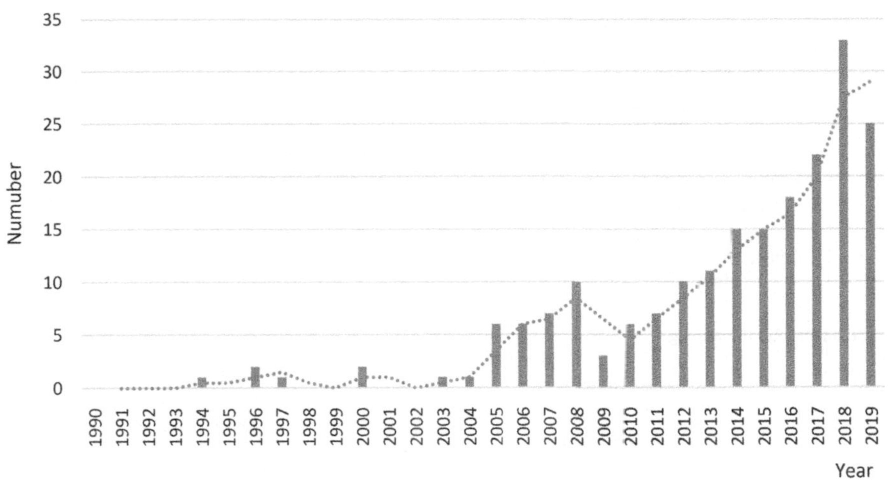

Fig. 10.5 Yearly figures showing upward trend of literature relating to healthy city planning. Source: Drawn by author

10.4.2.2 Research Focus

The research focus is analysed using keyword co-occurrence based on the software. Its basic principle is to count the frequency of a pair of words co-occurring in the same group of articles and to measure the strength of association between them (Jie and Chaomei 2016). The research focus includes built environment (11 times), landscape architecture (10 times), physical activity (10 times), health impact assessment (6 times), and urban construction (4 times), appearing with the basic keywords of healthy city, health, urban planning and urban design (Fig. 10.6, Table 10.2). Community, assessment indicators, urban renewal, and urbanisation have also come to the attention of researchers.

In the network of research focus, 'healthy cities' and 'health' are primary nodes, while 'urban planning,' 'landscape architecture,' 'built environment,' and 'physical activity' are important secondary nodes (Table 10.2). They are not only important research topics per se, but also function as a pathway to connect health with other research fields.

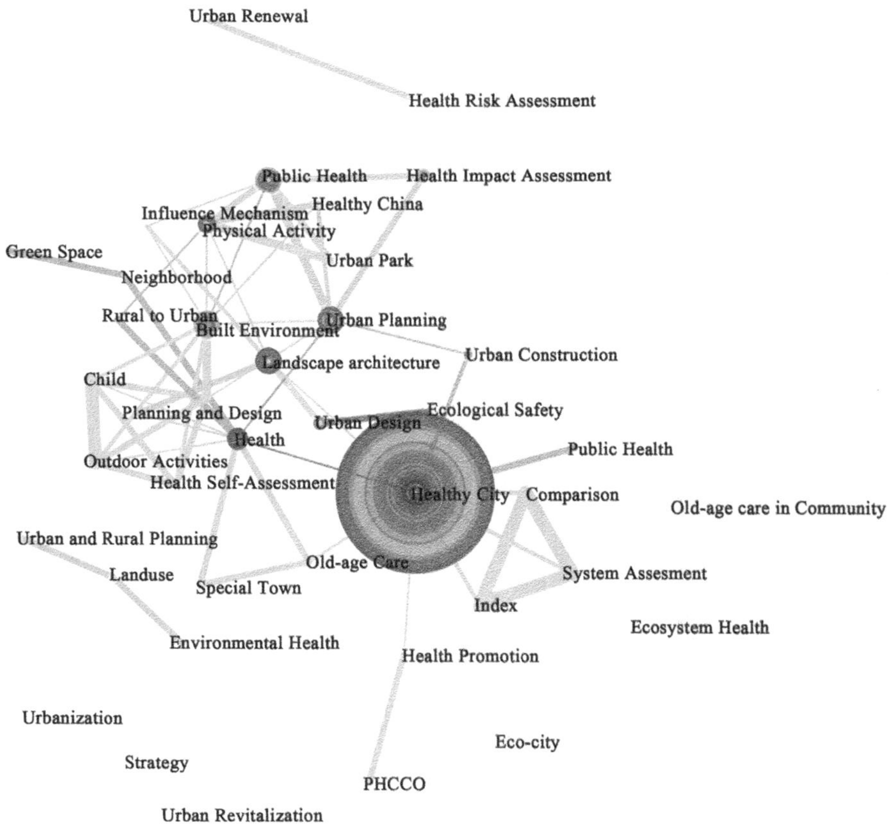

Fig. 10.6 Research focus of healthy city planning. Source: Drawn by author

Table 10.2 centrality and co-occurrence frequency of research focuses

No.	Keyword	Centrality	Co-occurrence frequency	No.	Keyword	Centrality	Occurrence frequency
1	Healthy city	0.44	57	22	City park	0	2
2	Health	0.41	16	23	Comparison	0	2
3	Urban planning	0.17	15	24	Old-age care in community	0	2
4	Public health	0.05	12	25	Neighbourhood	0.07	2
5	Built environment	0.1	11	26	Urban and rural planning	0	2
6	Landscape architecture	0.25	10	27	Impact mechanism	0.03	2
7	Physical activity	0.12	10	28	Outdoor activity	0.04	2
8	Health impact assessment	0	6	29	Planning and design	0.02	2
9	Urban design	0.07	6	30	Self-assessment of health	0	2
10	Urban construction	0.04	4	31	Ecological safety	0	2
11	Land use	0	2	32	Urbanisation	0	2
12	Healthy China	0	2	33	Public health	0	2
13	Rural to urban	0.02	2	34	Old-age care	0.01	2
14	Old-town renewal	0	2	35	Indicator	0	2
15	Special town	0	2	36	Ecosystem health	0	2
16	PHCCO	0	2	37	Strategy	0	2
17	Environmental health	0	2	38	Health risk assessment	0	2
18	Systematic assessment	0	2	39	Eco-city	0	2
19	Green space	0	2	40	Shanghai	0.01	2
20	Children	0.04	2	41	Health promotion	0.07	2
21	Urban renewal	0	2				

Source: Drawn by author
Note: 'Centrality' is an index used to discover and measure the importance of literature. In the corresponding figure, purple circles are used to highlight such literature (or authors, journals, institutions, etc.) (Jie and Chaomei 2016)

Notes: The parameters of the CiteSpace software as: (1) 'Time Slicing' is set at '1990–2019' and 'Year Per Slice' is set at 2; (2) Node Types at Institution and Keyword successively; (3) using the Cosine algorithm for Links; (4) nodal data at TopN = 50; (5) no trimming for the network. 'Year Per Slice' is a parameter used to set the number of years as the time period unit for bibliometric analysis.

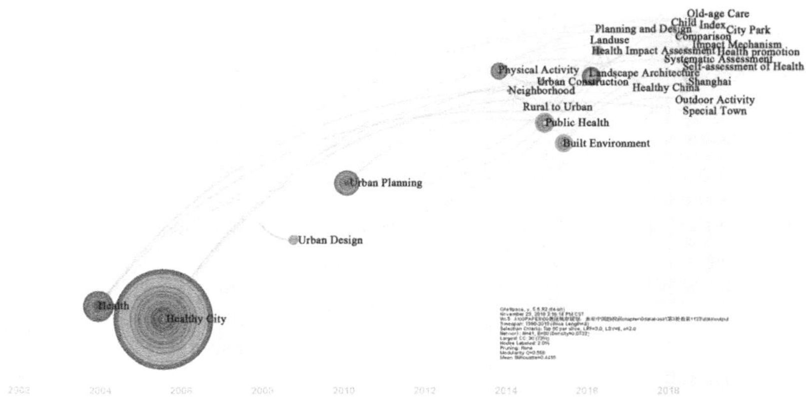

Fig. 10.7 Research trends of healthy city planning. Source: Drawn by author

The trend of research topics along the timeline is examined to capture overall change and identify areas for future research (Fig. 10.7). Visual analysis of Time Zone in CiteSpace reflects the evolution of knowledge based on two-dimensional orientation, including the time of research focus and its successive relationship expressed by the connecting lines (Yue et al. 2016). It is found that: (1) primarily, healthy city research in China initially focused only on health and healthy cities; (2) around 2010, urban planning and urban design became the main focus for research; (3) since 2014, the research focus has shifted from physical space (such as neighbourhood and built environment) and physical activity to more specific environmental factors (such as land use and city parks), and to particular population groups (such as children and old people), and to various assessments (such as self-assessment of health, HIA and assessment indicators). It also shows that research objects are becoming increasingly diversified and detailed.

10.5 Healthy City Planning Practice in China

Some cities (or districts) in China have already constructed pilot healthy cities with the support of the WHO (e.g. Beijing's Xicheng District, Shanghai's Jiading District, Suzhou and Dalian), and some have formulated action plans for healthy city development (e.g. Guiyang, Shanghai, Huangshan, Shanghai's Jiading District and Liuzhou's Liujiang District). However, these action plans are mainly supervised by health departments, focusing more on basic medical care, health management, elderly services and environmental governance rather than planning of physical space. One typical example is the '*Three-year Plan for the Healthy City Construction in Zhengzhou (2018-2020)*'. Its development goals are mainly reflected in the settings of per capita indicators, such as the area of green space, parks per capita and the area of sports fields per capita in the city, whereas reasonable allocation of health factors based on the features of spatiotemporal behaviours and population spatial distribution are overlooked.

Despite the fact that some cities have implemented healthy city planning and undertaken research such as, '*A Study of Spatial Planning for Healthy City Development in Hangzhou*' (2009), the practice of physical space planning for healthy city development is generally still in its early stages, requiring development of a working process to support healthy city planning in China.

10.5.1 Working Process of Healthy City Planning

The difficulty for healthy city planning lies in the challenge for practitioners to consider multiple health impacts at the same time. The healthy planning guidelines, therefore, need to identify key planning interventions, focusing on major health problems and optimal solutions (Forsyth 2018). With a focus on spatial planning in China, a working process is proposed to integrate health into city planning. Three steps are followed: Health-oriented Site Analysis, Health-oriented Planning and Design, and HIA (Fig. 10.8).

Health-oriented Site Analysis calls for inclusion of more content regarding public health within normal site analysis for planning. At least two types of site analyses are proposed: Health Demand Analysis and Exiting Health Risk Analysis. Health demand refers to the requirement of local residents for health resources based on

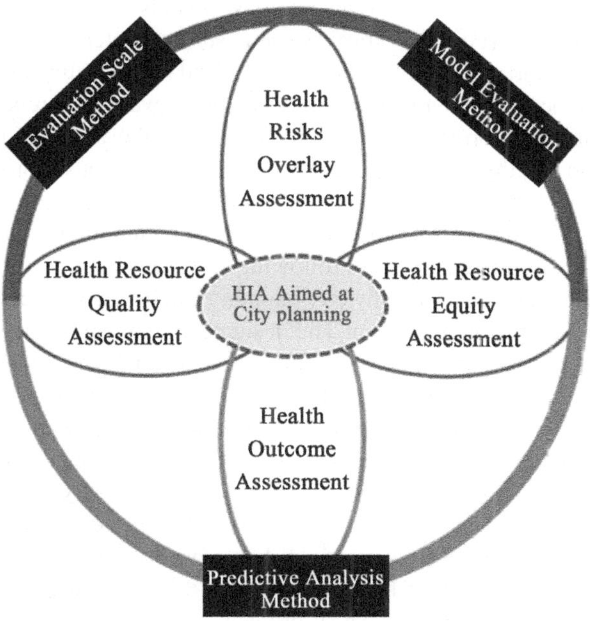

Fig. 10.8 Framework for the practice of healthy city planning. Source: Drawn by author

their evaluation of current health support within and around the site. This could provide information for planners allocating facilities and amenities to promote physical activity and social interaction. The Exiting Health Risk Analysis focuses on identification of potential risks to public health within and around the site before rezoning or redevelopment. These analyses would help planners and designers to understand the current health status of the site, and thus propose health promotion based on spatial arrangement.

In *Health-oriented Planning and Design*, existing planning principles can be optimised, based on relevant research of healthy cities, and specific planning content can be added. It is proposed that pollution source control and guidance planning, health resource planning, integration planning for a slow traffic system and service facilities are added. There inevitably exist both point and linear sources of pollution in cities, including garbage disposal sites, large-flow motor vehicle roads, etc. The distance between these pollution sources and densely populated areas, such as residential zones and public facilities, should be carefully controlled (e.g. it is improper to set up schools and residential buildings within 100 metres of an expressway). Detailed designs should be put forward to reduce the risk of human exposure to pollution (e.g. the crossroad area for cyclists waiting for the green light could be designed to avoid motor vehicle exhaust). Comprehensive planning of health resources will be based on accurate spatial distribution data of population. Various types of medical service facilities, fitness facilities and venues, and green parks will be planned according to the needs of different age groups. The integration planning of slow traffic systems and service facilities, which is based on Point of Interests (POIs), emphasises consideration of the route selection of pedestrian and cycling systems, together with all types of destinations. It is also important to provide service facilities to support the slow traffic system, such as water fountains and shower rooms. These are part of healthy urban planning, and more health-oriented planning and design will be further supplemented, based on relevant researches.

For the planning proposal and its implementation, the HIA can be used as the final step of the healthy city planning process, to simultaneously provide feedback on the impact on public health. HIA provides the tool for decision makers to select multiple planning scenarios and to evaluate its implementation for later adjustment. The HIA is discussed in detail in the following section.

Based on such a process, healthy city planning can be carried out on three levels within the existing planning system. The design elements and concerns of healthy city planning differ on a multiple geographic scale. On a macro-scale, overall planning focuses on the systematic health factors of the whole city, such as slow traffic systems, the layout of healthcare facilities, pollution source control and guidance. On a meso-scale, the urban design and regulatory plan, focusing on neighbourhoods and blocks, emphasises the quality of spatial healthy elements at a local level, the demand of different groups within the neighbourhood and the vitality of public space. On a micro-scale, health-oriented urban regeneration in communities could support spatial intervention for health promotion. By integrating health at different levels of planning and multiple types of projects, city planning may be able to provide essential solutions for the intervention and improvement of public health.

10.5.2 HIA in City Planning

Health Impact Assessment (HIA) has been adopted by the WHO, Asian Development Bank and cities in the United States and Europe to evaluate the impact of their plans, programmes and projects on the health of affected populations. According to the WHO, HIA is a series of complementary procedures, methods and tools for assessing the potential impact of a policy, programme or project on the health of a particular group, as well as for the distribution of such impacts (WHO 2003).

The introduction of HIA into city planning provides a useful policy tool to make decisions on spatial resource allocation from the perspective of public health. A theoretical framework is proposed to implement HIA within the context of the Chinese planning system, and to discuss its current status and potential application.

Compared with that of Western countries, China's HIA is still in its infancy and more often part of the Environmental Impact Assessment (EIA). China developed an EIA system in 1979 and, in 2003, released the *Environmental Impact Assessment Law of the People's Republic of China*, which established the legal status of EIA. In 2005, the *Guidelines for Environmental Impact Assessment Techniques—Human Health* confirms Human Health Assessment as part of the EIA for projects and programmes. As an informal HIA in China, it highlights a series of assessment methods (both qualitative and quantitative) for appraising, predicting and assessing the impacts (both positive and negative) of proposed projects on the health of specific populations within the involved area. With the recognition of 'Healthy China' as an important national strategy and the promotion of 'HiAP', HIA in China has witnessed progress in both theory and practice in recent years.

As China has not been engaged in HIA for a long time, it lags behind European and American countries in theories, methods, tools, practice and institutional design. In terms of the assessment content, most Chinese HIAs focus on assessment of the health impact of air, water and acoustic environments (Zhen 2011), instead of the built environment, which could be adjusted by urban planning. The most common assessment methods in current HIA are qualitative, such as the expert consultation method and the literature research method. Quantitative analysis methods are rarely adopted because of the lack of data and baselines.

Certain cities have carried out HIA for their policies; for instance, the implementation plan of the 'Ecological Citizens' project of Xiling District, Yichang, Hubei Province (Center China Health Education 2019). HIA for plans and mega-projects are still lagging behind. The Healthy City WLAB Lab at Tongji University conducted an experimental HIA for the construction of the Minhang section of the Shanghai Metro Line 15. Overall, however, relevant theories and techniques are still immature and in the exploratory phase.

Currently, there is insufficient evidence to support a feasible and reliable HIA. The exact assessment content should be further defined and specified within the context of Chinese cities. The exploration and development of methods and tools are top of the agenda. The institutional design to integrate HIA in the planning

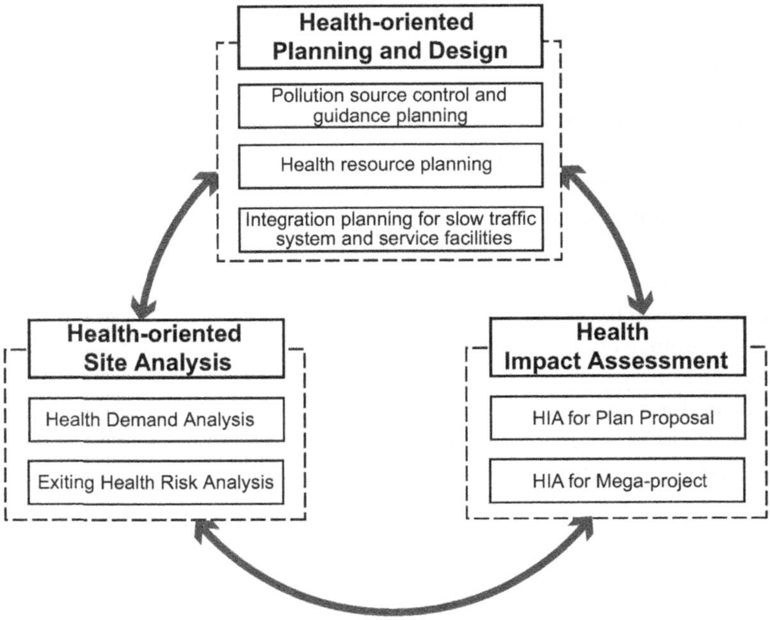

Fig. 10.9 Theoretical model of HIA oriented towards city planning. Source: Drawn by author

process is also significant. In this chapter, a theoretical framework of assessment content is proposed.

The purpose of HIA in city planning is to identify potential positive and negative impacts of spatial planning on public health, as well as its effect on health equity. Based on this understanding, we propose a comprehensive HIA theoretical framework of assessment content, including four assessment modules. They are titled Health Risks Overlay Assessment, Health Resource Quality Assessment, Health Resource Equity Assessment, and Health Outcome Assessment (Fig. 10.9). The combination of these four modules could evaluate the comprehensive effects of a plan and a project on public health.

Within this framework, the assessment module for health risks is designed to identify negative health impacts, while the two assessments for the quality health resources mainly focus on elements with positive effects. The assessment module for health outcome aims to predict the mortality and morbidity of certain diseases. All four modules together constitute the main content of HIA for city planning.

10.5.2.1 Health Risks Overlay Assessment

It is important to identify the distribution of various sources of pollution and overlay it with environmental factors that affect the impact of pollution on human bodies (Lan et al. 2018). A systematic analysis would simulate and identify impacts of health risk factors (e.g. major transportation facilities, polluting public facilities, urban trunk roads, exits to ramps, industrial land, polluted water) within the urban settings (e.g. wind fields, heat islands, hydrology and sound pressure) with the consideration of social and economic factors (e.g. the spatial distribution of elderly population). Areas with high health risks could be identified to support decision-making and plan making for new development.

10.5.2.2 Health Resource Quality Assessment

This module aims to evaluate the positive effects of a plan as spatial intervention for health promotion (Lan et al. 2018). Health resource refers to spatial elements that could promote physical activity and social interaction—such as urban green space—and healthcare facilities—such as hospitals and community nursing homes. The quality assessment of health resource evaluates the level of service these existing and planned facilities could provide. Planning and design, in terms of diversity and attraction of certain resources, may be improved following the HIA.

10.5.2.3 Health Resource Equity Assessment

Health equity is important for healthy city planning and development (Lindheim and Syme 1983; Corburn 2013). For instance, fairness in the use of health services is a key part of health equity (Jianmin and Jinjuan 2014). The spatial layout of health resources, therefore, should consider the spatial heterogeneity of population distribution, and give priority to vulnerable groups, such as children, the elderly and low-income groups. This module aims to evaluate the impact of plans and projects on different groups according to population structure.

10.5.2.4 Health Outcome Assessment

The assessment of health outcome attempts to predict the mortality and morbidity of certain illnesses, such as diabetes, cardiovascular diseases and respiratory system diseases. It provides a straightforward picture of the potential impacts that a plan may have. This module is either a predictive model built on local historical data or a predictive model that references similar areas. Two main methods may be considered: (1) using an existing model and taking two cities with similar background and environmental features into account to ensure the accuracy of analysis results; (2) developing a local prediction model and applying it to predict the future. The latter one tends to generate more accurate results, but it entails the support of rich data on local population health.

10.6 Inspirations and Experiences

Healthy city development in China has been continuously forging ahead. China has made significant improvements in the past two decades but must rise to new challenges. Despite the positive experiences gained with respect to theoretical and empirical studies, relevant theories have yet to be tested in practice and a collaborative platform involving multiple parties must be established. This chapter has systematically analysed multiple aspects of healthy city planning and development in China, including the status quo, the progress of relevant research, HIA and healthy city planning practice. The experiences gained by China could provide inspiration for other cities around the world. An emphasis on city planning, interdisciplinary research and multiple collaboration is essential. The theoretical framework of HIA and the working process of healthy city planning are expected to support better practice in the future.

It is important to recognise the role of city planning in leading and guiding healthy city development in terms of providing health facilities and promoting health equity. Today, Chinese cities have started to integrate healthy city ideas into their spatial planning system. Hangzhou, for instance, conducted 'A Study of Spatial Planning for Healthy City Development in Hangzhou' in 2009. Because the planning system differs among countries, it is necessary to understand the significant spatial factors for public health and their arrangement in different types of plans. The theoretical framework and working process proposed for healthy city planning in this chapter is expected to enrich the practice.

Meanwhile, interdisciplinary research can provide theoretical support for the practice of healthy city planning. Since 2013, research on healthy city planning in China has involved an increasingly wide range of health topics, many of which focus on the impacts of spatial factors on health and human behaviour. The research outcomes provide guidance for planning as spatial intervention. Additionally, the theoretical framework and working process for healthy city planning adapted to the features of China's space planning has taken shape. Although China has not yet formed a comprehensive collaborative network for the research, cooperation among institutions and scholars in different fields has been increasing steadily, which exerts a positive influence on innovative research and healthy city development.

Multiple collaboration is an important basis for reaching a consensus on healthy city development and promoting the efficiency of implementation. Health departments and commissions play a leading role in these efforts, which are equal in status to other relevant departments of local governments. They should boost interaction and cooperation among them. Differences in the focus of concern and authority between departments lead to difficulties in seeking collaborative efforts to build hard and soft infrastructures for healthy cities. Some cities have set up healthy city development or promotion teams and work together to create inter-departmental collaborative action teams. Inter-departmental barriers have been torn down gradually in China. The newly established collaboration and communication platforms and research institutions, as well as exchange meetings, have also played important roles in promoting the planning and construction of healthy cities in China. All have contributed greatly to healthy city development, such as the Healthy City Academic Committee of Chinese Society for Urban Studies (established in 2018), Healthy City WLAN Lab (Tongji University), Healthy City Research Centre (Tsinghua University), the Global Health Promotion Conferences (organised by the WHO in 2016), the national-level China Healthy City Construction Leaders' Forum (held in 2018), and the regional-level International Symposium on Healthy City (held in Shanghai in 2018 and 2019).

Furthermore, the concept of healthy city covers a wide range of areas, and involves numerous aspects of urban development. However, various disciplines differ from each other in knowledge system, terminology and basic principles, resulting in barriers to research and entailing a platform for communication and interaction. To cite an example, no courses on the planning principles, techniques and methods for public health and healthy cities are available in the curricula of urban and rural planning, architecture and other relevant disciplines. Students majoring in public health or health education are not taught ideas and knowledge about city planning and design. Furthermore, different disciplines address different objects. Urban and rural planning and architecture mainly focus on space, while medicine focuses on people. There is a pressing need to connect space with people by bridging the two disciplines and exploring the impact of space on human health to develop the overall framework of healthy city construction. Hence, the Healthy City Academic Committee of Chinese Society for Urban Studies, the first Chinese academic commission with healthy city in its name, has been established to bring together experts and practitioners from multiple disciplines, such as medicine, urban and rural planning, geography and public management, and to promote research on, and practice of, healthy cities.

Another urgent mission would be to localise knowledge on healthy cities. Vigorous efforts should be put into research into the phases and characteristics of urban development in China. Evidence-based practice aims to connect research with practice, setting high requirements for targeted research. It is necessary to take into account the background conditions, problems and characteristics of practice objects in research. Chinese cities, which are undergoing transformation of industrial structures, feature high density of both buildings and population, as well as unique spatial elements and health problems. Relevant foreign research may serve as a reference, but it is essential to collect data and analyse socio-economic and policy conditions, thus developing practices applicable to specific conditions. Meanwhile, multiple principles and standards for health planning and design must be established with meticulous attention to detail in research. More endeavours should also be made to promote application of academic research results, so that health research can exert greater influence on the formulation, management and decision-making of urban planning.

Healthy cities are receiving more attention during urban development in China, but challenges still exist in achieving this target, including support mechanisms and knowledge base. Notwithstanding the fact that the practice of healthy city planning is not yet in full swing in China, the outlook is promising thanks to the active guidance of national policies, the continuous accumulation of achievements in theoretical and fundamental research, and the constant explorations of Chinese cities.

Acknowledgement This study has been carried out with the financial support of the National Natural Science Foundation of China through research project 41871359 and 52078349.

References

Barton, H. et al. (2009) 'Healthy urban planning in European cities', *Health Promotion International*, 24(suppl_1), pp. i91–i99. doi: https://doi.org/10.1093/heapro/dap059.

Center China Health Education (2019) *Theoretical and Practical Research on Health Impact Assessment*. China Environmental Publishing Group.

Corburn, J. (2013) *Healthy city planning: From neighbourhood to national health equity*. Routledge.

Duhl, L. J. and Sanchez, A. K. (1999) *World Health Organization. Healthy cities and the city planning process: a background document on links between health and urban planning*.

Forsyth, A. (2018) 'Evidence-based Practice', in Beatley, T., Jones, C. L., and Rainey, R. (eds) *Healthy Environments, Healing Spaces: Practices and Directions in Health, Planning, and Design*. University of Virginia Press.

Giles-Corti, B. et al. (2016) 'City planning and population health: a global challenge', *The Lancet*, 388(10062), pp. 2912–2924. doi: https://doi.org/10.1016/S0140-6736(16)30066-6.

Jianmin, G. and Jinjuan, Y. (2014) 'Overview on Health Fair', *Health Economics Research*, 10, pp. 51–54.

Jie, L. and Chaomei, C. (2016) *CiteSpace Text Mining and Visualization in Scientific Literature*. Capital University of Economics and Business Press.

Lan, W. (2018) 'Healthy Urban Planning: Return and Improvement', in Shiwen, S. (ed.) *Academic Achievement of China Urban Planning Society: Quality Planning*. China Construction Industry Press.

Lan, W. (2019) 'Construction of the core content of a healthy city plan promoted by the 'Healthy China' national policy', in *Annual Report on Healthy City Construction in China(2019)*. Chinese Academy of Social Sciences Literature Press.

Lan, W., Wenyao, S. and Jiayu, G. (2018) 'The Methodological Development of Health-oriented Urban Design and Its Practical Exploration: A Case Study of Huangpu District, Shanghai', *Urban Planning Forum*, (05), pp. 71–79.

Lawlor, D. A. et al. (2003) 'The challenges of evaluating environmental interventions to increase population levels of physical activity: the case of the UK National Cycle Network', *Journal of epidemiology and community health*. BMJ Group, 57(2), pp. 96–101. doi: https://doi.org/10.1136/jech.57.2.96.

Lindheim, R. and Syme, S. L. (1983) 'Environments, people, and health', *Annual review of public health*. Annual Reviews 4139 El Camino Way, PO Box 10139, Palo Alto, CA 94303-0139, USA, 4(1), pp. 335–359.

McCarthy M (1999) 'Transport and health', in Marmot M, Wilkinson, R. (ed.) *Social determinants of health*.

Neira, M. (2018) *Health must be the number one priority for urban planners*. Available at: https://www.who.int/news-room/commentaries/detail/.

Shi Qi (2019) 'The Application of Health in All Policies in healthy cities initiative in China', *Chinese Journal of Health Education*, 35(2), pp. 186–191.

WHO (1998) *Health Promotion Glossary*. Geneva.

WHO (2003) *Health Impact Assessment (HIA)*. Available at: https://www.who.int/hia/about/en/.
WHO EURO (2003) *Healthy Cities around the world: An overview of the Healthy Cities movement in the six WHO regions*. Belfast.
Yue, C., Chaomei, C. and Zhigang, H. (2016) *Principles and Applications of Analyzing a Citation Space*.
Zhen, H. (2011) *Problems and Countermeasures of Health Impact Assessment of Construction Projects in China*. Huazhong University of Science and Technology.

Chapter 11
From Green Building to Green City: The Practice in Jiangsu Province, China

Dengyun Wang

Abstract This chapter focuses on Jiangsu's experience in promoting green buildings and green eco-city, including green ecological special planning, policy mechanisms, fundraising, technical standards systems, and promotion. Taking the demonstration zones of Nanjing Hexi New Town as an example, the practical experience of the demonstration area in special planning guidance, policy mechanism formulation, green building development, and urban green infrastructure construction was introduced.

This not only has great reference significance for the development of China's new cities but also has certain reference significance for the development of new cities in the world and the renewal of old cities.

Keywords Green building · Green eco-city · Jiangsu province · Sustainable development

11.1 Overview of Jiangsu

Since the Tang and Song dynasties, Jiangsu has been recognized as one of the ideal places to live all over the nation, with the rich heritage of traditional architecture. One proverb said that Paradise in Heaven; Suzhou and Hangzhou on Earth. By the end of 2018, the urbanization rate in Jiangsu had reached 69.6%,[1] and the residential population was 80.507 million. Jiangsu's GDP reached 92,595.4 trillion yuan in 2018, with per capita GDP, regional Development and Life Index (DLI) ranking first among all the provinces, and its economic and social development has entered the level of middle-and-upper-class developed countries. As an economically leading area, urban and rural construction level in Jiangsu is in a leading position in the

[1] Official website of Jiangsu provincial people's government

D. Wang (✉)
School of Architecture, Southeast University, Nanjing, China
e-mail: email:wxdy1603@163.com

L. Bian et al. (eds.), *Chinese Urban Planning and Construction*,
Strategies for Sustainability, https://doi.org/10.1007/978-3-030-65562-4_11

country. In 2018, the total output value of the construction industry was 308.467 billion yuan (Mou et al., 2019). Four cities (counties) won the "United Nations Habitat Environment Award" and 15 cities (counties) won the "China Human Settlements Award." The number of awarded cities remained the first in the country and Jiangsu has the most national garden cities (counties) in the country. The Yangtze River Delta urban agglomeration consists of Jiangsu, Shanghai, and Zhejiang has become one of the six world-class urban agglomerations.

While economic and social development and urban construction have gained achievements, Jiangsu is facing an increasingly severe situation due to resource and environmental constraints. With a population density of 750 persons/km^2, Jiangsu now has the highest population density in the country. The province's arable land area is 68.70 million mu but the per capita arable land is only 0.9 mu, which is less than two-thirds of the national average (Zhang, 2010). Jiangsu's energy self-sufficiency rate is low. The province's proven oil recoverable reserves account for only 0.2% of the nation's total. Hydropower resources possess only 0.3‰ of the country's total hydropower development. Per capita share of coal resources is only 6.8% of the national average (Ni, 2011). Therefore, taking the road of sustainable development and continuously implementing green development are the inevitable choices to deal with the bottleneck of development.

11.2 Top-Level Design

The Jiangsu Provincial Party Committee and the provincial government have attached great importance to and actively promoted the construction of ecological civilization. In the urban and rural construction aspects, they seized the opportunity that occurred during rapid urbanization to actively promoted green development, and the transformation to a greener, more economical and ecological urban and rural construction development pattern (Implementation in Jiangsu, 2015). From building energy conservation, renewable energy building applications, green buildings, resource-efficient urban and rural construction to green eco-city areas, from pilot demonstrations to comprehensive deployment, from explorations of individual technology application to regional and city-wide integration practices, the green construction in Jiangsu is developing as its connotation constantly improved. The content is increasingly integrated, the practice is gradually deepened, and the forms are diversified. It has embarked on a road of green development from the concept to practice, along with continuous exploration and innovation. Green development has increasingly become an industry consensus.

11.2.1 Improving the System of Policies and Regulations

In recent years, Jiangsu has continuously improved the green building development policy and regulation system. To make specific deployments to promote the development of green buildings, the Provincial Party Committee and the provincial government have successively issued *The Notice on the promotion of resource-efficient urban and rural construction*, *The Implementation Plan for the Construction of Better Urban-Rural Areas Within the Provincial*, and *The Implementation Plan for green building action in Jiangsu Province* and other documents. The provincial government has established a joint meeting system for green building actions, and the joint meeting office is located in the Department of Housing and Urban-Rural Construction of Jiangsu Province (hereinafter referred to as the Department of Housing and Urban-Rural Construction), which has effectively guaranteed the promotion of green building work.

In 2015, Jiangsu released and implemented the country's first local green building regulations, *the Regulations on the Development of Green Buildings in Jiangsu Province* (*Implementation in Jiangsu*, 2015), which initiated the process of comprehensively promoting green buildings under the guarantee of the legal system. *The Regulations on the Development of Green Buildings in Jiangsu Province* stipulated that new civil buildings in cities and towns in the whole province shall be fully planned, designed, and constructed under the green building standards; large public buildings using state capital investment or state-financed shall be planned, designed and constructed compliance with the two-star green building standards or higher.

11.2.2 Setting up Provincial Special Funds

In 2008, the Provincial Department of Finance and the Department of Housing and Urban-Rural Construction jointly established a provincial-level building energy-saving special guidance fund (hereinafter referred to as the "provincial special fund") (Hao, 2008), which is used as financial support for green buildings, green eco-city areas, green building demonstration cities, renewable energy integrated building application, energy-saving retrofit of existing buildings, energy performance contracting and ultra-low energy consumption passive buildings. Through the guidance of provincial special funds, it has promoted the exploration and practice of building energy conservation, green buildings and green urban areas in various places. Since the establishment of provincial special funds 11 years ago (2008–2018), Jiangsu has established a total of 831 demonstration projects of various types and has accumulatively allocated 2.18 billion yuan of financial funds (average of about 200 million yuan/year). A total of 72 provincial-level green ecological urban areas were established with a planned area of about 12000.0 km^2, achieving full coverage in designated cities and expanding to counties (cities, districts). The province's green building identification projects and the scale of energy-saving buildings have maintained a national leading level for a long time. The number of projects that have won the National Green Building Innovation Award is among the top in all provinces. At the same time, significant results have been achieved in the efficient use of green energy, the development of green transportation, the comprehensive utilization of water resources, the recycling utilization of solid waste and other aspects.

With the strong support of the provincial special funds, various places have carried out active practices of adapting strategies according to local conditions, and completed several individual and regional integrated demonstration projects. This has played a positive role in promoting local innovation practices and improving the level of green development in urban and rural construction.

11.2.3 Sticking to Special Planning Guidance

Since 2010, Jiangsu has deployed provincial special funds to encourage and guide local governments to target green buildings scaling up and key projects of resource-efficient urban and rural construction, and to start the establishment of green eco-city in small-scale urban areas. In the process of promoting the construction of green eco-cities, the Department of Housing and Urban-Rural Construction adhered to the concept of "planning as the direction" and guided all regions to develop and compile special plans based on the overall urban planning and detailed regulatory planning. A technical system has been gradually built including green buildings, building energy, comprehensive utilization of water resources, green transportation, and urban solid waste utilization. What's more, a special planning system of green eco-city was also built, covering the whole spectrum of subsystems of urban planning and construction (No author, 2015; No author, n.d.-a). While the Department of Housing and Urban-Rural Construction has continuously improved the theoretical basis and technical guidelines of the special planning system for green eco-city urban areas, each of the green eco-city urban areas has organized more than 300 special plans to scientifically guide the construction work.

It is stipulated in the *Regulations on the Development of Green Buildings in Jiangsu Province* that relevant departments of the people's governments at or above the county level shall organize and prepare the green building development plans for their administrative areas and special plans for green buildings, comprehensive energy use, comprehensive utilization of water resources, comprehensive utilization of solid waste, and green transportation and other aspects. All the relevant requirements in the special plan shall be incorporated into the detailed regulatory plan. Each district cities have actively explored and continuously improved the compilation methods and implementation approaches of green building development plans and green ecological special plans. The Suzhou Municipal Government has issued *The Green Building Layout Plan for Suzhou Urban Area*, which requires that indexes of two-star and above green buildings are required to be implemented in the construction land planning. Therefore, the construction of high-quality green buildings can be promoted from the very beginning. Based on special planning and relevant researches, Wuxi, Nantong, Xuzhou, and other cities have required indexes related to the green building should be clarified in the land assignment contracts and planning conditions of new projects, which has ensured the popularization of green buildings.

11.2.4 Building a Closed-Loop Supervision Mechanism

To implement the requirements that new urban civil buildings shall be designed and built-in full compliance with green building standards in *The Regulations on Green Building Development in Jiangsu Province*, Jiangsu Province has formulated *The In-depth Regulations for the Preparation of Green Design Documents for Civil Construction Drawings in Jiangsu Province* and *The Technical Review Highlights of Green Design Documents for Civil Construction Drawings in Jiangsu Province*. Before the construction projects apply for planning permission, examination requirements of green building planning schemes are set up, thus realizing the effective connection between the green building design plan and technical requirements of the evaluation standards, and ensuring that the projects through investigation at least meet the one-star standard for green buildings. *The Green Building Design Standards of Jiangsu Province* (DGJ32/J173-2014), *Green Building Construction Quality Acceptance Specifications* (DGJ32/J19–2015), *Green Building Indoor Environmental Testing Technical Standards* (DGJ32/TJ194-2015) and other local standards have been issued and implemented, which improved the green building standard system in Jiangsu. Under the support of various management regulations and technical standards, Jiangsu has established a long-term supervision mechanism that covers the planning program review, construction drawing review and completion acceptance inspection in administrative level and green building planning and design, construction, inspection and evaluation in the technical level.

Starting from 2012, to effectively implement the overall development goals of green building and building energy conservation, the Department of Housing and Urban-Rural Construction has annually formulated and released *The Provincial Building Energy Conservation and Green Building Work Task Decomposition program*, with green building and building energy-saving tasks to be decomposed and allocated to the construction departments of each district and city. It requires clear responsibility, measurements to promote implementation, and completion of tasks. Meanwhile, data collection, screening, statistics, and summary reporting need to be done well. Also, they shall organize assessments, research, and evaluation at the end of each year. Through the spot checks and on-site inspections, problems and deficiencies could be identified timely. The assessment results were used as one basis for the provincial government's total energy consumption, in-depth *Dual Control* assessment, and ecological civilization assessment. Since 2015, task decomposition indicators have been further optimized and improved, with the addition of ultra-low energy consumption (passive) buildings, the green retrofit of existing buildings and other indicators. The implementation of this system has effectively promoted the rapid development of green buildings and building energy conservation in Jiangsu. In addition, Jiangsu also incorporated green building development indicators into the provincial ecological civilization construction planning, provincial habitat environment award creation indicators, air pollution prevention and control plan and other systems, which further improved the closed-loop supervision mechanism for green buildings.

11.2.5 Launching Publicity Training and Promotion

From 2008 to 2018, Jiangsu has successively held 11 *Green Building Development Conference in Jiangsu Province* (*No author,* 2018), with a total of more than 50 sub-forums on various thematic forums set up, more than 500 academic reports and speeches organized. Participants came from three continents including Asia, America, and Europe, with the total number of attendees exceeded 10,000.

Past conferences have actively promoted exchanges and communication with other countries in the world. International experts and scholars from the United States, Canada, the United Kingdom, France, Germany, Japan, and other countries were invited many times to speak at meetings. During the tenth Jiangsu Green Building Development Conference, establishing *The International Green Building Alliance* was proposed, which aims to promote the high-quality development of green buildings and build an international platform for green building cooperation and innovation. The conference attached great importance to information sharing on green buildings in various fields among government, industry, university, and research. The invited speakers included both industry experts and representatives from management departments, scientific research institutes, and project builders. This encouraged the exchange of work in various fields including green building policy and mechanism, scientific research innovation, project practice, technology transfer, and other aspects. The development of green buildings from shallow to deep was effectively promoted, industry consensus was enhanced, and social impact was expanded.

In 2012, Jiangsu built the national first green building and eco-smart city demonstration center, which systematically showcased green ecological concepts, technologies, and lifestyles. It was the nation's first *Green Building and Eco-Smart City Demonstration Education Base* by the Ministry of Housing and Urban-Rural Development. The base was open to the public for free, which effectively improved the publicity effect and social influence of green concepts.

Jiangsu cooperates with the National Energy Conservation Publicity Week every year and launches green building publicity at the same time (No author, n.d.-b). By setting up publicity booths in public places, distributing green building brochures, and releasing scientific knowledge propagation in various media about energy-saving emission reduction and green building-related, Jiangsu is publicizing the green and low-carbon concept, guiding the whole society to participate in energy conservation and consumption reduction activities and promoting the green lifestyle and production manners.

11.3 Green Eco-City

11.3.1 Introduction of Progress

Since 2010, based on provincial *Demonstration Zones of Building Energy Conservation and Green Building* (hereinafter referred to as the Demonstration Zone), Jiangsu has started to build provincial green eco-city in small-scale urban areas, including various types of green building regional demonstrations supported by provincial special funds. Motivated by promoting scientific development and building a better Jiangsu, local governments are encouraged to target their goals, which is the construction of key projects of large-scale green buildings and resource-efficient cities and villages. Based on this, the local governments have carried out diversified innovative practices such as the implementation of green ecological concepts, formulation of policies and mechanisms, research and development of new technologies, and construction of demonstration projects (Zhu & Wang, 2014). As a result, a greener and more ecological model of city construction and development is being explored. During guiding the construction of demonstration projects, Jiangsu is continuously optimizing the target, formulating a persistent effective mechanism, and promoting technological advance, so that we can gradually expand the core concepts of resource conservation, environmental friendliness, and ecological livability to urban and rural planning, construction, operations, and other steps (Zhou, 2011). By the end of 2014, provincial green eco-cities have been established in 13 provincial districts and cities in Jiangsu, and these demonstrations have effectively promoted the practice of constructing green buildings and resource-efficient cities and villages over Jiangsu.

In 2015, with the issuance and implementation of the *Regulations on the Development of Green Buildings in Jiangsu Province*, the construction of a green eco-city had entered a full-scale development stage under the guarantee of the legal system.

In 2016, Jiangsu launched the compilation of *Development Report of Green Eco-city in Jiangsu Province*. It systematically analyzed and summarized the achievements of the construction of the green eco-cities over Jiangsu. This report was officially published in 2018.

In 2018, the Department of Housing and Urban-Rural Development issued the *Technical Guidelines for the Special Planning of Green Eco-urban Districts in Jiangsu Province* to guide the compilation of the special planning for green eco-city over Jiangsu. In the same year, the compilation of the local standard, *Green City Planning and Construction Standards*, was officially launched. This standard aimed to provide management and technical guidance of the green city during planning, construction, and operation. It focused on the supervision and control of important steps during implementation and promotion of the key technologies meanwhile.

By the end of 2018, Jiangsu had established 72 provincial green eco-city areas. From 2015 to 2018, 18 new provincial green eco-cities were established, including 6 green building demonstration cities (districts or counties), 5 regional integrated demonstrations of green buildings and eco-city, 1 green building village and 6 cities that won incentive subsidies (Fig. 11.1).

11.3.1.1 Implementation of Policy and Establishment of Management Systems

To promote the development of green eco-city, explorations are actively carried out in different places. A system of policy, mechanism, and management has been gradually formed for planning, construction, and operation. These systems have covered

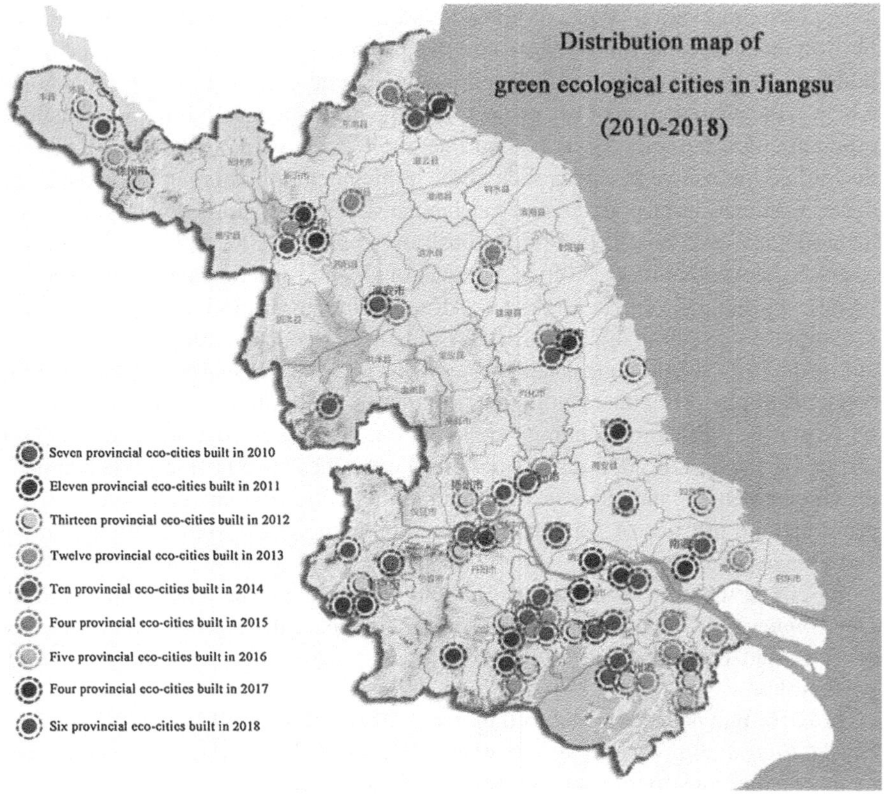

Fig. 11.1 Distribution map of green ecological cities in Jiangsu

the development of green buildings, retrofit of existing green buildings, energy-saving operation and management of public buildings, large-scale application of renewable energy, promoting the modernization of the construction industry, development of green building-related industry, and resource utilization of construction waste. The guidance of standard has effectively stimulated the orderly development of green ecological cities. Some places even combined green ecological cities with local characteristics so that creative achievements have been made in several aspects including acceptance inspection of green building and construction of green eco-city urban areas.

11.3.1.2 Compilation and Implementation of a Series of Special Plans

Jiangsu has not only achieved the full coverage of the detailed regulatory plan in green eco-city areas but also compiled more than 300 special plans and refined the planning system based on the concept of green ecology. The core contents include green buildings, urban building energy systems, comprehensive utilization of water

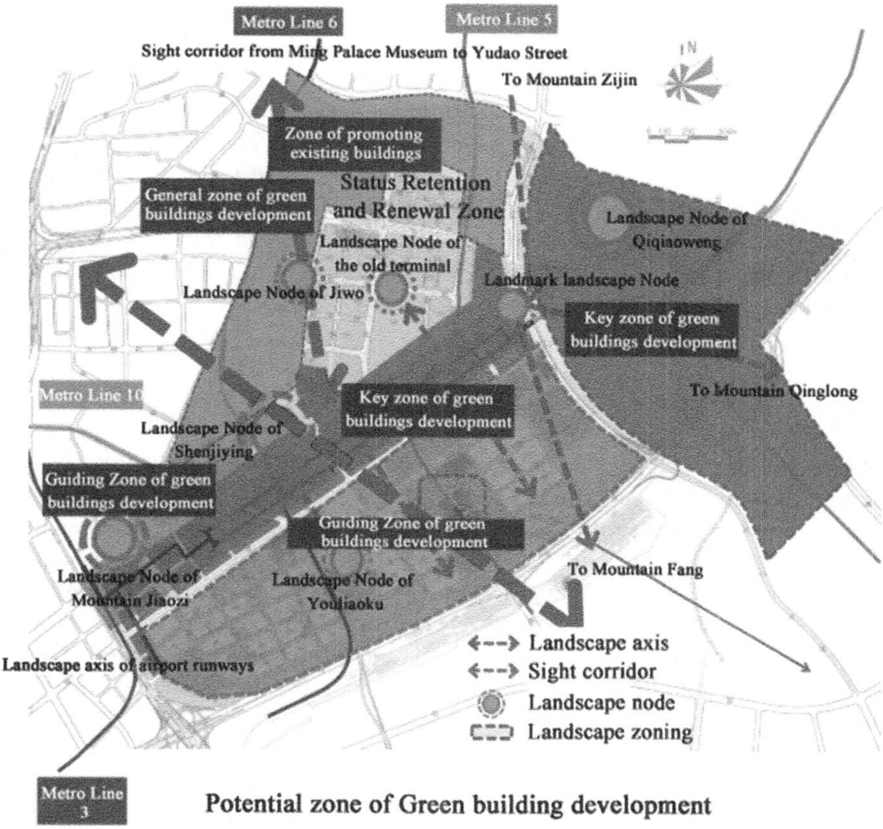

Fig. 11.2 Special plan for green buildings of Nanjing South New Town

resources, and green transportation. At the same time, implementations of special plans were promoted actively in different places and numbers of representative green eco-city have emerged including the Huaqiao International Business City of Kunshan, the Wuzhong Taihu New City of Suzhou, the Wujin District of Changzhou, and the Ecological New City of Huai'an. They have incorporated the results from special plans into regulatory plans to make sure the implementation of relevant contents, and achievement of relevant indicators (Fig. 11.2).

11.3.1.3 Research and Construction of Technology Support System

In recent years, the Department of Housing and Urban-Rural Development has successively organized a series of research projects, such as *Study on the Planning and Construction Index System of Building Energy Efficiency and Green Building Demonstration Zones in Jiangsu Province*, *Research on the Promotion Mechanism of Building Energy Efficiency and Green Building Demonstration Zone in Jiangsu Province*, and *Post-evaluation system of building energy-saving and green building demonstration area in Jiangsu Province*. Several technical documents were issued, such as *Technical Guide for Green Building Application in Jiangsu Province*, *Catalogue of Key Technologies for Building Energy Saving and Green Building Demonstration Zones in Jiangsu Province*, and *Technical Guidelines for Special Planning of Green Eco-city in Jiangsu Province*. *Green Building Standard System of Jiangsu*, *Development Report of Green Ecological City in Jiangsu Province* and other monographs were published. A relative complete technology and standard system of the green eco-city has been established.

11.3.1.4 Completion of the Green People-Benefiting Projects

By the end of 2018, the cumulative construction area of green eco-city in Jiangsu has reached about 120 million m^2. The total area of underground space under construction has reached about 16.803 million m^2. Thirty-six district energy stations have been built to serve about 11.666 million m^2 of building. 61.9 km planned urban utility tunnel and 53,000 m^2 ultra-low energy buildings were constructed. Applied energy-saving lamps and renewable energy streetlights reached about 486,000. Besides, 76 urban ecological wetlands were established, covering an area of about 1.7755 million m^2. The total annual utilization of construction waste achieved 130.0 million tons. It has been estimated that through the development of green eco-city, the total annual savings of standard coal was about 1.903 million tons, reducing CO_2 emissions by 500.0 million tons.

Each green eco-city closely has attached to the goal and development direction as a demonstration project, and they have completed several green people-benefiting projects with high technical integration, distinctive features, and remarkable benefits. For example, by introducing market funds, using domestic and foreign advanced equipment and mature technology and other measures, a utilization base of construction waste has been built and put into use in Wujin District. The comprehensive utilization rate of construction waste has reached over 95%. Permeable pavement, rainwater garden, ecological grass ditch, biological retention pond, and other green rainwater infrastructures were built in the Changzhou Financial Business District as a tentative of sponge city. Hexi New Town of Nanjing has built several green municipal infrastructures, such as the tram, the energy station of Youth Olympic Village, ecological wetlands, etc. The Suzhou Industrial Park and Wuxi New District have built projects such as District Energy Station, City Utility Tunnel, and Smart Park Management Platform. Xuzhou New District and Pei County have carried out ecological environment restoration and ecological wetland construction. They have achieved good results in applications of renewable energy in the building and green lighting. Featuring green concept in their education, Jiangsu Urban and Rural Construction College has created the country's first green campus which has become a leading model in China (Figs. 11.3, 11.4, and 11.5).

Fig. 11.3 Aerial view of Nanjing Hexi New Town

Fig. 11.4 Aerial view of Moon Bay in Suzhou Industrial Park

Fig. 11.5 Aerial view of Jiangsu Urban and Rural Construction College

11.4 Typical Case Introduction

11.4.1 Nanjing Hexi New Town (National Green Eco-City, Provincial Green Eco-City)

11.4.1.1 Overview

Nanjing Hexi New Town (hereinafter referred to as "Hexi New Town") is in the southwest of the main urban area of Nanjing. It is planned to be a sub-center of Nanjing with functions of business, commerce, sports, culture, etc. Its administrative area is 94.0 km^2, of which a total of 30.0 km^2 with the southern area and Jiangxinzhou Islet belongs to the provincial green eco-city. Hexi New Town always adheres to the concepts of compact land use, mixed layout, and ecological priority proposed in the general urban plan and regulatory detailed plan. On the core issue of how to organically diffuse the functions of the old urban district, green practices have been carried out in urban spatial layout, energy system, water resource system, transportation system and so on (Fig. 11.6).

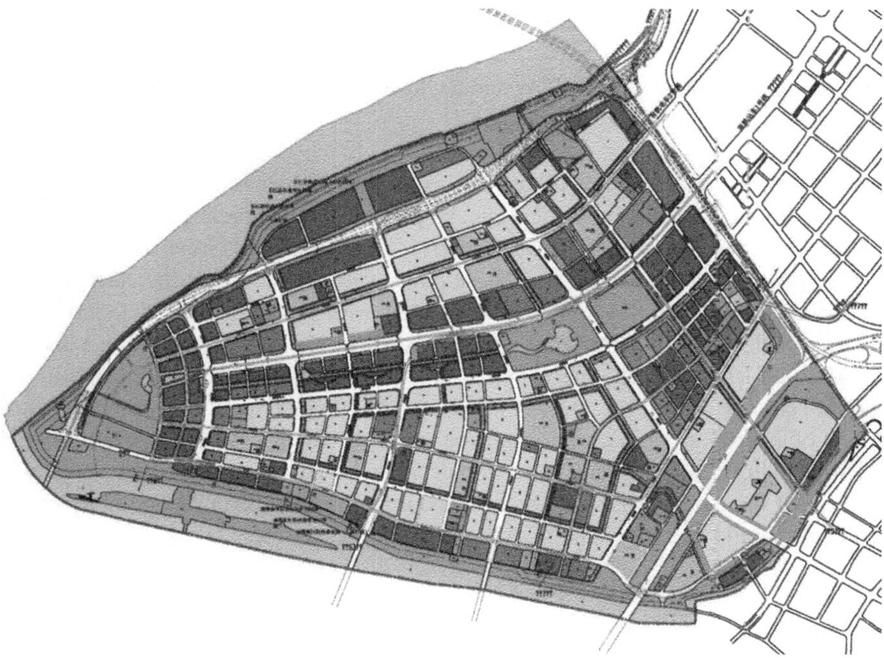

Fig. 11.6 Regulatory detailed plan for the southern part of Nanjing Hexi New Town

11.4.1.2 Case Innovations

As a carrier of diffusing the function of the old urban area and an expansion of city space, Nanjing Hexi New Town bears the important responsibility of evacuating the population of the old urban area. In the planning and construction process. It has always adhered to the people-oriented principle and closely focused on the concepts of "green, human, compact, intensive." Through a high standard of planning and design, it comprehensively coordinated the urban ground and underground space resources, ecological and environmental energy resources, and supporting resources for production and living. A new green city has been gradually built with a good living environment, compact and intensive layout, green ecology of building and diversified facilities. Through the construction of a high-quality living environment, high-end talents and high-tech industries were attracted and settled down here. An anti-magnetic center for evacuating urban functions has been formed and the development of local talents and innovation of industry have been promoted. Therefore, urban primacy ratio of Nanjing has been increased. According to a periodical evaluation, the existing population of Hexi New Town is about 460,000, and the index of jobs-housing balance in the built areas has reached 97.8% (Li, 2019).

11.4.1.3 Highlights in Hexi New Town

By Green Ecological Special Plan and a Closed-Loop Mechanism of Whole Process Management Was Established

Based on comprehensive planning systems such as the general city plan, Hexi New Town complied 8 green ecological special plans and 6 technical guidelines, including *Energy Utilization Plan of Nanjing Hexi New Town* and *Green Transportation Plan of Nanjing Hexi New Town*. These plans were embedded in the current urban planning management system and connected with the regulatory detailed plans. During making these plans, key indicators such as having green buildings and the application of renewable energy were integrated to guide the implementation of green ecological projects. During planning and implementation, dynamic evaluations of operation were carried out and tracked, which covered green buildings, district energy systems and some other items. The operational efficiency was improved since problems could be reported timely and corresponding improvement measures could be followed up.

A Compact Layout of Mixed Space with Economic and Efficient Production and Living Style

During the planning and construction, Hexi New Town focused on the jobs-housing balance. By focusing on the development of modern financial industry, information services, commerce, retail, and other industries, Hexi New Town is constantly attracting employment migration in the city and effectively reducing the pendulum traffic with the old urban area. According to the periodic operation evaluation, the index of jobs-housing balance in the built areas has reached 97.8%. Hexi New Town advocated mixed land-use such as commerce mixed with residence and commerce mixed with office buildings. Novel land-use patterns were explored with the successful formation of transportation complexes, municipal complexes, and community complexes. As a result, mixed land-use has reached a proportion of 15.9%. Meanwhile, Hexi New Town insisted on high-intensity land development and full-utilization of underground spaces, which effectively improved the land-use intensity. The net plot ratio has reached 1.6 and the average value reached 0.5 for underground space in the built areas.

Intensive Conservation of Energy and Resources, Livable and Ecological Environment

Hexi New Town payed attention to the overall coordination of regional energy, water resources, and transportation resources so that the consumption of existing resources can be reduced and the quality of production and living spaces can be improved. In the urban area of Hexi New Town, the proportion of renewable energy applications in buildings exceeded 70.0%, and 11 distributed energy stations were planned to be built as energy suppliers for the surrounding buildings. One hundred percent of new buildings have implemented rainwater reuse facilities, and the reuse rate has reached 5.2%. The renewable water supply network was planned to achieve full coverage in Jiangxinzhou Ecological Islet. The annual recycled water utilization rate reached 3.36 million m^3 upon its completion.

In Hexi New Town, a transportation complex was constructed, consisting of subways, trams, buses, public bicycles, etc. Green traffic has reached a ratio of 88.0% according to surveys and statistics. Besides, the original ecological bases such as water surface, wetland, and green space were fully preserved during the planning and construction. The per capita public green space in the built-up area was 16.9 m^2. Sponge ecological parks were constructed and have reached full coverage of parks and green spaces within a 500-meter radius of residential areas (Gu et al., 2013). Constructed urban utility tunnel has reached 8.9 km which protected the urban environment from a second underground excavation (Figs. 11.7 and 11.8).

11.4.1.4 Exemplary Significance

During promoting the planning and construction of the new urban area, Hexi New Town has fully considered the organic diffusion of the functions of the old urban areas. Compared to the older areas, Hexi New Town is more livable, the urban space is greener, the living facilities are more convenient, and the employment is more abundant. Better conditions in the new town are functioning like an anti-magnetic force of population migration, attracting more people to live and work here, which

Fig. 11.7 Tramcars in Hexi New Town

Fig. 11.8 Urban utility tunnel

diffused functions of the old city. In addition, through the development and construction of new city areas, Nanjing has promoted the development of new industries such as finance, technology, and information along with the attraction of high-end talents. This has become the driving force of innovation and creativity in the new town. Coordinated and harmonious development of new and old city areas is therefore realized.

References

Gu ZH, Han DQ, Ronald Weinstein, 2013. Spatial planning strategy for low-carbon and energy-saving cities: A case study of southern Hexi New City in Nanjing. *Urban Development Research*, 20(02), 94-104.

Hao R, 2008. Jiangsu Province's provincial energy-saving emission reduction special guidance funds increased to 500 million yuan. *Jiangsu Building Materials*, 2008(03), 13.

Implementation in Jiangsu, 2015. Regulations of Jiangsu Province on Green Building Development. *Urban Planning Newsletter*, 2015(14), 12.

Li XL, 2019. Evaluation of Phased Operation of Green Ecological City: Taking Nanjing Hexi New City Building Energy and Urban Lighting System as an Example. *China Urban Science Research Association*, 2019, 557-563.

Mou GY et al., 2019. Jiangsu Yearbook. *Jiangsu Yearbook Magazine*, 602.

Ni R, 2011. Research on Optimization of Energy Consumption Structure in Jiangsu Provinces Based on Environmental Constraints. *Reform and strategy*, 27(11), 119-122.

No author Ministry of Housing and Urban-Rural Development and Jiangsu jointly promote green building development. *Urban Planning Newsletter*, 2015(05), 4.

No author Jiangsu Green Building Development Conference Held 12 Projects Appraised as Provincial Green Building Innovation Projects. *Construction Technology*, 2018(22), 8.

No author Research on Strengthening Construction Industry Modernization in Jiangsu. *Engineering Quality*, 32(06), 77.

No author Jiangsu actively carries out the promotion activities of green building design standards. *Building Energy Efficiency*, n.d.-b 43(01), 125.

Zhang T, 2010. Research on Optimization of Jiangsu Energy Consumption Structure Based on Environmental Constraints. *China University of Mining and Technology*.

Zhou L, 2011. Green Practice of Urban and Rural Construction in Jiangsu Province. *Construction Technology*, 2011(22), 17-20.

Zhu YB, Wang DY, 2014. Carrying out regional demonstrations and promoting ecological civilization: A review of the work of demonstration zones for building energy efficiency and green buildings in Jiangsu Province. *Construction Technology*, 2014(06), 65-67.

Part IV
Modern Miracles Brought Forth by Technological Innovation and Economic Growth

Chapter 12
New Countryside in the Internet Age: The Development and Planning of E-Commerce Taobao Villages in China

Zhendong Luo and Yibo Qiao

Abstract With the progress of economic globalization and Internet informatization, Taobao Village become a special economic geographic phenomenon in Chinese rural region. Through analysis of the spatial distribution data of Taobao Villages from 2014 to 2018, this chapter points out that the spatial distribution of Chinese Taobao Villages shows a longitudinal "North-Middle-South" agglomeration structure and a latitudinal "East-Middle-West" diffusion trend, and a rapid fissile proliferation among Taobao Village industry clusters. In the Yangtze River Delta region, three major Taobao Village agglomeration areas are located in the "double-marginal" zone, that is, the urban margin and the margin of the metropolitan area. Then we propose a "diamond model" to explain the formative mechanism of Taobao Villages, in which grassroots entrepreneurs, main products, facilities support and governments support constitute the four key factors, and the village Party branch and village committee are the main agents for Taobao Villages to transform the rural governance pattern and accomplish beautiful village construction. In order to cope with the disorder of built environment which caused by weak regulation, the planning and construction of Taobao Villages should focus on the bottom-up and implementation-oriented construction of rural living environment.

Keywords Taobao village · E-commerce · Development characteristics · Formative mechanism · Panning · Governance

Z. Luo (✉)
Urban and Regional Planning Department, School of Architecture and Urban Planning, Nanjing University, Nanjing, China
e-mail: luozhendong@nju.edu.com

Y. Qiao
Human Geography and Spatial Planning Department, Faculty of Geosciences, Utrecht University, Utrecht, Netherlands

L. Bian et al. (eds.), *Chinese Urban Planning and Construction*,
Strategies for Sustainability, https://doi.org/10.1007/978-3-030-65562-4_12

12.1 Introduction

The influence of Internet on rural development has been the everlasting focus of academic research since the beginning of twenty-first century. Although as a technology designed in and for the very different situations prevailing in the developed economies (James 2010), many wonderful examples are well documented where Internet finds its suitable place and contributes much to the economic development and social governance in rural areas of developing economies, which is usually featured with low education and incomes, lack of skills and human capitals. Scholars have widely investigated the adoption and application of Internet in rural areas of developing countries, such as Iran (Jalali et al. 2011; Alibaygi et al. 2011; Alavion and Allahyari 2012), India (Ghosh 2011; Jamaluddin 2013), Bangladesh (Islam and Grönlund 2010) and South Africa (Emmanuel and Muyingi 2010; Dalvit et al. 2007), etc., while, in most of these existing researches, the application of Internet mainly involves supplying immediate and advanced information to the rural poor, such as agricultural production knowledge, computer literacy training, skills acquisition, purchasing commodities and services online, entertainment, health care, etc., and collecting information about rural regions to policy makers, such as the e-governance practice (Alibaygi et al. 2011). In this perspective, rural area is just a passive object waiting for being enlightened and modernized, and a unidirectional receiver of urban civilization and capitalist production. On the whole, the productive force and active aspect of rural areas still remain untouched and isolated from the Internet.

As China gradually integrated into the process of economic globalization and Internet informatization at the end of the last century, Chinese rural development is also affected by globalization and informatization process proportionately. With the rapid construction of ICT infrastructure and the popularization of mobile phones in the rural areas, accordingly, although the "rural-urban digital divide" (Hindman 2000) tends to increase (Fong 2009), still, the Internet development in Chinese rural area is rather amazing. The macro-level data in 43rd China Statistical Report on Internet Development published by China Internet Network Information Center (CNNIC) shows that, until December 2018, Chinese Internet penetration has reached 59.6% with 0.829 billion netizens in total, and the rural Internet penetration has risen to 38.4% with 0.222 billion netizens, and rural e-commerce sales volume has reached 1370 billion yuan (China Internet Network Information Center 2019). Many Chinese rural areas with disadvantaged location are involved in the national and even global production and transaction networks through Internet e-commerce, forming a new economic geographic phenomenon, that is, Taobao Village.

Taobao Village, according to the definition given by Alibaba, which is the biggest e-commerce platform in China, is an administrative village where there are more than 100 active online stores or 10% of the number of households owning online stores, and the annual e-commerce transaction volume is more than ten million yuan. The emergence of Chinese Taobao Villages has been well recorded in both Chinese and English literature since it appeared in journals in 2011 (Chen 2011). Existing researches on Taobao Village mainly focus on the following aspects: (1) Spatial distribution and influence factors of Taobao Villages. Some scholars use kernel density estimation, gyroradius method, clustering analysis, spatial analysis and statistical models to explore the spatial distribution of Taobao Villages and its corresponding influence factors from the national (Zhu et al. 2016; Shan and Luo 2017; Diao et al. 2017; Xu et al. 2017a, 2017b; Xin and Qiao 2018; Qi et al. 2018; Bai and Sun 2019), provincial (Zhao et al. 2017; Zhou et al. 2017), municipal (Hu and Liu 2016), and county (Xu et al. 2017a, 2017b) levels. Some scholars refine the evolution of the internal space of Taobao Village with detailed and in-depth case studies, and propose corresponding spatial planning methods based on the particularity of the e-commerce industry (Zhang and Qian 2015; Yang et al. 2016; Zhang et al. 2016; Dai and Zhang 2017). (2) The formative mechanism, industrial development and classification of Taobao Village. Some scholars use the theory of "space of flow," inclusive innovation or other theories to explore the deep mechanism and formation model of Taobao Village (Lei and Liu 2017; Wang et al. 2017; Yu et al. 2018; Fan et al. 2018; Chen and Zhang 2018). Some scholars regard Taobao Village and Taobao Town as industrial clusters from the perspective of industrial ecosystem (Shi et al. 2017; Zeng 2018; Zhou and Liu 2018; Qiao 2019), and classify Taobao Villages from the perspective of industry and space (Zhang 2015; Cai 2018; Mao 2018; Fu 2019). (3) Rural urbanization. Some scholars pointed out that the development of Taobao Village is a new bottom-up rural urbanization process driven by e-commerce, and proposes that Taobao Village is a breakthrough in the traditional rural urbanization dilemma. Compared with the urbanization of migrant workers and other urbanization types, the development of Taobao Villages and Taobao towns belongs to local urbanization (Luo and He 2017; Xu et al. 2015; Guo 2016; Fang 2016; Lou and Hu 2018). (4) Rural governance. Scholars pointed out that the development of Taobao Village and the return of developmental agents to Taobao Village led to the reconstruction of rural governance, and explored the role played by e-commerce practitioners, village committees, local governments and e-commerce associations (Chen et al. 2016; Li and Zhang 2015; Zeng and Guo 2016). (5) Rurality and rural culture. Some scholars point out that the development of Taobao Villages will lead to a re-understanding of the concept content of "rurality" and the reconstruction of rural culture (Chen 2016; Wu et al. 2015).

Based on the existing researches, this chapter exerts efforts to give a comprehensive overview of the development and planning of e-commerce Taobao Villages in China. In the following sections, we will investigate the development characteristics of Taobao Villages using the Chinese Taobao Villages data from 2014 to 2018, and explore its formative mechanism through many cases of excellent Taobao Villages, then propose some planning principles and methods to solve the governance dilemma of Taobao Villages.

12.2 Methodology and Data Sources

12.2.1 Methodology

12.2.1.1 Theory Construction Based on Case Study

Due to the unique characteristics and history of individual Taobao Village, case study is widely used in this chapter. Through field investigations of many Taobao Villages, and detailed interviews with relevant individuals who play a central role in the development of specific Taobao Village, a theoretical framework of the formation of Taobao Villages is proposed to explain the developmental mechanism of Chinese Taobao Villages.

12.2.1.2 Kernel Density Analysis

Kernel density analysis uses moving windows to calculate the density of point features (or line features) in surrounding fields, which will help to clearly understand the degree of heterogeneity of features and the agglomeration characteristics of their geographical distribution. This method takes the geographical coordinates of the selected point features as the center of the circle and sets a specific value as the radius. The attributes of the point features are distributed in this circular area according to a specific density function. The center point has the highest density and the radius is limited. This study performs kernel density analysis on the national and regional level to show the spatial distribution of Taobao Villages.

12.2.2 Data Sources

Since 2014, Ali Research has published the name list of Chinese Taobao Villages in Chinese Taobao Village Research Report annually, in which, the exact location of every Taobao Village could be identified. In this chapter, data from Chinese Taobao Village Research Report 2014, 2015, 2016, 2017, 2018 are used to display and analyze the patterns of the spatial distribution of Taobao Villages.

12.3 The Development Characteristics of Chinese Taobao Villages

12.3.1 Spatial Distribution of Chinese Taobao Villages

12.3.1.1 National Pattern: Highly Concentrated in the Eastern Coastal Areas

Since the discovery of three Taobao Villages[1] in 2009, nationwide Taobao Villages have experienced a rapid growth spurt, from the sporadic distribution of the initial coastal provinces to an economic geographical phenomenon with national influence. An in-depth analysis of the growth data of Taobao Villages in 2014–2018 shows that,[2] although the spatial distribution of Taobao Villages in China began to appear in 2017, which coincides with the economic geographical gradient of Chinese "East, West, and West" structure, Taobao Villages are still intensively distributed in the eastern coastal areas, in which the total proportion is always above 96%[3](shown in Fig. 12.1).[4]

On the provincial level, Chinese Taobao Villages present obvious characteristics of agglomeration along the eastern coastal areas. The number of Taobao Villages in Zhejiang Province is far ahead, accounting for 36.6% of the country in 2018, almost twice the number of those in Guangdong province in second place (Table 12.1). In addition, the top ten cities with the largest number of Taobao Villages[5] and the top ten Taobao Village clusters[6] were mostly located in Zhejiang Province in 2018. The provinces following Zhejiang Province are Guangdong Province, Jiangsu Province, Shandong Province, Fujian Province, and Hebei Province. The numbers of Taobao Villages in these provinces are more than 200, and they are all located in the eastern

---

[1] Qingyanliu Village in Yiwu, Zhejiang, Donggao Village in Qinghe, Hebei, Dongfeng Village in Suining, Jiangsu.

[2] Since the introduction of big data and other methods in the Ali Research Institute in 2014, the data of Taobao Village has been greatly improved, and the data comparability and research value are correspondingly higher.

[3] From 2014 to 2008, the proportion of Taobao Villages in the six provinces and two cities in the east was 98.11%, 97.69%, 97.71%, 96.32%, and 96.47%, respectively.

[4] All figures and tables in this chapter which are not marked with sources are made by the authors of the report.

[5] In 2018, they are Heze of Shandong Province (267 Taobao Villages); Wenzhou of Zhejiang Province (254); Jinhua of Zhejiang Province (245); Taizhou of Zhejiang Province (203); Quanzhou of Fujian Province (140); Suqian of Jiangsu Province (132); Ningbo of Zhejiang Province (131); Hangzhou of Zhejiang Province (126); Guangzhou of Guangdong Province (125); Jiaxing of Zhejiang Province (113).

[6] In 2018, they are Yiwu of Zhejiang Province (134 Taobao Villages); Cao of Shandong Province (113); Wenling of Zhejiang Province (97); Suining of Jiangsu Province (92); Yueqing of Zhejiang Province (83); Cixi of Zhejiang Province (78); Yongkang of Zhejiang Province (74); Ruiyan of Zhejiang Province (71); Puning of Zhejiang Province (64); Suqian Sucheng District of Jiangsu Province (61).

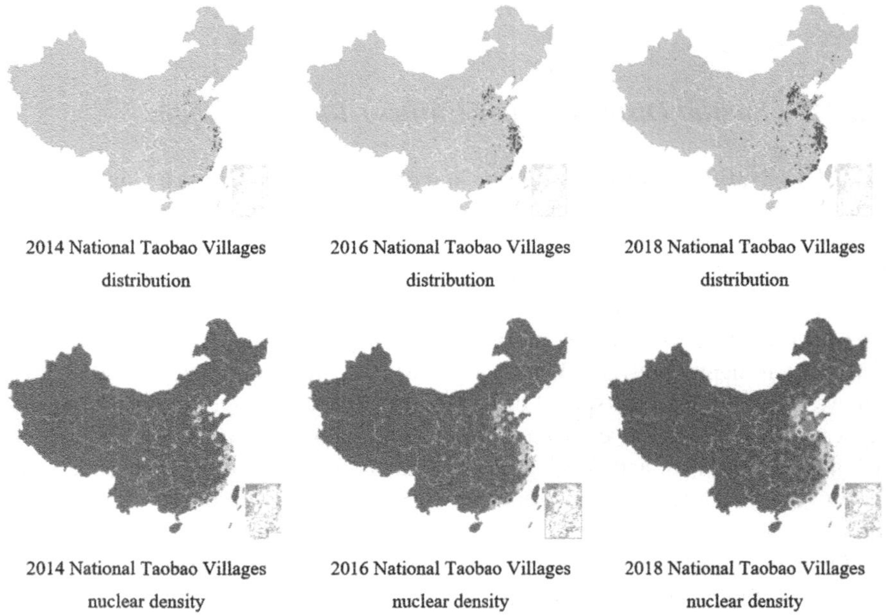

Fig. 12.1 Distribution changes of Taobao Village in China in 2014, 2016, and 2018

coastal areas. In Henan Province, which is located in the central part of China, the growth rate of the numbers of Taobao Villages has been extremely fast in recent years, and it has become the first and only province with more than 50 Taobao Villages except the six eastern provinces. In Jiangxi, Hubei, and Anhui provinces, and Liaoning Province, the development of Taobao Villages has begun to accelerate in the past two years. In addition, the number of Taobao Villages in the vast central and western provinces is generally small. The limitations of geographical location, traffic conditions, technical level, and government governance have made the level of rural e-commerce development in these areas lag significantly behind the eastern coastal areas.

On the county level, Taobao Villages manifest a cluster development characteristics in the eastern region, scattered distribution in the central and western regions, and the spatial distribution of Taobao Villages is quite similar to that observed at the provincial level, that is, mainly distributed in the southeastern coastal areas, especially in the form of large-scale contiguous development of the county in Jiangsu and Zhejiang provinces (see Fig. 12.2). Among the top ten industrial clusters of Taobao Villages in the county, Zhejiang Province occupies the seventh place (Yiwu City, Wenling City, Yueqing City, Cixi City, Yongkang City, Ruian City); Cao County of Shandong Province, Suining County of Jiangsu Province, and Puning County of Guangdong Province are also among the top ten Taobao Village clusters due to the specialization and large-scale development of the e-commerce industry. In addition, in the Yangtze River Delta, Pearl River Delta, North Jiangsu, Southeast

Table 12.1 Distribution of Taobao Villages and Taobao Towns in China from 2014 to 2018

Province	2014		2015		2016		2017		2018	
	Taobao Village	Taobao Town	Taobao Village	Taobao Town	Taobao Village	Taobao Town	Taobao Village	Taobao Town	Taobao Village	Taobao Town
Zhejiang Province	62	6	280	22	506	51	779	77	1172	128
Guangdong Province	54	5	157	20	262	32	411	54	614	74
Jiangsu province	25	2	126	11	201	17	262	29	452	50
Fujian Province	28	2	71	7	107	13	187	24	233	29
Shandong province	13		63	6	108	12	243	36	367	48
Hebei Province	25	2	59	5	91	8	146	16	229	27
Henan Province	1		4		13		34	2	50	3
Sichuan Province	2		2		3	1	4	1	5	
Hubei Province	1		1		1		4		10	
Tianjin	1		3		5		9	1	11	2
Liaoning Province			1		4	1	7	1	9	1
Jiangxi Province			3		4		8		12	
Hunan Province			3		1		3		4	
Yunnan Province			2		1		1		1	
Beijing			1		1		3	1	11	1
Jilin province			1		1		3		4	
Shanxi Province			1		1		2		2	
Anhui Province					1		6		8	
Guangxi Zhuang autonomous region							1		1	
Guizhou Province							1		1	
Shanxi Province							1		1	

(continued)

Table 12.1 (continued)

Province	2014		2015		2016		2017		2018	
	Taobao Village	Taobao Town	Taobao Village	Taobao Town	Taobao Village	Taobao Town	Taobao Village	Taobao Town	Taobao Village	Taobao Town
Ningxia Hui autonomous region							1		1	
Xinjiang Uygur autonomous region							1		1	
Chongqing							1		3	
Total	212	17	778	71	1311	135	2118	242	3202	363

Data sources: Ali Research, Chinese Taobao Village Research Report 2014, 2015, 2016, 2017, 2018

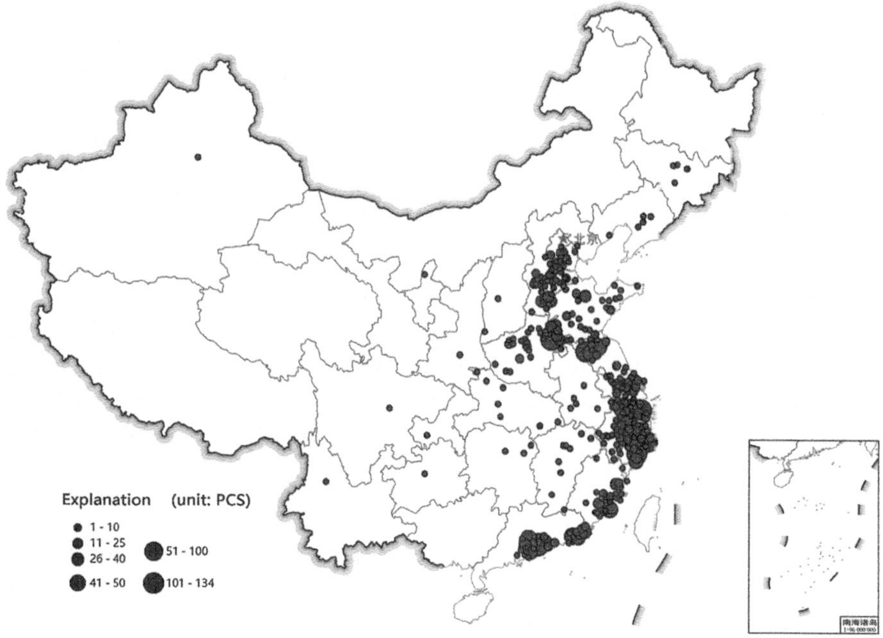

Fig. 12.2 2018 Chinese Taobao Villages county classification

Fujian, and Chaoshan region there are a large number of counties with more than 30 Taobao Villages. Compared with the high-density distribution pattern in the eastern region, the number of Taobao Villages in the central and western part and the north-eastern region of China are generally small, and are essentially in the low-level range below five. Xinzheng City in Henan Province and Luolong District in Luoyang City are among the few midwest counties with more than five Taobao Villages. Haicheng City, Liaoning Province is the only county in Northeast China with more than five Taobao Villages.

12.3.2 Spatial Gradient: Decreasing in "East-Middle-West" Structure

Because of the superior location conditions and relatively developed economy in the eastern coastal areas, the popularization of Internet and convenient transportation lay a good foundation for the emergence of Taobao Villages. In the six eastern coastal provinces and two cities (Hebei Province, Shandong Province, Jiangsu Province, Zhejiang Province, Fujian Province, Guangdong Province, Beijing, and Tianjin), the total number of Taobao Villages reached 3089 in 2018, accounting for 96.47% of the country, showing a cluster feature of rural e-commerce industry (see Fig. 12.2). In particular, the whole Zhejiang Province, Central and South part of

Hebei province, southwestern Shandong, northern Jiangsu, southern Jiangsu, southeastern Fujian, Chaoshan region, the Pearl River Delta, and other regions have shown the continuous development of Taobao Villages across several counties. The six provinces in the central region (Henan Province, Anhui Province, Jiangxi Province, Hubei Province, Hunan Province, and Shanxi Province) and the two northeastern provinces (Liaoning and Jilin Province) formed the second echelon of Taobao Villages; the total number of Taobao Villages reached 99 in 2018, accounting for 3.09%. The development of Taobao Villages in the central region depends more on the characteristic industries or on the upstream and downstream industries with the eastern region. The spatial distribution shows an island-like distribution feature that is connected across one or two adjacent counties. A typical region is in Henan Province, which include Luolong District (steel furniture) in Luoyang City, Mengjin County (Tang Sancai glazed ceramics, peony flower paintings), Yanshi City (drums), Changge City (bee products), Xinzheng City. In the western region, there have been four provinces, three autonomous regions, and one municipality (Sichuan, Guizhou, Yunnan, Shanxi, Guangxi Zhuang Autonomous Region, Ningxia Hui Autonomous Region, Xinjiang Uygur Autonomous Region, and Chongqing) which have Taobao Villages—the total number reached 14 in 2018—but the total proportion is still less than 0.5%. A Taobao Town was discovered in 2017 (Anjing Town, Pidu District, Chengdu City, Sichuan Province), however this Taobao Town disappeared with the disappearance of the corresponding Taobao Village in 2018. Because the formation of most Taobao Villages in western regions mainly relies on its own resource endowments and traditional handicrafts, so the industrial development has obvious local stickiness, and is difficult to be replaced or imitated by large-scale industrial production. However, the scale of the industry is relatively small and the diffusion effect is insufficient. Taobao Villages have a point-like distribution in space, such as Xinhua Village (traditional handmade silverware) in Heqing County, Dali, Yunnan, and Cangfanggou Village (high-altitude mountain flower nectar) in Shayibake District, Urumqi, Xinjiang, and Wangxian village (apple) in Wanrong County, Shanxi Province, and Libei Village (agricultural products), in Xifeng County, Guizhou province, and so forth.

12.3.2.1 Regional Agglomeration: "North-Central-South" Distribution

The northern gathering area covers a relatively large area, mainly including northern Jiangsu, southern Shandong, and central and southern Hebei (Fig. 12.3). In recent years, Taobao Villages have experienced explosive growth and the concentration degree has increased continuously. For example, in Heze City, Shandong Province, the number of Taobao Villages increased from 7 in 2014 to 269 in 2018. The northern gathering area belongs to the agricultural economic zone in the North China Plain, and its economic development level and industrial base lag far behind other areas along the eastern coast. Before the formation of Taobao Villages in most villages, the industrial base was very weak, and the production factors such as population and capital were in an outflow situation for a long time, and to a certain extent

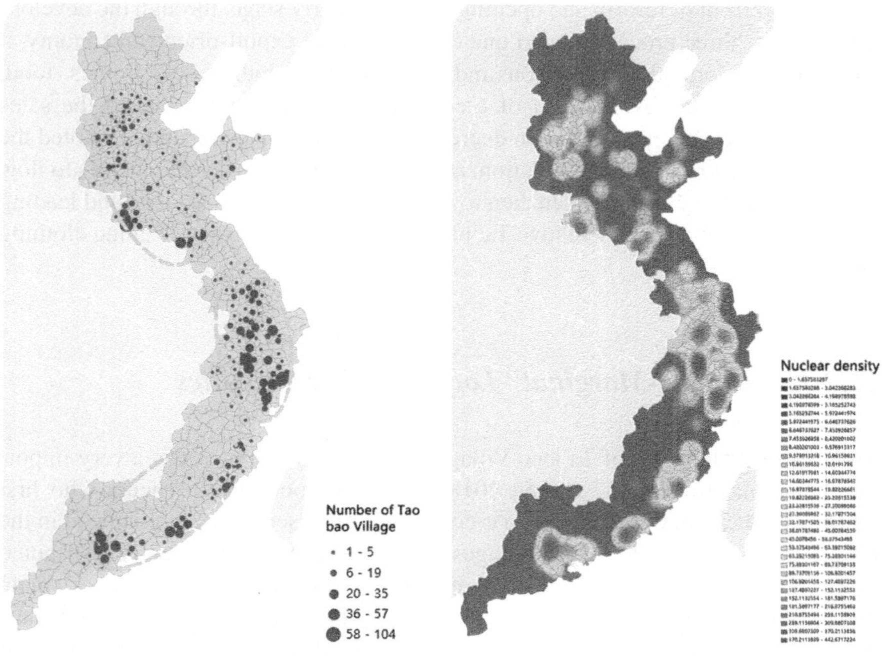

Fig. 12.3 County-level classification map of Taobao Villages in the eastern coastal area (left) and Taobao Village nuclear density (right)

belonged to decaying rural areas. The combination of the global link characteristics of e-commerce and the low-cost entrepreneurial environment in the countryside has opened and accelerated the local rural industrialization process, which has spawned clusters of emerging Taobao Villages in Cao County (performance costumes), Suining-Sucheng (furniture industry), and Shuyang County (flowers and plants), showing a new way for rural revitalization.

The central gathering area mainly includes the whole province of Zhejiang and the southern part of Jiangsu Province. It is the most concentrated area of Taobao Villages in the country, gathering more than 40% of Taobao Villages. In 2018, six of the top ten cities having Taobao Villages in the country were located in this area (Jinhua City, Wenzhou City, Taizhou City, Hangzhou City, Ningbo City, and Jiaxing City). This area is one of the most developed areas of Chinese township enterprises and private economy. It is also the birthplace of Chinese e-commerce and has the firmest industrial foundation. At the same time, because of its location in the Yangtze River Delta, the intensive transportation networks and the central location grant the products of this region significant comparative advantages in transportation efficiency and transportation costs, and can easily radiate to the most important consumption markets in the country.

The southern gathering area consists of three relatively independent gathering clusters in the Pearl River Delta, Chaoshan region, and Southeastern Fujian Province. Because of its proximity to Hong Kong, Macao, and Taiwan, this region is at the forefront of reform and opening-up. In the early stage, through the development of the "three processing and one compensation" export-oriented economy, it has gradually formed advantageous industries such as clothing, electronics, food, and toys, etc. The development of e-commerce has further broadened the sales channels, enhanced the integration degree of the industrial chain, and promoted the production factors, such as population, capital, technology, and other factors to flow back from Guangzhou and Shenzhen to the origin area of the products, and leading to the emergence of representative Taobao Village clusters, such as Puning clothing and Chenghai toys.

12.3.3 "Double-Marginal" Location Characteristics

The rapid development of Taobao Villages driven by e-commerce is a very important "marginal revolution" (Liang 2018). The grassroots entrepreneurs who first returned to the country to engage in e-commerce are essentially marginalized in the city. A large number of Taobao Villages and Towns are also essentially distributed in the marginal locations of the city-regions, meanwhile, Taobao Villages' main products are mostly Blue Ocean products. Combined with the marginal crowd, marginal location, and marginal products, the marginal attributes make the rise and prosperity of Taobao Villages full of challenges, and quickly become an important force to change the urban and rural economic geography.

We take the Yangtze River Delta region (Jiangsu, Zhejiang, and Shanghai) as an example, as Taobao Villages are most densely distributed there. The distribution of Taobao Villages in 2014–2018 is different from that of regional central cities (Shan and Luo 2017). Except Hangzhou, the central cities of the Yangtze River Delta region such as Nanjing and Shanghai do not have Taobao Villages around. Even in Hangzhou, Taobao Villages are only located at the marginal metropolitan area. Combined with the distribution of the newly added Taobao Villages in 2018, the dislocation of the distribution of Taobao Villages and the regional central cities are more obvious. It shows a "double-marginal" effect: the first is that Taobao Villages and clusters are mostly distributed on the margin of the regional central city, and the second is that Taobao Villages and their clusters, which are not on the margin of the regional center city, are mostly distributed on the edge of the Yangtze River Delta city region.

12.3.3.1 Marginal Effect on the Central City Level

The distribution of Taobao Villages around the three central cities of Shanghai, Nanjing, and Hangzhou shows obvious differences. According to the data of 2018, Taobao Villages do not exist in the jurisdiction district of Shanghai. Nine Taobao Villages are found within 30 kilometers outside the downtown. The number of Taobao Villages in the 30 to 80 kilometers ring area has increased to 146. However, there is still no Taobao Village within the 60 kilometers of Nanjing, and only six Taobao Villages are found within 80 kilometers outside the Nanjing downtown. The biggest difference between Hangzhou, Shanghai, and Nanjing is that more and similar Taobao Villages are located in the near-marginal ring of 10 to 30 km and the far marginal ring of 30 to 60 kilometers. Clearly, in Yangtze River Delta, the global city region, the multi-values of the suburban villages of the metropolitan areas are increasingly appearing. The top-down development and the degree of space control are relatively high, and the marginal conditions for the development of Taobao Villages are gradually lost. On the one hand, the demand for urban short-distance leisure vacations is enormous, and the resulting rural tourism has appeared in large numbers. Then the government and private capital are also more inclined to build villages by meeting urban needs. Therefore, beautiful rural construction has become the main mode of suburban rural development and can form a good demonstration and driving effects. For example, when Jiangning District in the suburbs of Nanjing City has gradually formed a pattern of rural tourism through the creation of "Five Golden Flowers" (five exemplar villages of rural tourism) in the beautiful countryside, Lishui, Gaochun, and Jurong in the far suburbs of Nanjing City have followed suit. On the other hand, the economic value of land in the suburbs of metropolitan areas is generally high. At the same time, the strong radiation and driving effects in the central urban areas make the rural population in the suburbs mostly multi-employed. The comprehensive income level mainly on rent collection is relatively high, therefore, rural residents lack motivation to start a business.

12.3.3.2 Marginal Effect on the Regional Level of the Yangtze River Delta

The spatial distribution of Taobao Villages has the second marginal effect in the regional scale of the Yangtze River Delta. From the spatial distribution of Taobao Villages in 2018, it can be seen that except the large number of Taobao Villages around the Taihu Lake and in Hangzhou-Shaoxing-Ningbo area, the central and southern areas of Zhejiang and Xuzhou-Suqian area in Jiangsu province are two other densely distributed areas. They are all located in the outer margin of the Shanghai- Nanjing-Hangzhou city belt, which is the core area of the Yangtze River Delta (Luo et al. 2015) (Fig. 12.4). The mechanisms for the emergence, development, and diffusion of Taobao Villages in these two regions are completely different from those at the central city scale. The central and southern regions of Zhejiang Province, which have inconvenient transportation and resource constraints, belong to the southern margin of the Yangtze River Delta. The emergence of Taobao Villages in these areas is mainly based on the production and trade base of the "block economy" formed by long-term development. The longstanding combination of business, entrepreneurial tradition, and e-commerce has rapidly spawned a large number of Taobao Villages and Towns. The "marginal temperament" in the Xuzhou-Suqian area in northern Jiangsu is more obvious. They are all relatively backward economic areas, and the industrial base is almost blank. Relying on the

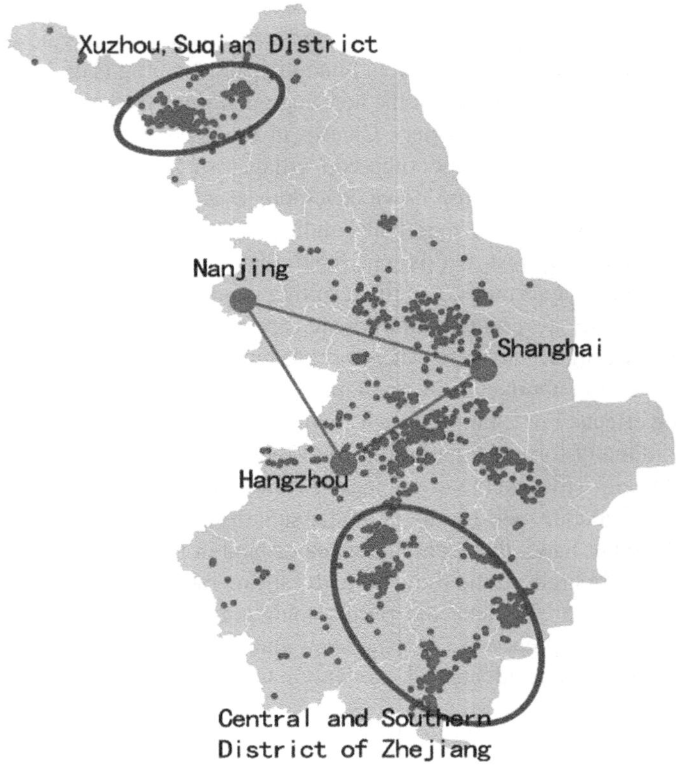

Fig. 12.4 Schematic diagram of the distribution space of Taobao Village in the Yangtze River Delta region

grassroots entrepreneurial spirit of the local people and the low-cost Internet entrepreneurial environment, it has been one of the three largest Taobao Village clusters in the Yangtze River Delta during the past decade.

12.3.4 Agglomeration and Fission of Taobao Villages

Taobao Villages and Towns are highly densely distributed in the eastern coastal areas. This remarkable spatial feature is closely related to the fissile growth model of the newly added Taobao Villages. Although the number of Taobao Villages and Towns has increased rapidly, there is no significant expansion in spatial distribution. It is more likely to be adjacent growth within a certain range (county), that is, most of the newly added Taobao Villages appear in the originally more concentrated places, as if the cells undergo fissile proliferation (Shan and Luo 2017). In 2014–2018, the number of provinces that had Taobao Villages increased from 10 to 24, but the total number of Taobao Villages in newly added provinces was only 59,

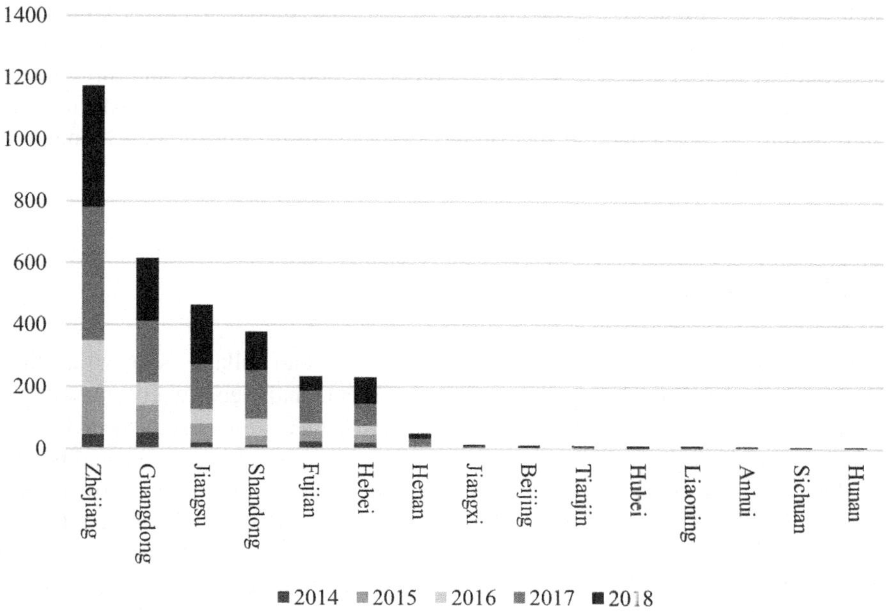

Fig. 12.5 The growth of Taobao Village in some provinces and cities in 2014–2018 (in 2014, the initial value). Data sources: Ali Research, Chinese Taobao Village Research Report 2014,2015,2016,2017,2018

accounting for only 2.0% of the new quantity. Taobao Villages are far from forming a situation of agglomeration and development in new areas. Compared with 2016, 94% of the newly added Taobao Villages in 2017 were still located in the six provinces and two cities on the eastern coast, and 63% were located in 59 Taobao Village clusters. Most of the newly added Taobao Villages are located in the six coastal provinces where the distribution was relatively concentrated before (Fig. 12.5), accounting for more than 95% of the newly added villages. From 2014 to 2018, the number of Taobao Villages increased from 212 to 3202, increasing by 14 times, while, the expansion of spatial distribution is far less than the expansion of Taobao Villages' number.

12.4 The Formative Mechanism of Chinese Taobao Villages

The formation and development of Taobao Villages is a process of the overall reconstruction of rural material space and social systems by e-commerce in the Internet era. The pioneering rural grassroots entrepreneurs, relying on the local industrial base and resource endowment to explore specialty products, with the link capability of e-commerce platforms, assisted by the government's policy support and financial assistance, have promoted the rapid development of the rural e-commerce industry, and make the ordinary villages gradually grow into Taobao Villages. Grassroots entrepreneurs, main products, supporting facilities, and governments support are the four key factors affecting the development of Taobao Villages, and rural governance is the key factor for Taobao Villages to achieve upgrading and transformation, thus forming the diamond mechanism model for the formation and development of Taobao Villages (Fig. 12.6).

12.4.1 Grassroots Entrepreneurs

The development process of a large number of Taobao Villages shows that they were first implemented by a number of entrepreneurs and then quickly spread after success. These grassroots entrepreneurs who have been exposed to advanced knowledge and cutting-edge technology are the rural elites of the Internet age and heroes of the time (Luo and He 2017). They are the core driving force for the formation of Taobao Villages, such as the famous "Shaji three musketeers," "Junpu twelve arhats," and "Taobao Village excellent leader" selected by "China Taobao Village Summit Forum" in 2017 and 2018. Entrepreneurship highlights the core personality of grassroots entrepreneurs (Zhuang 2005). They try and find excellent business opportunities with the low cost of living in rural areas and Internet. When an

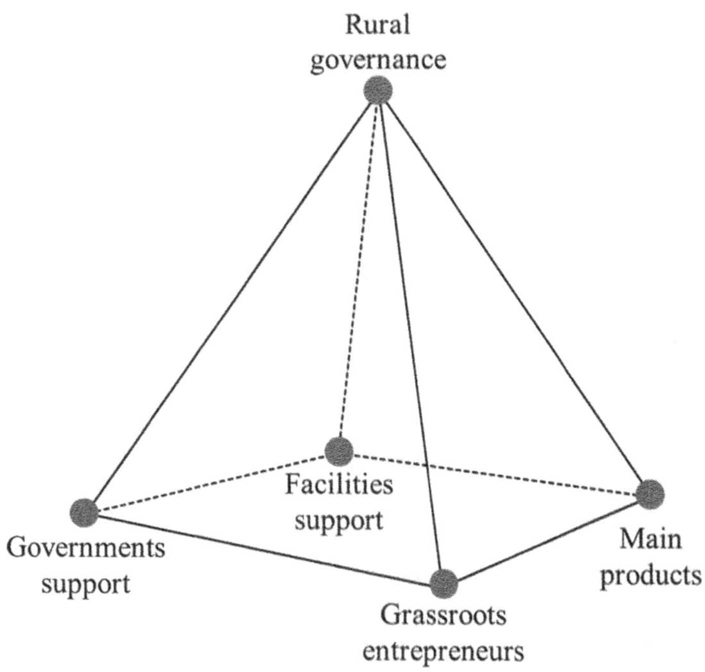

Fig. 12.6 The diamond mechanism model of the formation of Taobao Village.

e-commerce start-up succeeds, its dizzying wealth effect immediately triggers imitation behavior among villagers. As China's rural areas are essentially acquaintance society organized with blood, kinship, friendship, and other relationship ties, the social trust network is developed in a relatively closed space with obvious self-organization. Therefore, the spread of entrepreneurial information and tacit knowledge of grassroots entrepreneurs within the village are extremely fluent (Hagerstrand 1968). The entrepreneurship and the self-organization of rural social networks constitute the basic logic for the formation and development of most Taobao Villages.

12.4.2 Main Products

Finding products that are suitable for local production and that are uniquely competitive in the external market is critical to the formation of Taobao Villages. Synthesizing Taobao Villages across the country, we find that there are two main types of primary products. In regions with a firm manufacturing base, relying on the local physical market and related supporting facilities, it is easier to expand the sales channels of original products from physical stores to the Internet, thus bringing about the rapid development of Taobao Villages. In regions with weak manufacturing bases, due to the low maturity of production technology and the industrial chain, they must rely on local resource endowment and special technology to create Blue Ocean products through dislocation development so as to link local products to the broad external market through the Internet. Typical examples are children's costumes in Daji Town, Cao County, Shandong Province, handmade silverware in Xinhua Village, Heqing County, Yunnan Province, and Tang Sancai and peony paintings in Mengjin County, Luoyang City, Henan province. These are all Blue Ocean products with unique features in areas where the manufacturing industry is not well developed.

12.4.3 Facilities Support

The superimposed support of infrastructure has laid a solid hardware foundation for the rapid development of Taobao Villages. In recent years, with the increasing investment in agriculture, rural areas, and farmers, the construction of infrastructure in rural areas has significantly improved, which in turn has greatly improved the traffic and transportation efficiency of a stream of people and logistics. In addition, the rapid development of mobile Internet has further amplified the comprehensive utility of infrastructure. The massive online market created by e-commerce platform enterprises not only meets the huge and continuously upgraded consumer demand, but also effectively matches the demands of both the supply and demand sides, providing grassroots entrepreneurs with a low threshold for entrepreneurial channels. In addition to powerful transaction functions, major e-commerce platforms are also technical service providers for acquiring various information, learning production and operation skills, and conducting data analysis, which can effectively solve the difficulties of grassroots entrepreneurs' lack of knowledge (Hong and Cui 2017) and continuously promote the diffusion of Internet knowledge and the upgrading of the Taobao Village industry.

12.4.4 Governments Support

In the course of development, Taobao Villages have received strong support from different levels of government. In the early stage of the development of the rural e-commerce industry, because of its small scale and weak performance, it is usually difficult to attract the attention of superior governments, and it is mainly the local governments that exert administrative advantages to support it. For example, local governments provide entrepreneurs with assistance in online shop registration, start-up loans, technical training, which promote the rapid development of the rural e-commerce industry and result in significant performance. Typical examples include the Daji Town Government of Cao County, Heze City, the Jianchi Town Government of Yunxi County, Shiyan City, the Sucheng District Government of Suqian City, and the Jieyang Municipal Government. With the rapid growth of the rural e-commerce industry, Taobao Villages have gradually attracted the attention of all sectors of society and senior leaders. In November 2014, Premier Li Keqiang visited the "First Village of Online Shop," Qingyanliu Village, in Zhejiang Province,[7] and introduced this case to the world at the Davos Forum in January 2015.[8] The Prime Minister and the central government's recognition of poverty alleviation and rural revitalization driven by e-commerce began to encourage governments at all levels to participate in the development and cultivation of Taobao Villages actively. With the support of a large amount of funds and policies, various public service products began to enter the countryside continuously. Road traffic, land supply, communication facilities, and other aspects were constantly improved. The rural e-commerce industry began to spread from Taobao Villages to a larger space.

With the continuous development of e-commerce industry, the improvement of living facilities and the optimization of the living environment have become important demands for the further development of Taobao Villages. At present, the spatial transformation of a large number of Taobao Villages is limited to the production space, while the living space essentially stays in the original state, and some are even worse. Such spatial representation of unbalanced development is caused by the lack of coordination of the current governance system inside and outside of the

[7]In November 19, 2014, Premier Li visited the "First Village of Online Shop" (2014-11-20). Retrieved from http://www.gov.cn/zhuanti/2014-11/20/content_2781163.htm

[8]In January 1, 2015, Premier Li praised Taobao village in Davos and continued to promote mass entrepreneurship (2015-1-23). Retrieved from http://www.gov.cn/zhuanti/2015-01/23/content_2808977.htm

Taobao Villages. Compared with the "governance vacuum" status of ordinary villages, with the emergence of Taobao Villages in recent years, the attention that governments at all levels have paid to them is continuing to increase, and the national governance system is sinking down again. Additionally, due to the development of the industry, the village itself has activated the villagers' subjective consciousness and no longer hold an indifferent attitude toward village affairs (Chen, Luo and He 2017). In addition, the presence of upstream and downstream suppliers, express companies, financial credit, and other external market players enriched the participation of rural governance to a certain extent. At the same time, the development of Taobao Villages attracted extensive attention and deep participation from many news media outlets, experts, and scholars. So, the governance of Taobao Villages has currently formed a diverse group of participants, and rural construction has a very good foundation and opportunities. However, because of the differences in the power, responsibility, and profit of all parties, there are also differences in the capacity and vision of participating in governance. To achieve co-governance, it is necessary to promote the establishment of a modern governance mechanism. In the process of the modernization of rural management, the village Party branch and village committee are the key agents. Whether the village Party branch and village committee can take advantage of the situation and form an effective governance mechanism will directly determine the sustainable development of Taobao Villages. Active and promising village Party branches and village committees need to effectively integrate the input of governments at all levels with the forces within the village, so as to form a strong force to promote the continuous upgrading of Taobao Villages.

12.5 The Governance and Planning of Chinese Taobao Villages

12.5.1 Governance Dilemmas of Chinese Taobao Villages

In the process of transforming the rural economic and social structure, e-commerce Taobao Villages also promote the change of rural governance structure, which is visually manifested as the rapid disintegration of the pyramid structure with the township government as the top, the village political organization as the middle layer and the majority of the villagers as the bottom. The multi-subject governance pattern, consisting of local government, village committees, e-commerce elites and villagers, has been formed. The "vacuum" dilemma of traditional rural governance situation has been alleviated to some extent, which not only realizes the two-way reconstruction of the governance structure inside and outside the village, but also promotes middle governance agents which coordinate and communicate between the internal and external agents of the village. On the one hand, the development and labor division of e-commerce industry reshaped the economic relation between villagers, which represented as the disintegration of unitary and homogeneous economic structure based on traditional agricultural activities, and the formation of diverse and heterogeneous economic structure based on e-commerce industry. In pursuit of greater policy convenience and return of benefits, e-commerce practitioners start to participate in the village governance with greater enthusiasm, and express their own interest claims through political election, elite agent, media opinion, asylum expression, and resistance expression, etc. On the other hand, compared with urban areas, in the vast rural areas of those underdeveloped areas, both the supply of hard infrastructure, such as land, Internet, and transportation, and the soft infrastructure, such as human capitals, intellectual resources, financial support, and policy support, are often insufficient. These defects have become the main obstacles restricting the upgrading and sustainable development of Taobao Villages, and cannot be solved by individual village or even town. So, it is necessary to pull in higher level government and transform the focus and methods of external governance. In this process of transformation, the recognition of the remarkable poverty alleviation effect and wealth-enhancing performance of Taobao Villages from the central government of China further accelerated the restructuring of the service-oriented governance system led by the local government. In particular, the policy support and infrastructure construction by the all levels of government, including the government of city, county, town, and village, have completely changed the traditional characteristics of rural external governance. In a word, with the coexistence of the diverse subjects, the governance pattern of Taobao Village has undergone the transformation from traditional pyramid structure to diversified network structure.

The continuous development of e-commerce industry puts high demands on rural spatial construction, infrastructure, and public services, however, most of Taobao Villages failed to transform the governance pattern timely. The interests of various subjects are not well communicated and coordinated, then presenting a situation in which the subjects competed fiercely for their own interests respectively, while the village as a whole is caught in a collective action dilemma. The main phenomenon are as follows: (1) The "involution" of e-commerce industry development under excessive competition. Due to the lack of effective industrial governance measures, many Taobao Villages have fallen into the dilemma of "involution" under fierce and homogeneous competition, which means, although the economic aggregate of the e-commerce industry is still growing, but most e-commerce practitioners did not obtain any more profit from it, and even have the problem of gradually reducing revenue; (2) The problem of disordered spatial construction caused by weak control, such as the spontaneous expansion of the residential site by the villagers, the construction of the plant along the traffic road, and the inefficient construction of the e-commerce industrial park; (3) The transfer of rural public service functions under the imbalance of supply and demand. The absence of the public goods, which should be provided by the government and the village committee, makes Taobao Villages become places for the production, processing, sales and service functions of the e-commerce industry, while the service functions related to living are gradually transferred due to the mismatch between supply and demand. As the income of the villagers increases, so does the need for their children to receive high-quality education. Due to the lack of high-quality schools and good educational facilities in towns and villages, most villagers choose to send their children to the county and even to the city for education, resulting in a "reversed left-behind" phenomenon, that is, the labor force stays in the countryside to carry on economic activities, while the elderly and children are "left-behind" in the city (Chen, Luo and He 2017). In order to cope with these governance dilemmas, the comprehensive and effective planning of Chinese Taobao Villages, Taobao Towns and Taobao Village industry clusters is necessary.

12.5.2 *Planning of Chinese Taobao Villages*

12.5.2.1 Planning Transition of Taobao Villages

The rapid development of e-commerce industry of Taobao Villages reshaped the rural spatial structure dramatically, which on the one hand, manifests the huge vitality of rural development, while, on the other hand, it also leads to serious rural problems, which mainly is the problem of disordered spatial construction caused by weak control. Meanwhile, different Taobao Villages with specific industries, products and process procedures have different spatial requirements. So, it is necessary to implement different planning guidance and strategies for different kinds of Taobao Villages.

For the planning of scale-industrialized Taobao Villages, it is necessary not only to meet the spatial requirements of land expansion and centralized production in an individual Taobao Village, but also to maintain the order of rural development and optimize the rural residential environment. The important issue for the planning of this type of Taobao Village is to deal with the spatial relation of production, living and ecological function, and to rationally update and replace the stock land and incremental land of the village. For advantaged villages, further consideration should point out how to gather development resources and form an industrial spatial organization that integrates functions of production, innovation, sales and after sales service. Then, the planning of the scale-industrialized Taobao Village should also pay attention to the effective supply of living service facilities, which could optimize the rural living environment and promote the sustainable development of the village through high-quality education, medical, cultural and entertainment facilities.

For the planning of the light industry Taobao Village, the spatial demand of which is mainly based on existing land, then, there will not be a large amount of incremental land demand, but it is still necessary to pay attention to the impact of industrial development on the functional structure of villages and small towns. Firstly, rural planning must address the land requirement for production agglomeration and the corresponding supporting service. Usually, the spatial distribution of production of such Taobao Village is highly scattered, lacking the agglomeration effect and scale effect, and this could be solved to some extent through the agglomeration of production space. Furthermore, a series of e-commerce supporting services, such as innovative design, quality inspection, incubation, etc., can be implanted to promote industrial transformation and upgrading. Secondly, rural planning should pay attention to the improvement of rural residential settlements and the beautiful village construction. The improvement of the rural built environment can satisfy the villagers' pursuit of high-quality life, and enable the village to truly become a decent living place. Lastly, it should be noted that different light industries have different spatial requirements, and at the same time, Taobao Villages with different ecological bases and different locations also need to be considered respectively in specific planning practice.

For the planning of Taobao Villages that do not involve industrialization, firstly, it is necessary to improve the transportation and logistics systems between villages and towns, because this type of Taobao Villages usually has very close industrial links with towns and urban areas, so convenient transportation and logistics systems are the basis for rural e-commerce development. Secondly, the planning should leave blank land for production and storage. Although this kind of village does not involve processing in the initial stage, the demand for production space is relatively low. However, as the scale of the industry expands, more and more production and storage functions need to be satisfied. Therefore, rural planning should appropriately arrange the corresponding reserved land. Finally, similar to the above two types of Taobao Village, rural planning must pay attention to the supply of public services and the improvement of environment, and improve the quality of rural life. On the basis of improving the basic public service functions of the village, village planning should further develop the beautiful villages and characteristic towns, and realize the transformation of rural functions, and promote the development of the countryside into a settlement with distinctive characteristics and vitality.

As a bottom-up rural development path, Taobao Villages are characterized by villagers' awakening of their own interests, and fierce competition among them. In this developmental context, interest coordination, solving problems and implementation are the inevitable objectives of the Taobao Village planning, which means the planning g methods and expressions must be transformed. In terms of planning methods, it is necessary to adopt a bottom-up planning approach, and carry out collaborative planning based on village participation, and make full use of information technology to implement dynamic and accompanied planning. In terms of the planning expressions, traditional professional expressions need to be transformed into concise and implementation-oriented expressions. And also, the planning outcomes should be institutionalized and legalized to guide the development of Taobao Villages, which means the content of planning should be reflected in the public decision-making process, and a consensus on rural governance should be formed for rural planning to effectively play a public policy role.

12.5.2.2 Planning Principle and Method of Taobao Town

According to the definition of Ali Research Institute Taobao Town is a town with three or more than three Taobao Villages or achieves more than 30 million yuan sales a year on the Alibaba platform, and possesses more than 300 active online stores. Taobao Town should become a platform to promote the upgrading of Taobao Villages, and transform the current small and low-tech e-commerce industry into a specialized and high-tech one. At the same time, for the e-commerce industry with typical bottom-up development characteristics, the huge e-commerce practitioners are the main practitioners of Taobao Town. Their active participation in planning is an important guarantee for the operation and development of Taobao Town, so, the communication and collaboration between top and down must be implemented throughout the planning process. In addition, the Taobao Town is also an important carrier to show the regional cultural characteristics. Therefore, Taobao Town planning should also stress the following three corresponding aspects. First of all, Taobao Town planning should be based on the region to build a highly integrated industrial system combining online and offline, and make clear the industrial development ideas from the perspective of regional and local development, and fully consider the space demand and service function of larger industrial upgrading, and build a relatively complete and radiating industrial service system to attract talents and resources to gather in Taobao Town. Secondly, Taobao Town planning should focus on bottom-up, demand-oriented spatial customization. By dividing the target people and demand analysis, Taobao Town planning should develop a spatial function division and service matching plan, and deepen the micro-space refinement. Finally, Taobao Town planning needs to integrate the cultural and ecological characteristics of Taobao Town. On the one hand, the planning should protect, refine and display the local culture including both material cultural heritage (ancient architecture, cultural sites, cultural relics, etc.) and intangible cultural heritage (traditional performing arts, folk activities, etiquette and festivals, traditional handicraft skills, etc.) to create the cultural soul of Taobao Town. On the other hand, it should make full use of the local ecological landscape resources to create a characteristic ecological region.

12.5.2.3 Interregional Coordination and Planning of Taobao Village Industrial Clusters

The rapid development of e-commerce in the rural region has brought about industrial clusters across administrative regions, which inevitably leads to the need for interregional coordination and planning. The well-known furniture e-commerce industry cluster, Shaji-Gengche region, is a typical one. The rapid development of e-commerce industry has completely changed the development track of these two ordinary towns like the magician's point stone, and Internet informatization has driven local industrialization and urbanization to create a dazzling wealth myth. However, this development model that relies on market opportunities has brought a series of severe challenges in recent years due to large-scale replication and lack of regional coordination. The regional development lacks unified planning guidance and control, and industrial space and living space are staggered, and there is a lack of systematic integration of the transportation system and infrastructure. Fierce homogeneous competition leads to lower and lower profits, and the network business community is in a "multiple, small, scattered, and weak" situation, lacking a brand with strong appeal, and the overall development dilemma of the industry is prominent. The regional comprehensive service facilities are scattered, and the lag of public services has caused obvious "reversed left-behind" phenomenon in Taobao Villages. In a word, the Shaji-Gengche region is facing not only the upgrading and transformation challenges of individual Taobao Village and Taobao Town, but also the coordinated development challenges of the region as a whole. Obviously, such a challenge exists not only in the Shaji-Gengche region, but also in other areas where Taobao Villages and Taobao Towns are agglomerated.

The main content of interregional coordination and planning is divided into the following four aspects. (1) Defining the region of coordination. In order to define the coordination region, research on the development characteristics and trends of urban functional region is necessary. Qualitative research can make overall judgments through the characteristics and trends of adjacent industrial development and the network characteristics of regional productive service industries, and quantitative research requires further investigation about the spatial distribution and linkage characteristics of various flows carrying functional links, such as online store transaction volume, logistics express delivery, division of labor among enterprises, and the impact of major transportation infrastructure on business communication capabilities. On the basis of functional regional studies, the final definition of the region of coordination needs to fully consider the operability of institutional design and select appropriate spatial scales according to the level and status of the coordination agents. (2) Division of coordination units. As a relatively complete spatial unit, the coordination region based on functional linkages is likely to have multiple cross-border situations, which means, there exist different level sub-regions. Based on the principle of adapting to local conditions, interregional coordination and planning must focus on these sub-regions. Therefore, it is necessary to divide the coordination units according to the characteristics and trends of the development of different cross-border areas within the coordination region, and to carry out various planning designs and spatial arrangements with the coordination unit as the smallest unit. (3) Designing coordination mechanism. According to the development characteristics and trends of coordination region and coordination unit, interregional coordination and planning should take the coordination unit as the main carrier and make some important arrangements about long-term, structural and deterministic regional development strategies. The design of coordination mechanism should not only fully understand the industrial and spatial development characteristics and trends of the coordination unit, but also pay attention to local and private institutional innovations, and respect the "native methods" explored by grassroots entrepreneurs and excellent promoters. So that, the bottom-up power and top-down planning control could form an organic combination. (4) Formulating spatial strategies. Interregional coordination and planning not only requires institutional design, but also should guide spatial development effectively, especially the interconnection and sharing of regional infrastructure. Broadly defined infrastructure refers to the sectors and industries in the national economic system that provide general support for social production and reproduction, including transportation, post and telecommunications, water supply and power supply, commercial services, scientific research and technical services, landscaping, environmental protection, cultural education, and health services. Due to the characteristics of the development of e-commerce industry, logistics infrastructure is usually the focus and difficulty of interregional coordination planning.

12.6 Conclusion and Discussion

12.6.1 Conclusion

The main conclusions of this chapter are as follows: (1) The spatial distribution of Chinese Taobao Villages is highly concentrated in the eastern coastal areas on the national level, and decreases in "East-Middle-West" spatial gradient. On the regional level, the spatial distribution manifests the "North-Central-South" agglomeration pattern. Furthermore, Chinese Taobao Villages reveal "double-marginal" locational characteristics, that is, marginal effect on the central city level and marginal effect on the regional level of the Yangtze River Delta. In addition, the growth of Chinese Taobao Villages features fissile proliferation. (2) The pioneering rural grassroots entrepreneurs, relying on the local industrial base and resource endowment to explore specialty products, with the link capability of e-commerce platforms, assisted by the government's policy support and financial assistance, have promoted the rapid development of the rural e-commerce industry, and make the ordinary villages gradually grow into Taobao Villages. Grassroots entrepreneurship, main products, supporting facilities, and governments support are the four key factors affecting the development of Taobao Villages. Rural governance is the key factor for Taobao Villages to achieve upgrading and transformation, thus forming the diamond mechanism model for the formation and development of Taobao Villages. (3) The governance dilemmas of Chinese Taobao Villages mainly present as, the "involution" of e-commerce industry development under excessive competition, the problem of disordered spatial construction caused by weak control, and the transfer of rural public service functions under the imbalance of supply and demand. In order to cope with these governance dilemmas, the comprehensive and effective planning of Chinese Taobao Villages, Taobao Towns and Taobao Village industry clusters is necessary.

12.6.2 Discussion

As a rural development path driven by the bottom-up e-commerce industry in rural areas, Taobao Village has proved its great development potential and wide applicability in practice, and realized the partial return of urban talents, technology, capital and other development factors to rural areas, thus activating a new rural urbanization model, which provides a new possibility for rural revitalization. While, in order to investigate any economic geographic phenomenon, it is necessary to prepose the corresponding social, institutional and economic context. In the 1980s and 1990s, due to the social background of shortage economy, weak government regulation, and low production costs, enterprises in some villages and towns undergone great prosperity and development, and formed an equally impressive industrialization and urbanization process, which was also regarded as the hope of rural revitalization and a new way of modernization. However, with the large-scale transfer of global capital, technology and industry to China, the development of village and town industries were entangled into internal and external difficulties due to inherent shortcomings, such as blurry property right allocation, backward technology and management, inferior location, and then gradually became faded. After that, what remains in the rural area are sewage, soil pollution, damaged ecology and land (Luo 2020). To some extent, the development of Taobao Villages shows many similarities, which also arises in rural areas with weak government control and low cost, and causes deterioration of rural ecology and living environment. The difference is that the Internet e-commerce has brought the villages into a global division of labor and cooperation network at a greater and deeper level. As to the social background, the coexistence of the largest luxury consumption market in the world and the low-cost, low-quality industrial products of Taobao Villages is the reflection of the polarized distribution structure of Chinese society. That Taobao Village is a phased phenomenon or a long-term trend depends on the transformation of social distribution and consumption structure. If there is no substantial improvement in social distribution and consumption structure, low-cost and low-quality Taobao Villages will be a long-term economic geographic phenomenon. As to the future of different types of Taobao Villages, Taobao Villages which offer special agricultural products, traditional handicraft products and other special products, have strong rural embeddedness and then long-term vitality. However, Taobao Villages which produce low-cost and low-quality industrial products, will lack human capitals and skills for further upgrading, and have weak rural embeddedness. Most of them will gradually die out with the increasing rationality of Chinese social distribution and consumption structure. A few successful Taobao Villages will be upgraded to characteristic industrial towns and lose their rural style. Therefore, we should look upon the development of Taobao Villages with deliberate optimism.

References

Alavion, S. J., & Allahyari, M. S., 2012. E-readiness of rural ICT offices for rice e-marketing in Rasht Township, Iran. *Acta Agriculturae Slovenica*, 99(2), 175-183.

Alibaygi, A., Karamidehkordi, M., & Karamidehkordi, E., 2011. Effectiveness of rural ICT centers: A perspective from west of Iran. *Procedia Computer Science*, 3, 1184-1188.

Bai, D., & Sun, Z., 2019. The spatial organization and geographic embeddedness of Taobao Village in China. *World Regional Studies*, 28(01), 121-129. [in Chinese]

Cai, X., 2018. Research on the spatial characteristics of Taobao Village. (Master's thesis). *Guangdong University of Technology*, Guangzhou, China. [in Chinese]

Chen, F., Luo, Z., & He, H., 2016. Study on the pluralistic reconstitution of rural governance driven by e-commerce: A case study of Daji Town of Cao County, Shandong Province. *Modern Urban Research*, 10, 22-29. [in Chinese]

Chen, H., & Zhang, J., 2018. Taobao Villages: Rural development transformation driven by the space of flow. *City Planning Review*, 42(09), 99-107. [in Chinese]

Chen, R., 2016. Self-consciousness and rebuilding of rural areas: social analysis of the phenomenon of Taobao Village. *Journal of Huazhong Agricultural University (Social Sciences Edition)*, 3,74-81+134-135. [in Chinese]

Chen, T., 2011. From Rubbish Village to Taobao Village. *For the Public Good3*, 54-57. [in Chinese]

China Internet Network Information Center., 2019. "The 43rd China Statistical Report on Internet Development". available at: http://www.cnnic.net.cn/hlwfzyj/hlwxzbg/hlwtjbg/201902/P020190318523029756345.pdf [Accessed 10 Sep. 2019]. [in Chinese]

Dai, Y., & Zhang, X., 2017. The spatial planning innovation of rural e-commerce town: A case of Shaji Town, Xuzhou City. *Urban and Rural Development7*,40-41. [in Chinese]

Dalvit, L., Muyingi, H., Terzoli, A., & Thinyane, M., 2007. The deployment of an e-commerce platform and related projects in a rural area in South Africa. *Strengthening the Role of ICT in Development*, 27.

Diao, B., Chen, K., & Ding, L., 2017. Spatial pattern and influence factors of Taobao Village in 2013 and 2014. *Tropical Geography*, 37(1), 56-65. [in Chinese]

Emmanuel, E. A., & Muyingi, H. N., 2010. A mobile commerce application for rural economy development: a case study for Dwesa. *In Proceedings of the 2010 Annual Research Conference of the South African Institute of Computer Scientists and Information Technologists*. 58-66.

Fan, Y., Yao, M., & Wu, W., 2018. Research on the inclusive innovation model and mechanism of Taobao Villages in China. *Issues in Agricultural Economy*, 12, 118-127. [in Chinese]

Fang, G., 2016. Taobao Village and its possibility in breaking through the dilemma of urbanization in rural China: From the perspective of urban sociology. *China Rural Survey*, 3, 71-81+96-97. [in Chinese]

Fong, M. W., 2009. Digital divide between urban and rural regions in China. *The Electronic Journal of Information Systems in Developing Countries*, 36(1), 1-12.

Fu, Z., 2019. Study on the classification of development modes of the Taobao Village in China (Master's thesis). *Nanjing University*, Nanjing, China. [in Chinese]

Ghosh, A., 2011. Initiatives in ICT for Rural Development: An Indian Perspective. *Global Media Journal: Indian Edition*, 2(2).

Guo, X., 2016. Study on rural in-situ urbanization under the e-commerce: Take Daji as an example. (Master's thesis). *Nanjing University*, Nanjing, China. [in Chinese]

Hagerstrand, T., 1968. Innovation diffusion as a spatial process. Chicago, IL:University of Chicago Press.

Hindman, D. B., 2000. The rural-urban digital divide. *Journalism & Mass Communication Quarterly*, 77(3), 549-560.

Hong, W., & Cui, P., 2017. An empirical study on the relationship between trading platform, special knowledge and flexible production based on the investigation of Cao County, the Taobao village. *China Business and Market*, 31(01), 122-128.

Hu, Y., & Liu, L., 2016. Spatial distribution and influencing factors of Taobao Village in Guangzhou. *Planners*, 32(12), 109-114. [in Chinese]
Islam, M. S., & Grönlund, A., 2010. An agricultural market information service (AMIS) in Bangladesh: evaluating a mobile phone based e-service in a rural context. *Information Development*, 26(4), 289-302.
Jalali, A. A., Okhovvat, M. R., & Okhovvat, M., 2011. A new applicable model of Iran rural e-commerce development. *Procedia Computer Science*, 3, 1157-1163.
Jamaluddin, N., 2013. Adoption of e-commerce practices among the Indian farmers, a survey of Trichy district in the state of Tamilnadu, India. *Procedia economics and finance*, 7, 140-149.
James, J., 2010. Mechanisms of access to the Internet in rural areas of developing countries. *Telematics and Informatics*, 27(4), 370-376.
Lei, B. & Liu, M., 2017. Formation mechanism of e-commerce industrial cluster in rural area: Based on cases of typical Taobao Villages. *Science and Technology Management Research*, 37(11), 177-184. [in Chinese]
Li, Y., & Zhang, Y., 2015. Analysis of the role of local government in the development in the Taobao Village: Taking Junpu Village in Guangdong Province as an example. *Science and Technology Management Research*, 333(11), 174-178. [in Chinese]
Liang, C., 2018. Reform and opening up and the Internet are all marginal revolutions. Retrieved from http://www.aliresearch.com/blog/article/detail/id/21517.html. [Accessed September 21, 2019] [in Chinese]
Lou, J., & Hu, D., 2018. Taobao Village, simultaneous urbanization and new urbanization practice. Academic Research, 5, 58-62. [in Chinese]
Luo, Z. & He, H., 2017. New urbanization from below in China: Rural urbanization driven by e-commerce. *City Planning Review*, 3, 31-40. [in Chinese]
Luo, Z., 2020. New bottom-up urbanization: The development and governance of Chinese Taobao Villages. Nanjing, China: *Southeast University Press*. [in Chinese]
Luo, Z., Zhu, C., & Xue W., 2015. Re-examination of the spatial structure of the Yangtze River Delta based on high-speed rail passenger flow. *Shanghai Urban Planning*, 4, 74-80. [in Chinese]
Mao, P., 2018. The development characteristics and mechanism of Taobao Villages in the "Internet +" context. *Journal of Commercial Economics*, 18, 131-133. [in Chinese]
Qi, J., Zheng, X., & Guo, H., 2018. The formation of Taobao villages in China. *China Economic Review*106-127.
Qiao, Y., 2019. The diffusion and upgrade of creative knowledge of Taobao Villages: A proximity perspective (Master's thesis). Nanjing University, Nanjing, China. [in Chinese]
Shan J., & Luo Z., 2017. Agglomeration and fission: Spatial distribution characteristics and evolution trends of Taobao Villages and Towns. *Shanghai Urban Planning Review*, 2, 98-104. [in Chinese]
Shi, X., Wei, T., & Liu, Q., 2017. Research on the development model and spatial difference of rural e-commerce industrial cluster. *Modern Economic Research*, 11, 124-131. [in Chinese]
Wang, L., Yun, Y., & Ni, J., 2017. The geo-space of Taobao Village in China: Theoretical framework based on space of flows. *City Planning Review*, 41(6), 27-34. [in Chinese]
Wu, X., Yuan, Z., & Zhu, H., 2015. Global information network and the social and cultural construction of rurality: A case based on Taobao Village in Guangzhou. *Journal of South China Normal University (Natural Science Edition)*, 47(02), 115-123. [in Chinese]
Xin, X., & Qiao, J., 2018. Formation mechanism and spatial-temporal evolution of Taobao Villages agglomeration. *Areal Research and Development*, 37(1), 11-15. [in Chinese]
Xu, C., Lv, B., & Wen, T., 2015. New patterns of county in-situ urbanization and rural development based on e-commerce. *Urban Planning International*, 30(1), 14-21. [in Chinese]
Xu, J. Luo, Z., He, H., & Zhou, Y., 2017b. Research on the spatial characteristics and influence factors of county e-commerce development in China. *Shanghai Urban Planning Review*, 2, 90-97. [in Chinese]
Xu, Z., Wang, Z., Zhou, L., & Wang, H., 2017a. The spatial distribution characteristics and driving factors of Taobao Village in China. *Economic Geography*, 37(1), 107-114. [in Chinese]

Yang, S., Li, Y., Wie, Z., & Chen, T., 2016. Spatial transition and restructuring of Taobao Villages in Internet+ Era. *Planners*, 32(5), 117-123. [in Chinese]

Yu, H., Wang, C., & Zhao, Z., 2018. Research on the motivation and mechanism of rural e-commerce entrepreneurship agglomeration: The case study of Shuyang Taobao Village. *Business Management Journal*, 40(12), 39-54. [in Chinese]

Zeng, Y., 2018. The formation of agricultural Taobao Village cluster and its impact on small holders' income (Doctoral dissertation). *Zhejiang University*, Hangzhou, China. [in Chinese]

Zeng, Y., & Guo, H., 2016. The development mechanism of e-commerce association's promoting effect on Taobao Villages: A case study of Junpu village, Jieyang City, Guangdong Province. *Chinese Rural Economy*, 6, 51-60. [in Chinese]

Zhang, J., & Qian, X., 2015. Research of the spatial transformation of Taobao Village in Information Age. *Urban Development Studies*, 22(10), 81-84. [in Chinese]

Zhang, J., Qian, X., Chen, Y., & Jiang, Y., 2016. The spatial change of Taobao Village in Lirengdong, Guangzhou in the perspective of spatial production. *Economic Geography*, 36(01), 120-126. [in Chinese]

Zhang, S., 2015. Research on Internet mechanism between e-commerce and regional development: A case study of Lishui Prefecture Zhejiang Province (Master's thesis). *East China Normal University*, Shanghai, China. [in Chinese]

Zhao, J., Ding, J., & Wang, X., 2017. Spatial and temporal distribution and evolution characteristics of Taobao Villages in China. *World Regional Studies*, 26(6), 73-82. [in Chinese]

Zhou, J., Yang, Z., & Gao, W., 2017. The development characteristics and mechanism analysis of Taobao Villages in Jiangsu Province under e-commerce economy. *Urban Development Studies*, 24(2), 9-14[in Chinese]

Zhou, Y. & Liu, C., 2018. Analysis on the aggregation of farmers' e-commerce and influencing factors in "China's Taobao Villages": A survey of Shaji Town and Yanji Town in Jiangsu Province. *South China Journal of Economics*, 37(1), 62-84. [in Chinese]

Zhu, Y., Song, Y., Li, G., & Yu T., 2016. Spatial aggregation pattern and influencing factors of Taobao Village in China under the C2C e-commerce mode. *Economic Geography*, 36(4), 92-98. [in Chinese]

Zhuang, Z., 2005. Entrepreneurship spirit, continuous technological innovation and the micro-mechanism of long-run economic growth. *Journal of World Economy*, 12, 32-43[in Chinese]

Chapter 13
Innovation Districts in Beijing: Evolution, Distribution, and Development Mechanisms

Yuan Xiaohui, Ren Junyu, Gu Chaolin, Shi Xiaodong, Liu Xiyu, Chen Jun, and Wang Liang

Abstract Innovation acts as a crucial role in urban economy development. As the technology and education center in China, Beijing has developed diverse types of innovation districts in the past few years, which have supported all kinds of innovation activities giving new impetus into urban growth. We reviewed the forming process of innovation districts in Beijing to reveal the development characteristics of four stages. We also show the distribution of innovation districts in Beijing recently, including six main types varied from science and technology park, university park, science city, entrepreneurship community, characteristic town to science and technology building. The newly developed entrepreneurship community has several new features. Based on the generalization for development mechanisms of innovation districts, planning policies for all kinds of innovation districts have been discussed. In general, the innovation districts in Beijing are experiencing the transformation from traditional science and technology parks to innovation districts with enterprises as its core, from government leading to enterprise and community leading, from spatial agglomeration to spatial decentralization.

Keywords Innovation districts · Innovation space · Science park · Entrepreneurship community

---

Y. Xiaohui (✉)
Beijing QuantUrban Tech Co. Ltd., Beijing, China
e-mail: yuanxiaohui@quanturban.com

R. Junyu
Peking University Shenzhen Graduate School, Shenzhen, China

G. Chaolin
Architecture School, Tsinghua University, Beijing, China

S. Xiaodong · C. Jun · W. Liang
Beijing Municipal Institute of Urban Planning & Design, Beijing, China

L. Xiyu
Beijing Tsinghua Tongheng Urban Planning & Design Institute, Beijing, China

L. Bian et al. (eds.), *Chinese Urban Planning and Construction*,
Strategies for Sustainability, https://doi.org/10.1007/978-3-030-65562-4_13

13.1 Introduction

According to China National Bureau of Statistics[1], China's GDP reaches $13.2 trillion in 2018, second in the world following USA. During decades of rapid economic growth, China invests a lot in research and development (R&D) and related areas to support innovation. In 2018, China is listed in the top 20 most innovative countries in the world according to Global Innovation Index (2018),[2] as the only Asian country in the top 20.

As the capital of China, Beijing has the leading innovation resources and capabilities. According to the Beijing Municipal Commission of Science and Technology[3](BMCSTC, in short), Beijing has more than 90 universities, more than 1000 research institutes, 120 state key laboratories, 68 national engineering research centers, and more than 20,000 state-level high-tech enterprises by 2018, all ranking first in China. Meanwhile, Beijing has 82 unicorn enterprises, accounting for half of the country. In 2018, The annual number of patent applications and grants in Beijing is 211,000 and 123,000 respectively also leading the country. According to the research of Global Innovation Index (2018), Beijing is among the top five leading science and technology innovation clusters and plays an important role in worldwide innovation ecosystem (Dutta et al. 2018).

The government has pledged to transform Beijing into a global powerhouse in innovation by 2050. According to Xinhua News Agency, the Communist Party's Central Committee and the State Council published a guideline for innovation in the world's second-largest economy, which is also a blueprint to develop a growth model based on quality, rather than quantity. It aims to be among "the front-runner countries in innovation" by 2030.

The innovation development of Beijing has been studied by scholars from different aspects, including education and research institutions (Guan et al. 2005; Chen and Kenney 2007), enterprise clusters (Zeng et al. 2006), innovation capability (Yam et al. 2004). Some papers emphasize the role of government policy in innovation development and economic development (Guan 2002; Segal 2003). Meanwhile, some studies pay attention to the intrinsic relationship between the process of space production in innovation development and the effect on urban development (Guan et al. 2009), such as the development of Science park (Tan 2006), High-Tech Industrial Parks (Wang and Leng 2012) in Beijing.

Before 2000, the development of innovation districts in Beijing were mainly influenced by the government. In the early 1990s, Chinese government started to establish science and technology parks in 53 major Chinese cities under the Torch Program and "863" Plan which is a science and technology initiative to promote technology transfer and innovation. The objective was to build within these parks a concentration of high-tech companies through policy incentives such as tax

[1] http://www.stats.gov.cn/

[2] https://www.globalinnovationindex.org/Home

[3] http://kw.beijing.gov.cn/

holidays (DeBresson and Hu 1996). Preferential government policies raise the private return to investment in the parks and therefore lead to rapid growth through expeditious resource accumulation (Hu 2007). The development of Zhongguancun Science Park in Beijing is the representative achievement for the Torch Program. Encouraged by all kinds of policies, a number of high-tech companies were born in the park with technologies transferred from the leading universities and research institute nearby.

Over the past two decades, the world has witnessed China's transformation from a planned economy into a more market-driven economy. The urban space supporting innovation activities is also gradually changing, from the previous government-led, planned and centralized development to more market-oriented, bottom-up, centralized-decentralized development. A number of scholars have studied the mechanism of innovation development in Beijing, however, there lacks an overall generalization and deep exploration on the space which supports innovation activities. What's more, some existing studies have well explained the government-led model of science parks and industrial parks, but they didn't cover the emerging types of innovation districts which are more market-driven and bottom-up. As urban space plays a more and more crucial role in promoting innovation, not only as a place where innovation resources aggregate, but also as a mechanism to inspire innovation activities, it is of great value to study the process, characteristic and mechanism of the development of innovation space. This paper will study the innovation districts in Beijing, including the formation process, characteristic of different development modes, and development mechanisms. It will not only help us better understand recent economic development of Beijing as a capital city, but also can contribute to the theory of innovation and space.

13.2 Definition and Method

13.2.1 Definition of Innovation Districts

The concept of innovation districts attracts a series of discussions these years as a new model for urban innovation. It was firstly proposed in the research for 22@ Barcelona by Leon (2008a, 2008b) who regards it as the new knowledge hub of the city through a regeneration plan. According to the researcher, through the transformation of the physical infrastructure, industry clusters are created which bring together industries, universities, research centers and technology transfer programs that provide all kinds of support for startups. Brookings Institute depicts the rise of innovation districts in America as a new urban model for supporting innovation compared to the suburban model such as Silicon Valley. They define the innovation districts as "geographic areas where leading-edge anchor institutions and companies cluster and connect with start-ups, business incubators, and accelerators." And also, this new urban model is "the manifestation of mega-trends altering the location preferences of people and firms, and in the process, re-conceiving the very link between economy shaping, place making and social networking" (Katz and Wagner 2014).

Some scholars consider innovation districts as a kind of economic development strategy which is non-participative top-down (Morisson and Bevilacqua 2018), whereas Esmaeilpoorarabi et al. (2019) emphasize the nexus of knowledge-based development where public and private actors work together. Baily and Montalbano (2018) proposes the specific place characteristics of innovation districts to foster, attract and retain talent and investment, and consequently accelerate their socioeconomic performance. Lawrence et al. (2019) focus on the influence of innovation district and indicate "the innovation district has impacts beyond the boundaries of the district through knowledge spillover in the labor market and the value chain."

We want to address innovation district as a broader concept rather than a specific model of urban area innovative development such as an urban regeneration plan or a top-down strategy. According to Kevin Lynch, district is a relatively large urban area with certain common characteristics, people can enter and leave freely and recognize its features; externally, it can be treated as a whole (Lynch 1960). Therefore, the innovation district can be understood as an urban area that can inspire, cultivate and support innovation no matter the scale of the urban space. It is the area that people in the city recognized as innovation-driven where diversified innovation participators and innovation activities aggregates or the area that people think has the great potential to be innovative in the near future.

13.2.2 Data and Method

13.2.2.1 Data Sources

To reveal the spatial distribution relationship of innovation participators and innovation districts, we acquire the data of innovation participators from BMCST. The data includes the name lists including Beijing 2018 high-tech companies accredited by BMCST, Beijing engineering tech centers, Beijing key laboratories, Beijing company level R&D centers, Beijing maker spaces, Beijing tech company incubators and Beijing international tech cooperation bases, etc. We further match these name lists with the company registration information database to get the detailed address for related innovation contributors, in order to illustrate the spatial distributions.

We also get some boundaries of the innovation districts from Beijing Municipal Commission of Urban Planning and Natural Resources including three science cities, Yizhuang economic and development zone and Zhongguancun science park to depict the spatial distributions of the innovation districts in Beijing.

13.2.2.2 Literature Research

Since we try to explore the influence of urban planning on the formation of innovation districts in Beijing, we select all the previous versions of Beijing comprehensive planning as research materials. Based on these literatures, we can tell the determined power shaping the innovation districts and historical evolution process of the innovation districts in Beijing.

13.2.2.3 Case Study

To get the newest progress about innovation districts in Beijing, we conduct some detailed case study on different types of innovation districts such as science cities, entrepreneurship communities and office buildings for tech companies. We carry out some field investigations and interview some managers for the innovation districts to help us clarify the internal mechanisms and leading power for innovation district development.

13.3 Formation of Innovation Districts in Beijing

The innovation districts in Beijing have a long history evolving which can be divided as four stages.

13.3.1 The First Stage: Science and Education Park (1949–1982)

From 1949 to 1982, the cultural and educational area in Beijing has been constructed in the northwest area and the scientific and educational park began to take shape during that time.

In May 1949, Beijing municipal government, in the preparation of Beijing comprehensive planning research, agreed that: "In addition to the administration center, Beijing shall not only be the cultural, scientific and artistic city but the big industrial city." It proposed that "colleges and universities shall be located in the northwest and adjacent to the scenic rest area." In Nov. 1953, Beijing municipal government reported the Reconstruction and Extension of Beijing Draft Plan to the central government and proposed that "build Beijing as China's strong center of science and technology." In Apr. 1955, Beijing municipal government has formulated the Primary Scheme for Beijing Comprehensive Plan (Draft). It proposed that "to build Beijing as the center of science and technology and make it stand at the forefront of China's technological and cultural revolution."

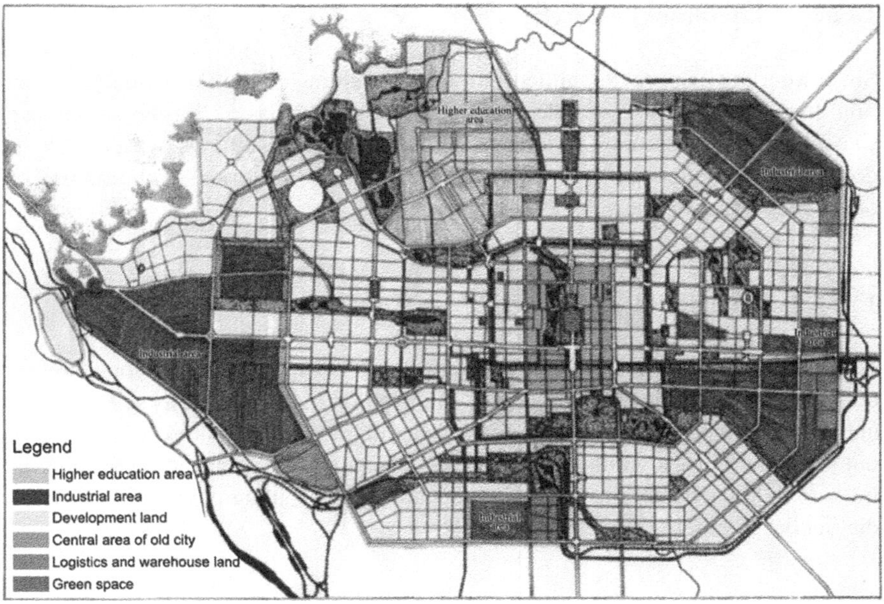

Fig. 13.1 Beijing Scheme of Comprehensive Planning (Revised Draft in 1954)
Source: Beijing Municipal Institute of Urban Planning & Design

With the development of Beijing scientific research and university construction, the fast development of scientific and educational district in Beijing laid the foundation of the subsequent evolution and innovation (Fig. 13.1).

13.3.2 The Second Stage: High-Tech Production Bases (1982–2003)

From 1982 to 2003, a number of high-tech tech industrial districts have been built to explore the economy suitable for the capital.

In Jul. 1983, the Scheme for Beijing Comprehensive Plan (Draft) proposed that "Beijing shall become China's best city of science and technology in the future." In Oct. 1993, official apply of Beijing Comprehensive Plan (1992–2010) proposed that "arrange necessary space for the development of high-tech industry." With the continuous improvement of the capital economic system, the "production space" supporting new industry has become the theme of innovative development in this period. Through nearly two decades of construction, the economic and development zones and science parks have become the new space of urban innovation in Beijing which expanded from urban areas to suburbs (Table 13.1).

Table 13.1 New areas of innovation set up in Beijing during 1991–1994

New areas of innovation set up in Beijing	Level	Time
Beijing Yizhuang economic development zone	National level	1991
Beijing Shilong industrial development zone	City level	1992
Beijing Liangxiang industrial development zone	City level	1992
Beijing Yanqing economic development zone	City level	1992
Beijing Shilong economic development zone	City level	1992
Beijing Daxing industrial development zone	City level	1992
Beijing Yanqi industrial development zone	City level	1992
Beijing Badachu High-Tech Park	City level	1992
Yongle economic development zone	City level	1992
Beijing Linhe industrial development zone	City level	1993
Beijing Xinggu economic development zone	City level	1993
Beijing Miyun industrial development zone	City level	1993
Beijing Tongzhou industrial development zone	City level	1994
Beijing Tianzhu airport industrial zone	City level	1994
Beijing Badaling economic development zone	City level	1994
Finance street	–	1992
Business district in Liangmahe and Yansha	–	1994

13.3.3 The Third Stage: R&D Parks (2004–2012)

From 2004 to 2012, urban R&D parks gradually integrated production, research and education functions with Zhongguancun as its core.

In 2005, the state council has issued the Official Reply of Beijing Comprehensive Plan proposed that Beijing is the "gathering place of national level universities and scientific research institution" and the "high-tech base for innovation, R&D, and production," which marks the new development stage of urban innovation space. Since 2000, Zhongguancun has been formed "1 zone 16 parks" (one demonstration zone and 16 science and technology parks) and built into the most intensive region of intelligent resource in our country and even the world. In Mar. 2009, the State Council has issued the Official Reply of Agreeing to Support the Construction of

National Independent Innovative Demonstration Zone in Zhongguancun Science and Technology Park and pinpointed the establishment of Zhongguancun as the center of scientific and technological innovation with global influence.

13.3.4 The Fourth Stage: Innovation Districts (2012-Present)

After 2012, The diversified innovation districts are formed under the new state policy of "innovation and entrepreneurship.

In Mar. 2013, Beijing proposed to build "one science city and three high-tech serving streets." In Jun. 2014, Zhongguancun entrepreneurial street is officially opened, gathering Garage Coffee, 3 W Coffee, IC Coffee, Hackerspace and other entrepreneurial "dream factories." In Sep. 2016, the State Council has issued the Overall Plan for Strengthening the Construction of National Science and Technology Innovation Center in Beijing and pinpointed the positioning of national science and technology innovation center in Beijing: the global leader in scientific and technological innovation, the growth pole for high-end economy and preferred location for innovative talents, the pilot area of cultural innovation and the exemplary city of ecological construction. Beijing Comprehensive Planning (2016–2035) further proposes the "Three Science Cites and One Economic Development Area" as the main platforms to support the development of national science and technology center, and based on that, diversified innovation districts are also encouraged to develop. At this new stage, the innovation districts integrate diversified elements including government, industry, university, research, investment, trade and media.

13.4 Distribution and Characteristic of Innovation Districts in Beijing

13.4.1 General Characteristic of Innovation Participators and Innovation Districts in Beijing

Innovation participators are more diversified. Compared with the years before 2012, there are more different participators in the innovation arena, no longer confined to the universities, state-owned enterprises, national or city sponsored research institute or that affiliated to the ministries and commissions under the State Council. Private companies, foreign investment companies and even individual entrepreneurs are joining the competition for innovation and they contribute a lot in promoting the innovation-based economic development. Among the 21,606 high-tech companies accredited by BMCST in late 2018, only around 2% of them are state-owned or collective owned enterprises and the rest 98% are all private companies who show strong capacity in innovation. Figure 13.2 shows the distribution of some

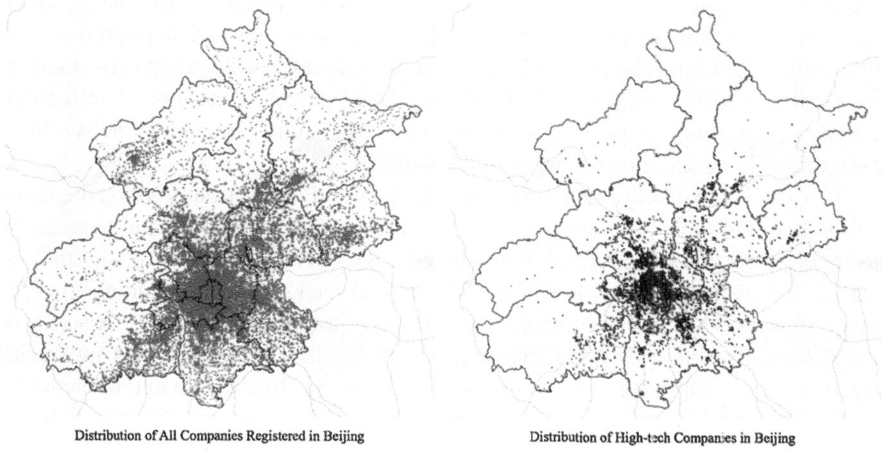

Fig. 13.2 Distribution of all companies and high-tech companies in Beijing

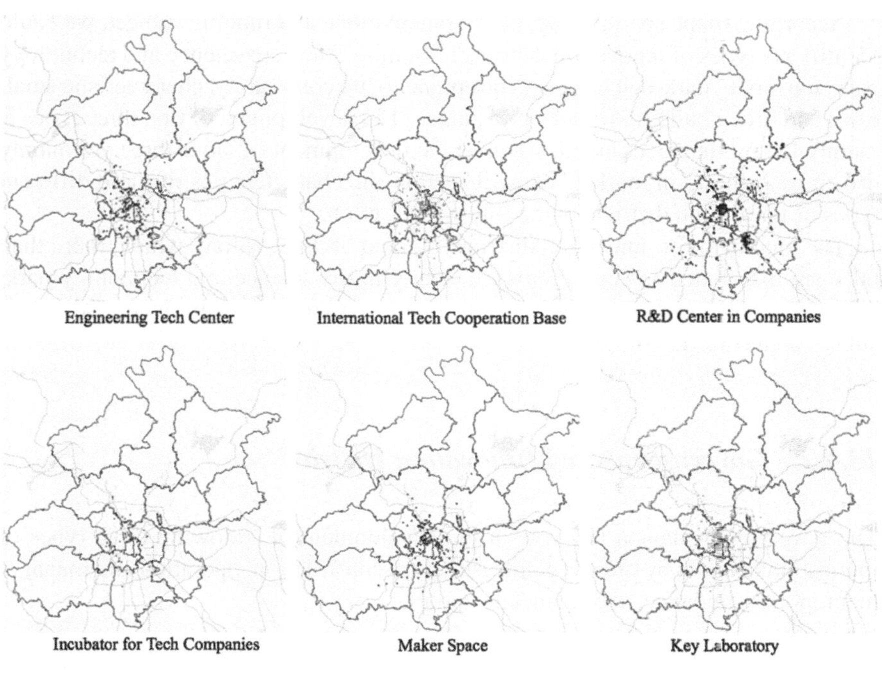

Fig. 13.3 Distribution of innovation participators

representatives of innovation participators, including the engineering tech centers, international tech cooperation bases, R&D centers in companies, incubators for tech companies, maker spaces, key laboratories (Fig. 13.3).

Innovation behaviors are more active ranging from science and technology innovation to business model innovation. There emerging specialized services covering the whole life cycle of innovation, research and development, fund raising, talent recruitment, prototype production, growth hack, marketing and networking. The innovation and entrepreneurship ecosystem gradually take shape in market. Especially with the background of Internet development, a number of influential companies are born in Beijing, bringing innovation achievements to the world in both technology realm or business model realm.

As a spatial projection of innovation behaviors and innovation process, the innovation districts show the transformation trend from government-led and aggregated to market-driven and separated. The subjects who are responsible for or running certain innovation districts have more diversified backgrounds, including government, universities, private companies and foreign investment institutes. A series of innovation districts become more open and dynamic which integrated more and more resources supporting innovation and entrepreneurship. Compared to the traditional designated science park development mode led by government, the distribution of innovation districts is more separated due to the increasing various subjects who will follow the market opportunity to decide the location of the innovation districts.

According to the spatial scale, development mode and running subject, we could identify six types of innovation districts in Beijing. They are science and technology park, university park, science city, entrepreneurship community, characteristic small town and office building for tech companies. The development of first three types is mainly led by the government, whereas the development of latter three are mainly driven by subjects in market. We will specify the characteristics of these different types of innovation districts in the following section.

The total land for innovation districts is about 38 km^2, among which, more than 60% are distributed in central city, especially in the science and technology park. The rest of them are located in new town and small town in suburb area and outer suburb area (Fig. 13.4).

13.4.2 Government-Led Innovation District

The government plays a key role in the development of following three types of innovation districts by planning, infrastructure construction, operation and management, policy guidance, and financial support.

13.4.2.1 Science and Technology Park

The science and technology parks, especially Zhongguancun science parks constitute a major part of innovation districts and contributes a lot in supporting all kinds of innovation activities in Beijing. It takes up 30% of construction land of the whole city and accounts for 70% of the total industrial land where 90% of the high-tech

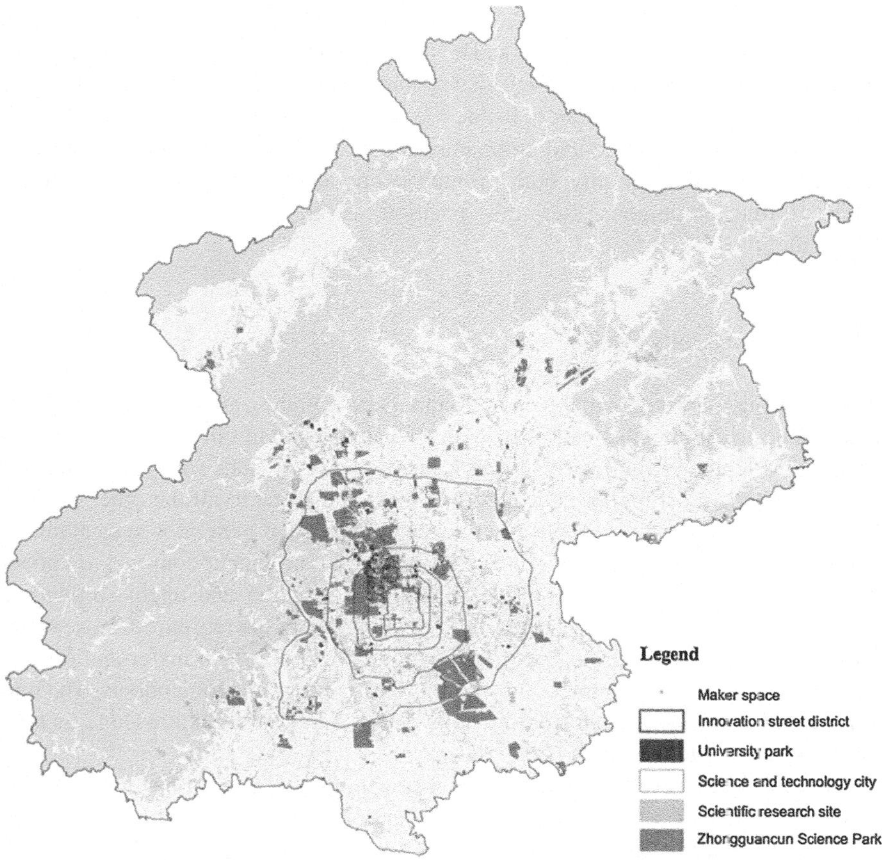

Fig. 13.4 Innovation districts layout in Beijing

companies accredited by BMCST locate. The economic development of Zhongguancun science park attributes to 42% of the economic growth of Beijing. More than 60% granted patents are born in it.

The spatial structure of Zhongguancun science park has evolved to "one zone and sixteen parks." One zone means one National Independent Innovation Demonstration Zone and sixteen parks stand for the sciences parks in the sixteen administrative districts. Among all the parks, Haidian, Yizhuang and Chaoyang parks are the first three parks who contributes most innovation achievements and GDP.

The development of Zhongguancun science park has long been led by government, and these years more and more private participators join the construction and development of the science park. The model of gated park with single industrial function gradually transformed into the model of open park with various functions integrated, including R&D, business, exhibition, entertainment, communication and even residence, which provide a creative environment and innovation ecosystem for the workers and residents in the park. The former industrial areas turn into the micro innovation centers of the city which also reshape the spatial structure of Beijing and complete the urban unit function. Inside the park, the connections between the universities, research institute and companies are strengthened, and the international innovation networks are also built through cooperation in academic and industry which further strengthen Beijing's position as an international communication center.

13.4.2.2 University Park

The university park refers to the campus and its surrounding area where some affiliated science parks are also included. With the background of innovation-driven economic development, the university park become a more active player in Beijing. Beijing has the most higher education institutes compared to all the other cities in China. Ninety-two universities and colleges are the great generator and incubator for innovation. The key mission for the universities is no longer confined to knowledge generation, science exploration, technology research and talent cultivation, they start to play a more active economy-boosting role in the region, such as setting up university-affiliated companies to speed up the technology transfer and providing incubation space, resources and startup training programs for students. The university park has transformed from the knowledge warehouse to knowledge factory, and then to knowledge center.

The development of university park is following the mode of "one university one zone." A series of private incubator, accelerator and innovative companies are emerging around the campus to take the advantage of knowledge spillover and talents resources. The area around Tsinghua University and Peking University becomes hotspot for startups.

Beijing Municipal Bureau of Finance provides the financial support for the construction of the innovation centers and key laboratory in university, such as Peking University, Tsinghua University, Beihang University and Beijing University of Technology. They offer 50 to 100-million-yuan fund to strengthen the innovation center within a five-year cycle. Until present, universities in Beijing have established 20 university innovation centers.

Unver the policy of "Evacuating the Noncapital Functions to Beijing-Tianjin-Hebei Region," some of the universities are required to move to the suburb area of Beijing. Figure 13.5 shows the location of the university parks in suburb. This strategy provides adequate space for the further development of some universities, and will also drive the development of suburb area by adding the innovation generator into it (Fig. 13.6).

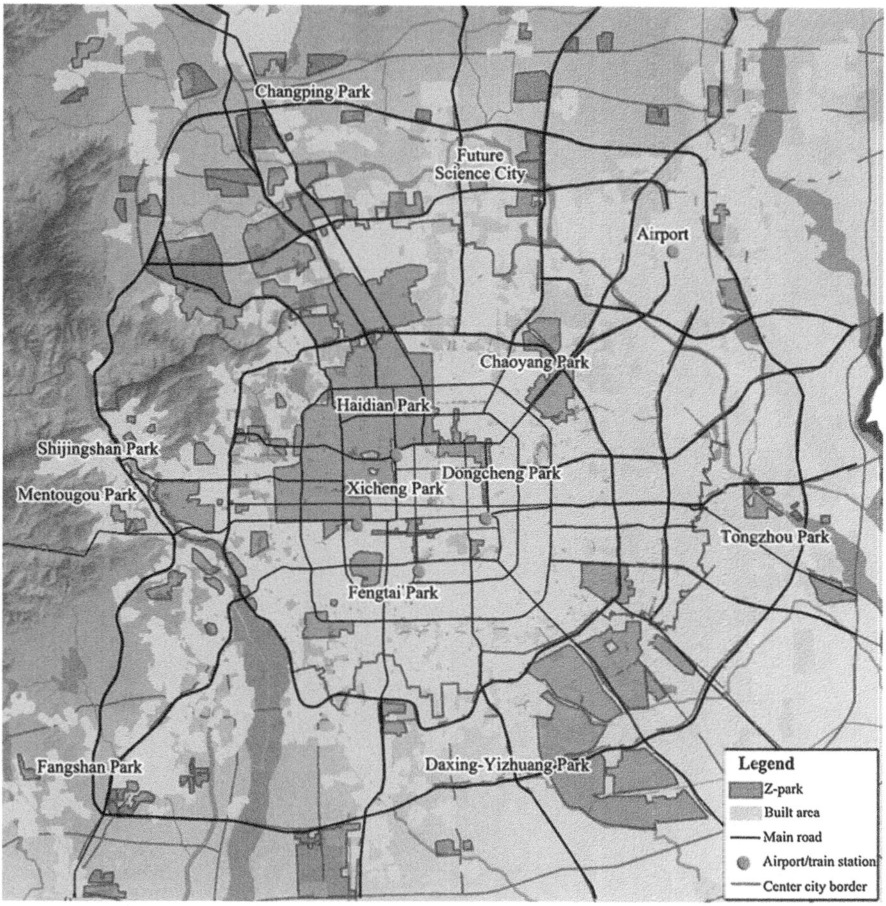

Fig. 13.5 Zhongguancun science parks layout

13.4.2.3 Science City

There are three science cities designated by the government, zhongguancun science city, future science city and Huairou science city. In the Overall Plan for Strengthening the Construction of National Science and Technology Innovation Center in Beijing, the three science cities and Yizhuang Economic Development zone are planned as the main platform for promoting the innovation capacity of Beijing. Based on universities, national level research institutes and central enterprise affiliated R&D institutions, this type of innovation districts develop in the mode of new town and are planned as a full-functional city from the beginning.

The government selects a certain goal for each of the three science cities and aggregates related innovation resources in order to build up the innovation ecosystem to support great innovation breakthrough.

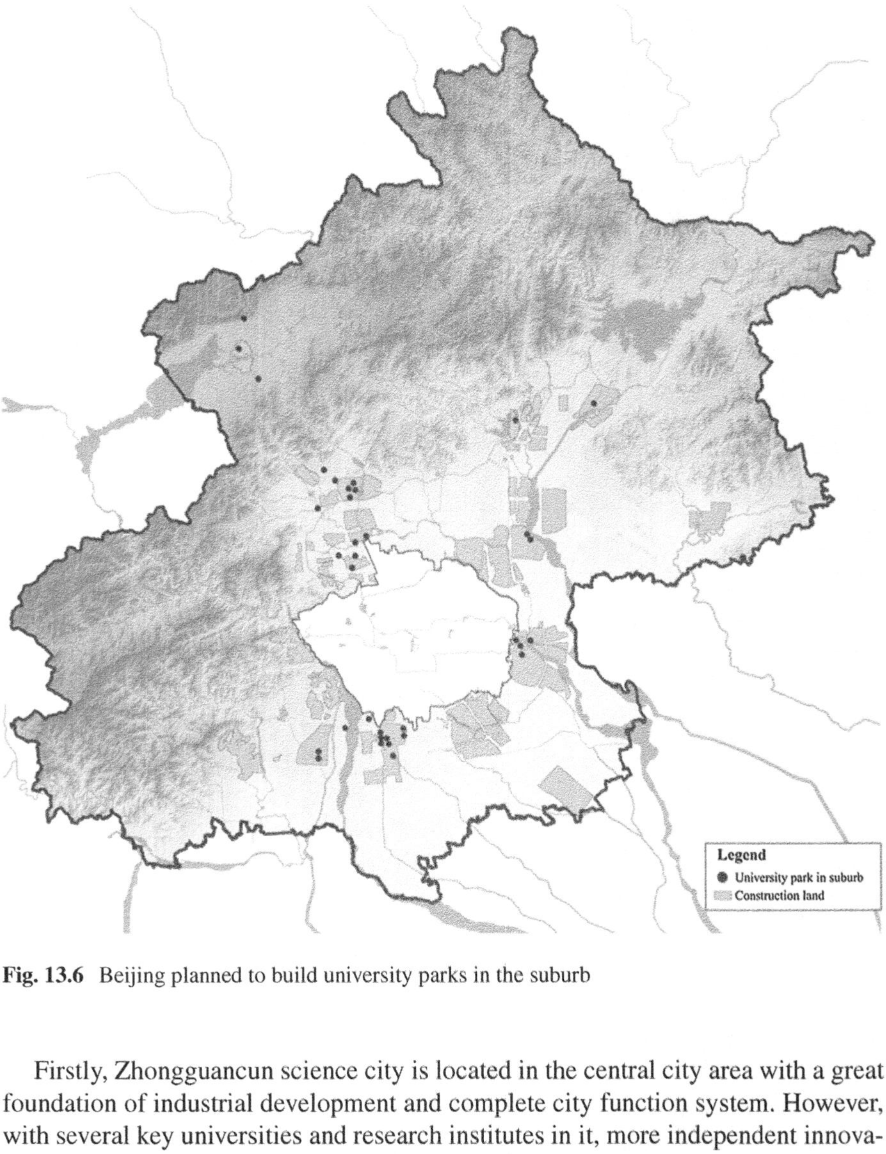

Fig. 13.6 Beijing planned to build university parks in the suburb

Firstly, Zhongguancun science city is located in the central city area with a great foundation of industrial development and complete city function system. However, with several key universities and research institutes in it, more independent innovation and technology transfer achievement are expected. The government tries to upgrade the innovation ecosystem through urban renewal plan in which excavates the non-capital functions and introduce more high-tech resources into the science city.

Secondly, Beijing future science city is planned as the innovation and entrepreneurship platform for national enterprise where aggregate some research institutes of national enterprises such as Shenhua Group and State Grid Corporation of China. It is the aim to promote the innovation capacity of national enterprises and provide a great environment for talents to accumulate innovation achievement to serve for the key industry in China. But right now, located in the suburb of the city, the future science city is still on the stage of starting. Some institutes of national enterprises have moved in, but there still lack of many service infrastructures due to the remote location. What's more, the national enterprises all adopt the gated park model and build up their own science parks separated from each other by the wall. The open atmosphere and innovation synergy which are expected in the planning are not realized, at least for this stage. The new planning for three science cities and one zone has realized this problem and try to propose related strategy.

Thirdly, Huairou science city is designated as the national science base and it is the newest science city which is still under construction. Its development is based on the new campus of Chinese Academy of Sciences. Through the construction of large-scale science facilities and technology service platform, the government expects Huairou science city undertakes the mission of leading the development of science and technology in China. Far from the central city and just starting its construction, Huairou science city also has a long way to go to form the innovation environment which can attracts talents.

13.4.3 Market-Led Innovation Districts

13.4.3.1 Entrepreneurship Community

The entrepreneurship community is the spatial form integrating the functions of innovation and residence. Such space provides flexible office service and reduces the commuting time and rental cost which is crucial for startups in early stage. Generally, there are three types of entrepreneurship community differentiated on the forming process. The first is called Local Silicon Valley, where startup groups spontaneously aggregates in the traditional residential districts. The second is Startup Complex, where the residence and incubation functions are implanted into traditional office building by private company. The third is Community with Maker Space, means some maker spaces are implanted in traditional residential districts to stimulate nearby talents to start up their own business (Fig. 13.7).

One typical case of the Folk Silicon Valley is Huaqing Jiayuan. Being the residential area near Tsinghua University, Peking University and Wudaokou subway, it had the reputation of "Local Silicon Valley" in Beijing. A number of famous Internet

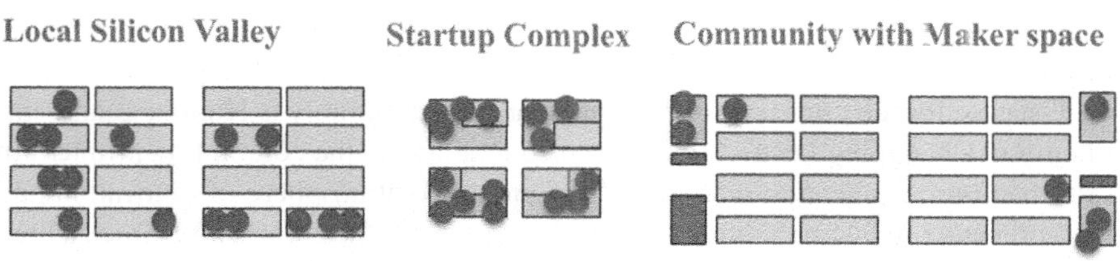

Fig. 13.7 Spatial pattern of three types of entrepreneurship community

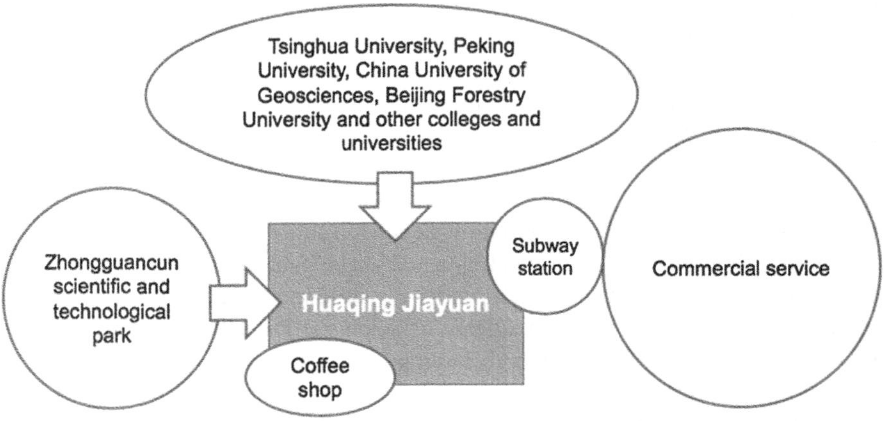

Fig. 13.8 Resources around Huaqing Jiayuan

companies are founded here such as Xiaonei, Meituan, Baofengyingyin, Zhuaxia, Kuwo, Meilishuo, etc. The unique location condition gives Huaqing Jiayuan the advantages of convenience in recruiting, working and communication thanks to the short distance to Tsinghua University and Peking University. At the same time, only small amount of rent is needed compared with that for the offices in surrounding area. The high resource density and low rent cost attracts youngsters to pursue their startup dreams. Most enterprises move away quickly after gaining certain development and they always move to other formal science park or office building to adapt the companies to new development phase. This also provides the opportunity for more latter entrepreneurs to take advantage of the entrepreneurship community. The community becomes an incubator for the youngsters who have some ideas and are willing to try. However, during these two years, with the rise of house rent, Huaqing Jiayuan is gradually losing its advantage of low cost for entrepreneurs (Fig. 13.8).

One case of Startup Complex is Xiaoyang Community with its aim to serving young entrepreneurs based on the entrepreneurship ecosystem. A standardized "Xiaoyang community" includes: entrepreneurial coffee shop, maker space, coworking space, quick office, entrepreneurial lecture hall, experiential business, apartment, and other space service. The first Xiaoyang Community is located on Flat A in Zhongguancun Dinghao building (Fig. 13.9). It is running by a private company which transform the former office building into the mixed-used complex catering for entrepreneurs. Due to the careful site selection and integrated service provided, the startup complexes satisfy a number of entrepreneurs' demand to start their business, including working, residence, recreation, social networking, and learning.

The case for the third type of entrepreneurship community is Huilongguan community in Changping District. It is designed to alleviate the commuting problem in suburban large residential districts and promote the atmosphere of innovation and entrepreneurship by implanting some maker space into the community. Since a high portion of the residents in Huilongguan Community has a background in Internet, biomedicine or other high-tech industries, the maker space will provide low-cost

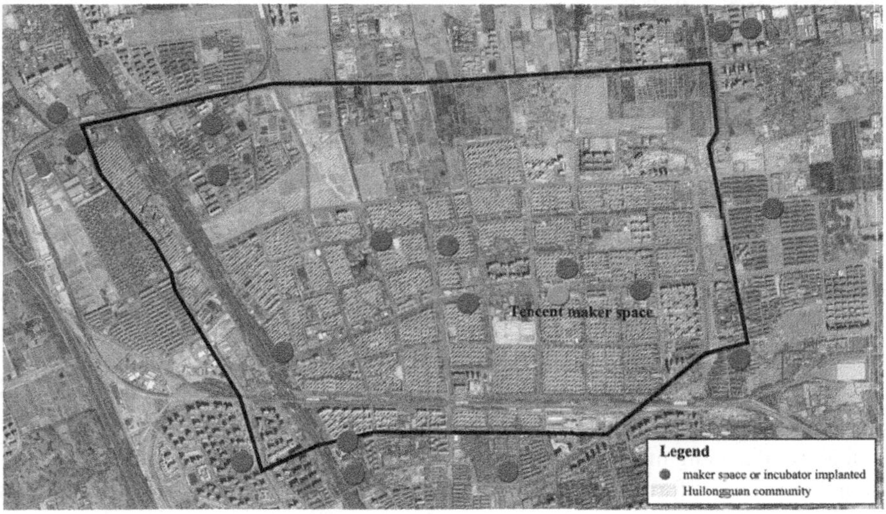

Fig. 13.9 Maker space implanted in the residential community

working space service and incubation service to the residents who have the desire to start their own business near their home. This idea of building the community into an entrepreneurship community is proposed by a company called ChangFazhan which is a subordinate company to Changping District Government. It will be responsible for the building up of the entrepreneurship ecosystem in the community. It is also a major work to cooperate with some private companies such as Tencent who is running the maker space. Hundreds of startups are born in the entrepreneurship community and 20 of them get the title of high-tech companies.

13.4.3.2 Characteristic Small Town

Characteristic small town is defined by Chinese government as the town with specific development characteristics. It is a way to explore the bottom-up and various development mode in the small towns in the suburb of cities. Due to the policy guidance, many private participators join the arena to take advantage of the space, resources in the small town. Some large companies invest a lot into certain towns, not only providing financial support, but also introduce more resources and opportunities. Some small towns in Beijing choose the direction of developing technology or creative industries such as Qianglong lake characteristic small town and Shiyan technology farm. The former selects culture and tourism as its feature and encourage the creative economy whose operator is a real estate company, and the latter is led by a postdoc in Tsinghua University focusing on the technology and ideas about farming. Combined with the special endowment of the town and the new resources and fresh ideas brought by extraneous private companies and entrepreneurs, some innovation, especially the social innovation occurs. As a result, the characteristic small town becomes a unique innovation district in Beijing.

13.4.3.3 Office Building for Tech Companies

The office buildings for tech companies are supporting diversified innovation activities, including office, R&D, business, art, and exhibition. They influence and support the innovation behavior by providing physical space for resource matching, and improving the efficiency of innovation activities.

These years more diversified operator appears with different background ranging from real estate company, property operation company, tech company to government. They adopt more intelligence building management systems to provide better service to tech companies. At the same time, they are exploring the method of setting up the innovation and entrepreneurship ecosystem by introducing incubators, accelerators, investment institutes to the office building.

The buildings in Tsinghua science park and the ones in Parkview Green are two typical cases. The first is developed and operated by Tsinghua science park landholders and leasing mainly for science and technology enterprises. A micro innovation ecosystem is formed around the area where a number of high-tech companies are supported. The second is developed and operated by a company who chooses the strategy of mixing the business, culture, exhibition and office functions and becomes a great success. It attracts some high-tech companies who are enjoying the diversified and mixed atmosphere. The deep understanding of the high-tech industry and the innovation mechanism helps these office building operators lead the development of this kind of innovation districts consciously.

13.5 Development Mechanisms of Innovation Districts in Beijing

The development mechanisms of innovation districts in Beijing will be explained in microscale and mesoscale. In microscale, it is the creation and innovation process and entrepreneurship behavior which drive the emergence and development of innovation district. In mesoscale, the incubation ecosystem and industrial cluster become the determinant mechanism to evolution of innovation district of Beijing.

13.5.1 Creation and Innovation Process

In microscale, the innovation process can be divided into three modes and each mode determines the specific logic of innovation district formation. The three innovation modes are science-technology-innovation, doing-using-interaction, and social innovation. The three modes correspond to different knowledge types and knowledge innovation sources and will prefer distinct location. For STI mode, the innovation district should be around or have easy access to the knowledge and innovation source to get cutting-edge technology in time; for DUI mode, the innovation district should be near to the users to help tech companies to see the feedbacks from users more quickly; and for social innovation mode, the innovation district will be around the centralized region of social issue.

13.5.2 Business Behavior of Entrepreneur

In microscale, the ambition of entrepreneur is the driving force of entrepreneurship and the aggregation of business behaviors of entrepreneurs becomes the fundamental impetus for the forming of innovation district. The key merits of entrepreneur are entrepreneurial motivation, opportunity identification and resource integration and they will choose the place where can inspire their motivation, provide more opportunity and integrate more useful resources. At the same time, the cost for office space and human resource, the channel to get investment and technology, and the full usage of policies are under their consideration. Generally, they will pursue the convenience to acquire all kinds of resources and at the same time control the cost including office rent and personnel wage. If there is an area can satisfy the above requirement, it will become an innovation district where the entrepreneurs aggregate spontaneously.

The optimization of innovation districts lies in the stimulation of entrepreneurial motivation, docking of entrepreneurial resource, reduction of entrepreneurial cost and formation of entrepreneurial ecology (Fig. 13.10).

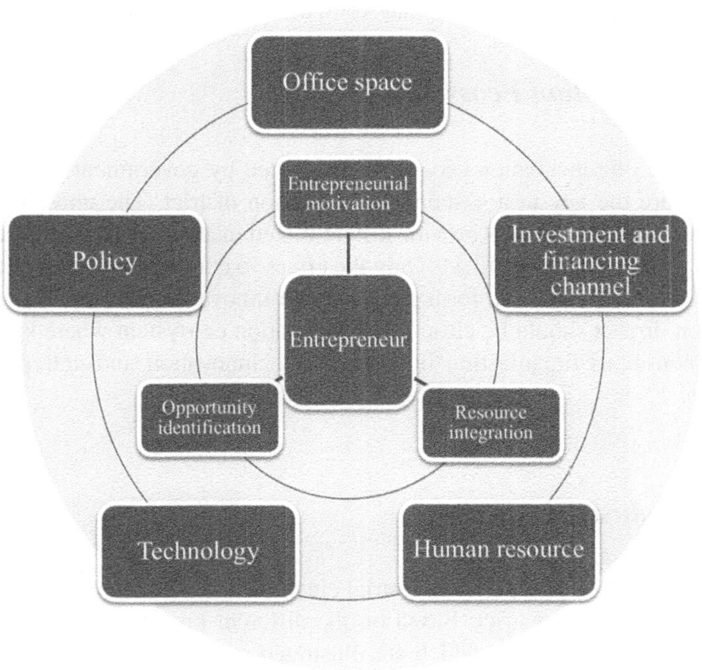

Fig. 13.10 Elements supporting entrepreneurship behavior

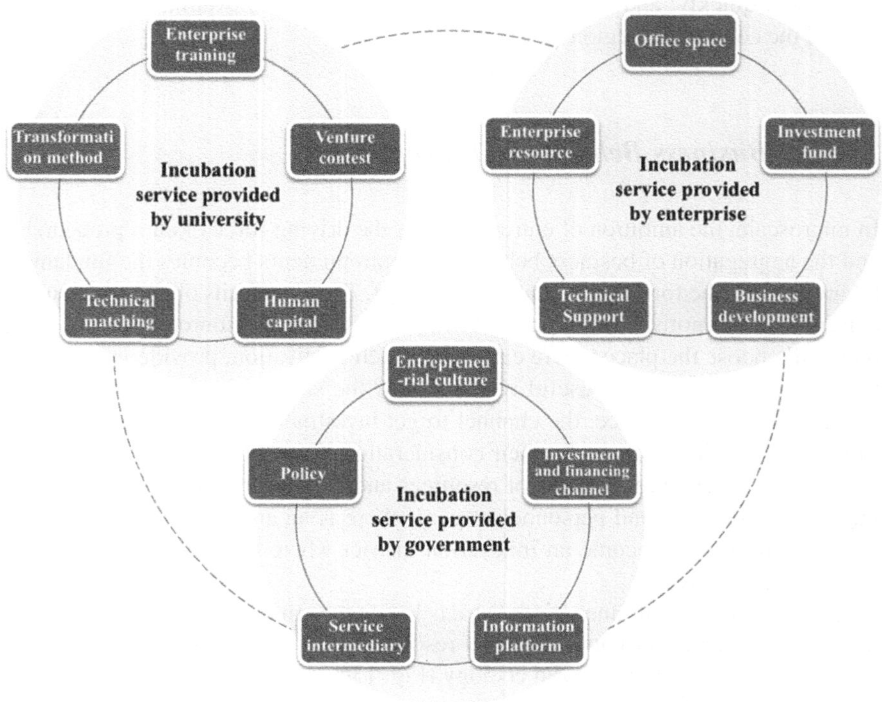

Fig. 13.11 Incubation ecosystem supporting innovation and entrepreneurship

13.5.3 Incubation Ecosystem

In mesoscale, the incubation ecosystem supported by government, university and enterprise are the key to a sustainable innovation district. The universities, large enterprises and governments provide differentiated incubation service for the entrepreneurs as showed in Fig. 13.11. Only the effective cooperation among these three types of subjects, can there form a sustainable innovation district. Therefore, the innovation district should be close to the incubation ecosystem where key subjects make a conscious organization of most of the innovation and entrepreneurship resources.

13.5.4 Industrial Cluster

In mesoscale, different types of industrial clusters have a great influence in the formation of innovation district. Based on the different knowledge type, innovation type and innovation process which are illustrated in Table 13.2, industrial clusters have distinct location selection rules and environment requirement (Tables 13.3 and 13.4). For the analytical or science based industrial cluster, the cluster could be

Table 13.2 Characteristics of three main type of entrepreneurship communities

Main mode	Folk silicon alley	Startup complex	Maker space in the community
Main impetus	Bottom-up spontaneous development under the influence of surrounding environment	Market leading. Developed and constructed by enterprises	Government leading. Provide planning lands and policy to attract enterprises to operate the space
Operator	Household autonomy	Real estate enterprise	Company subordinated to the government and financing enterprise
Space organization	Residential district	Integrate the residential area in the office and science and technology parks	Integrate renewable lands and buildings and implant the innovation space in traditional large residential districts
Site selection	Rich innovation resource (i.e., colleges and universities and science and technology park), public service and infrastructure in surrounding area	Near the city center; small mass; development and construction by means of "renovation"	In the traditional large residential districts; there are some innovation resources around; in the border areas of the city. Combine the reclaimable stock
Trend	With the increase of rent, it still maintains a certain scale of innovation activities	Rapid development scaling depending on local conditions	It has developed rapidly and received great attention from all parties. Huilongguan and Tiantongyuan have become the brands of entrepreneurship community
Typical case	Huaqing Jiayuan	Xiaoyang community	Huilongguan community

Table 13.3 formation mechanisms of innovation districts in three types of innovation process

Innovational mode	STI (science and technological innovation)	DUI (doing-using-interaction)	SNI (social need innovation)
Innovation process	Scientific research-technological invention-achievement transformation	Practice use-user feedback-interactive innovation	Social need-resource integration-social innovation
Knowledge type	Globalized explicit knowledge	Localized tacit knowledge	Social knowledge and value orientation
Knowledge content	Know-what; know-why	Know-how; know-who	Know-who; know-why
Knowledge sources	College and research institute; research and development in enterprise	User, competitor, supplier and institutions	College and institutes; research and in enterprise
Innovation driving forces	Paradigm of scientific and technological practice	Stimulation of user interaction in application environment	Culture and value guidance
Innovation district	Knowledge innovation areas	Users innovation areas	Local innovation areas
Cases	Technology transformation in university	Interaction based on user need	Local advantages

Table 13.4 Formation mechanisms of innovation districts in three types of innovation process

	Analytical (science based)	Synthesizing (engineering based)	Symbolic (arts based)
Knowledge type	Theory (know-why)	Skill (know-how)	Knowledge (know-who)
Innovation type	Developing new knowledge about natural sciences	Applying or combining existing knowledge in new ways	Creating meaning, desire, aesthetic qualities, affect, intangibles, symbols, images
Innovation process	Scientific knowledge, models, deductive	Problem solving, inductive, custom production	Creative process
Innovative interaction	Collaboration within and between research units	Interactive learning with customers and suppliers	Learning by doing, in studio, project teams
Knowledge sharing and diffusion	Patent, publication, network and science-oriented conference	Problem solving in the professional field and face-to-face communication	It is hard to share and be controlled by the key person
Site	Relatively constant between places	Varies substantially between places	Highly variable between places, class, and gender
Location selection	Relatively independent area of the city	Close to the customer and supplier	In area full of characteristics and aesthetic stimulation
Required environment	Quiet	Diversified and interactive	Diversified and interactive
Facilities	Research space and development space	Research space, development space and interactive space	Interactive space and leisure space
Industries	Bioengineering and pharmaceutical industry, new material industry and new energy industry	Information industry and smart manufacturing industry	Cultural creative industry

Source: Modified based on Asheim and Gertler (2005), Asheim et al. (2007), Gertler (2008), and Asheim and Hansen (2009)

located in a relatively independent area of the city, such as the future science city and Huairou science city; for the synthesizing or engineering based industrial cluster, the cluster should be closer to the customer and supplier, such as Yizhuang economic and development zone. For the symbolic or arts based industrial clusters, the cluster should be in the place where personality and aesthetic feeling could be stimulated.

13.6 Conclusion and Discussion

Beijing, China's technology and education center, has developed diverse types of innovation districts to support innovation which have injected new impetus into urban economic growth. The innovation districts in Beijing are experiencing the transformation from traditional science and technology parks to innovation districts with enterprises as its core, from government leading to enterprise and community leading, from spatial agglomeration to spatial decentralization.

Based on current characteristics and the development mechanisms of innovation districts in Beijing, we propose some planning policies for these innovation districts in two levels to better support innovation development, including the general development strategy and the guidance for six kinds of innovation districts.

There are three general development strategies in the city level. Firstly, in order to improve the regional innovation cooperation, it is necessary to establish an innovation ecosystem in Beijing-Tianjin-Hebei region to bring each districts advantage into full play. Especially the cooperation channels between Zhongguancun and other industrial clusters in the region should be built, in both of the physical form and virtual form. Secondly, to strengthen the spatial structure system of innovation districts to gather innovation resources. Three science cities and one development zone, one demonstration zone and sixteen science parks in Zhongguancun are key platforms for innovation agglomeration. The service provided by these innovation districts need to be further improved to attract more innovation resources. Thirdly, to increase the land supply for research and innovation. Innovation districts can provide low-cost space and services for innovation and entrepreneurship.

For the six types of innovation districts, there are also some suggestions for their development. As the regional engines and service centers, the science and technology parks should be more open and have more mixed-functions to attract more creative minds. University park has the responsibility to improve R&D transfer and at the same time cultivate the entrepreneurship atmosphere. Relying on the three science cities, Beijing aims to build a world-renowned science center and that will require loose and free academic environment to improve the interaction of diversified ideas. Government should provide loose planning regulation to encourage the bottom-up development of entrepreneurship community. Based on the natural endowment and cultural feature, the characteristic small town will attract companies and entrepreneurs to explore new way to creatively manage the innovation resources better than the government. The office buildings for tech companies will become the hotspot of innovation and the open and diversified cultural environment is the key which is conducive to mingle all kinds of talents, ideas, minds and other innovation resources.

Last but not least, the evolution of innovation districts is a projection of transformation of innovation activities. Only if there are more creative mind and risk-taking entrepreneurs, will there emerge a strong innovation center.

Acknowledgments The authors are grateful to Beijing Municipal Commission of Urban Planning and Natural Resources and Beijing Institute of Urban Planning & Design for providing funds through Beijing QuantUrban Tech Co. Ltd.

References

Asheim B T, Gertler M S. (2005). The Geography of Innovation//Fagerberg J, Mowery D, Nelson R. The Oxford Handbook of Innovation. New York: Oxford University Press. 291-317.

Asheim B, Coenen L, Vang J. (2007). Face-to-face, buzz, and knowledge bases: sociospatial implications for learning, innovation, and innovation policy. Environment and Planning C, 25(5):655.

Asheim B, Hansen H K. (2009). Knowledge Bases, Talents, and Contexts on the Usefulness of the Creative Class Approach in Sweden. Economic Geography, 85(4):425-442.

Baily, M. N., & Montalbano, N. (2018). Clusters and innovation districts: Lessons from the United States experience. Economic Studies at Brookings Institutions.

Chen, K., & Kenney, M. (2007). Universities/research institutes and regional innovation systems: the cases of Beijing and Shenzhen. World development, 35(6), 1056-1074.

DeBresson, Chris, Hu, Xiaoping. (1996). The localisation of clusters of innovative activity in Italy, France and China. In: Innovation, Patents and Technological Strategies. OECD, Paris, pp. 73–88.

Dutta, S., Lanvin, B., & Wunsch-Vincent, S. (2018). The Global Innovation Index, 2018.

Esmaeilpoorarabi, N., Yigitcanlar, T., Kamruzzaman, M., & Guaralda, M. (2019). Conceptual frameworks of innovation district place quality: An opinion paper. Land Use Policy, 104166.

Gertler M S. (2008). Buzz without being there? Communities of practice in context. Community, economic creativity, and organization, 1(9):203-227.

Guan, J. C., Richard, C. M., Tang, E. P., & Lau, A. K. (2009). Innovation strategy and performance during economic transition: Evidences in Beijing, China. Research Policy, 38(5), 802-812.

Guan, J. C., Yam, R. C., & Mok, C. K. (2005). Collaboration between industry and research institutes/universities on industrial innovation in Beijing, China. Technology Analysis & Strategic Management, 17(3), 339-353.

Guan, J.C., 2002. Comparison study of industrial innovation between China and some European countries. Production and Inventory Management Journal 43 (3/4),30–46.

Hu, A. G. (2007). Technology parks and regional economic growth in China. Research Policy, 36(1), 76-87.

Katz, B., & Wagner, J. (2014). The rise of innovation districts: A new geography of innovation in America. Washington: Brookings Institution.

Lawrence, S., Hogan, M. & Brown, E. (2019) Planning for an Innovation District: Questions for Practitioners to Consider. RTI Press. Retrieved from https://www.rti.org/sites/default/files/resources/rti-publication-file-5e2872d5-c00f-4e27-9b59-6644da6c091f.pdf

Leon, N. (2008a). Attract and connect: The 22@ Barcelona innovation district and the internationalisation of Barcelona business. Innovation, 10(2-3), 235-246.

Leon, N. (2008b). Attract and connect: The 22@ Barcelona innovation district and the internationalisation of Barcelona business. Innovation, 10(2-3), 235-246.

Lynch, K. (1960). The city image and its elements. MIT Press, Cambridge, 41, 73.

Morisson, A., & Bevilacqua, C. (2018). Beyond Innovation Districts: The Case of Medellinnovation District. In International Symposium on New Metropolitan Perspectives (pp. 3-11). Springer, Cham.

Segal, A. (2003). Digital dragon: High-technology enterprises in China. Ithaca, NY: Cornell University Press.

Tan, J. (2006). Growth of industry clusters and innovation: Lessons from Beijing Zhongguancun Science Park. Journal of Business Venturing, 21(6), 827-850.

Wang, J. H., & Leng, T. K. (2012). Production of space and space of production: High-tech industrial parks in Beijing and Shanghai. Cross-Currents: East Asian History and Culture Review, 1(1), 47-73.

Yam, R. C., Guan, J. C., Pun, K. F., & Tang, E. P. (2004). An audit of technological innovation capabilities in Chinese firms: some empirical findings in Beijing, China. Research policy, 33(8), 1123-1140.

Zeng, G., Liefner, I., & Fan, J. (2006). Comparison of regional innovation systems of high-tech enterprise clusters in Beijing and Shanghai. City Planning Review, 30(3), 2-8.

Chapter 14
Suzhou's "Growth Machine": Tracking the Driving Force Behind a Fast Growing Chinese Urban Region

Yifan Yang

Abstract Suzhou experienced fast economic growth in the past 30 years, which was accompanied by dramatic urban growth. During this process, the society benefitted from the economic boom and faced tough challenges and choices associated with diminishing land resources and environmental deterioration. By analyzing a wide range of literature, aerial maps and statistical data, and through interviews with scholars, government officials, stakeholders, this article traces the industrialization and urbanization process in Suzhou from 1984 to 2008 (between the Chinese Reform and Opening-up Policy and the World Financial Crisis, and also before XI Jinping's New Era). This analysis reveals how a pro-growth coalition involving state and market powers formed a "Growth Machine," which accelerated socio-economic changes. Because Suzhou is one of the fastest growing cities in China, other cities experiencing similar "Growth Machine" effects can gain insights regarding the implications of these strategies for community development and the evolution of urban form.

Keywords Growth machine · Suzhou · China · Industrialization · Urbanization · Regional development

14.1 Introduction

In the past more than two decades, Suzhou experienced fast economic growth. Gross regional product during the period 1985–2008 increased at an annual rate of 20.5% in Suzhou. Suzhou's Per Capita Disposable Income for Urban Households increased from $287 in 1985 to $3489 in 2008 (Investigation team of Suzhou Price

Y. Yang (✉)
China Urban Development Planning & Design Consulting Co., Ltd., Beijing, China

L. Bian et al. (eds.), *Chinese Urban Planning and Construction*,
Strategies for Sustainability, https://doi.org/10.1007/978-3-030-65562-4_14

Bureau, 2009). Ground vegetative coverage ratios decreased dramatically in the whole Grand Suzhou region. The quality of water in Tai Lake degraded one class approximately every decade: from 1960s' Grade I-II to 1990s' Grade IV. (V is the poorest grade.) After the city spent billions of dollars to clean up the pollution, the water quality stabilized at Grade IV but can hardly be regarded as improved. Developed and to-be-developed areas within Grand Suzhou grew 10 times in 20 years (from 264.45 km^2 in 1986 to 2691.38 km^2 in 2006). If urban growth continues at this rate and in this form, Suzhou's lands (including all vacant and farm land) will be exhausted before 2020 (China Academy of Urban Planning and Design, 2003).

The driving force behind the fast economic growth and accompanying urban growth is a Chinese version of a "Growth Machine" (Molotch 1976, 1984), the result of a coalition between the state and market interests. The term "state" consists of the central government and the local governments, but the latter plays a major role in this whole mechanism. This "Growth Machine" in Suzhou is similar to those identified in the U.S. and Europe in the 1930s through the 1990s (see, e.g., Bish, and Nourse 1975; Broadbent 1977; Cummings 1988; Clingermayer and Feiock 1990; Elkins, 1995; Logan and Crowder 1997; DiGaetano and Klemanski 1999), but different in organization, operation, cultural setting and political context.

The evolution of the "Growth Machine" in Suzhou can be divided into two phases: First, from 1984 to 1991, when the main players from the market side were locally based, small-scale enterprises, and the local governments acted more like regulators rather than direct game players. Industrialization mainly occurred in the county side, at town and village levels. During this period, the growth is commonly described by Chinese scholars as the "Sunan Moshi [South Jiangsu Pattern]" (Hu Xuwei, 1998; Wu and Zhen 1997; Cui Gonghao, 1999; Zhou et al. 2002; Gu Chaolin, 2002, etc.).

Second, from 1992 to present, the main players from the market side are international corporations and very large state-owned or private enterprises, with the local governments participating directly in the investments. They not only created local economic growth incentive policies, but also built government-owned companies to take part in new land development in hopes of attracting new private investments. In this period, cities become the main proponents for industrialization, and urban growth spread dramatically.

This chapter consists of four main sections: First, it will briefly introduce the basic characteristics of local natural resources and society, and the socio-economic development of Suzhou through history, as a broad context of the whole study. In the second and third sections, it will discuss the nature, operation and influences of the "Growth Machine" of the first phase (1984–1991), and the second phase (1992–2008). The fourth section will discuss some new trends for the evolution of the "Growth Machine," in which it rises to a broader regional level.

14.2 A Brief History of Socio-Economic Development in Suzhou

Suzhou City is a city with a history of more than 2500 years on the same site. Suzhou and Hangzhou together used to be called "the paradise in the human world" in China for their charming natural environment, rich culture, and commercial prosperity. Further, a series of natural and social factors made Suzhou the leading city in the Chinese history of industrialization (Fig. 14.1).

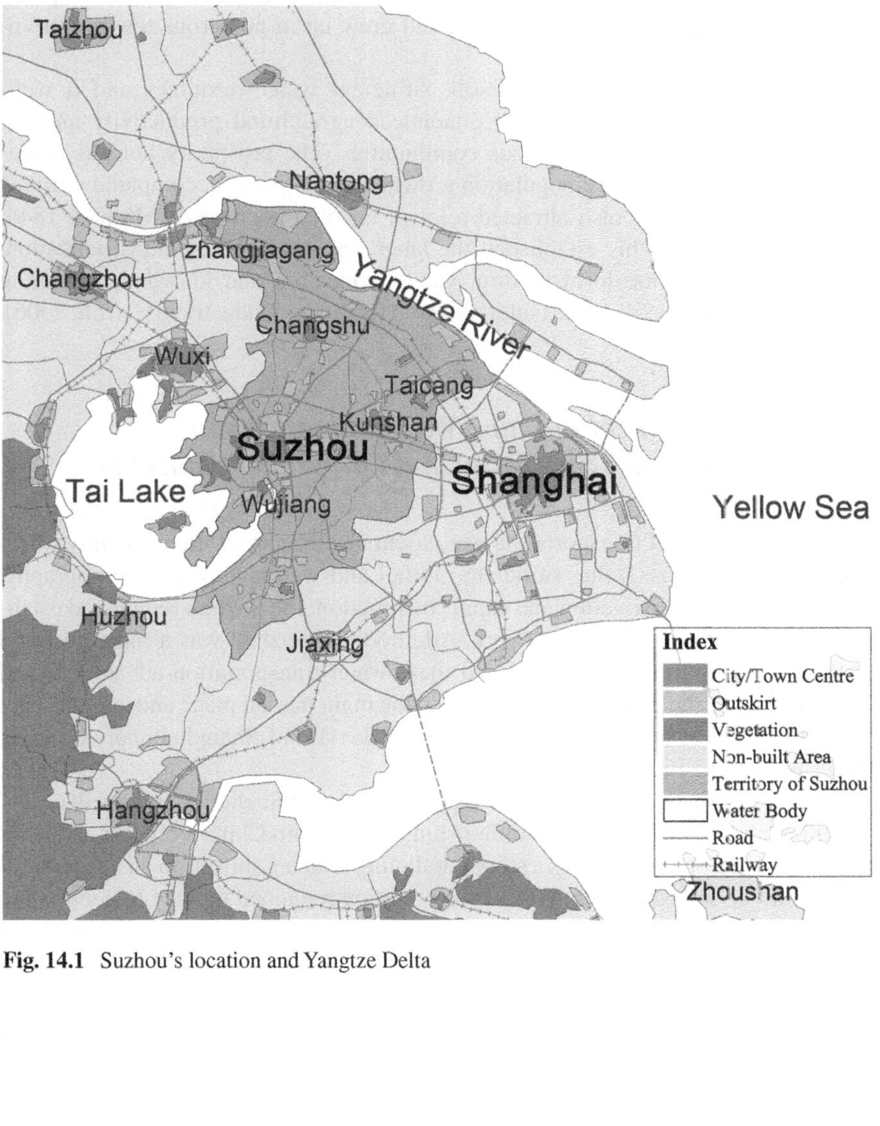

Fig. 14.1 Suzhou's location and Yangtze Delta

14.2.1 *Productive Agriculture Played a Fundamental Role for Socio-Economic Prosperity and the Preparation for Industrialization in Suzhou*

Productive agriculture was the foundation of prosperity in Suzhou, and led to economic prosperity and a skilled work force needed for industrialization. Before the early twentieth century, Suzhou was predominantly an agricultural society. Agriculture established the foundation for local social structure, culture, population distribution, regional landscape development, and rural countryside industrialization. Even before the Chinese Economic Reform in the late 1970s, most of the population and wealth was located in the countryside. Consequently, the first phase of the Growth Machine originated from and grew up in numerous towns and villages but not in the major cities.

The Suzhou region had fertile soil, sufficient water resources and a well-developed irrigation system, which guaranteed agricultural productivity and the foundation for relatively prosperous communities. The prosperity not only cultivated faster and more stable population growth within this area, compared with the rest of the country, but also attracted a lot of talented people (Zhu Kebao, 1844; Cheng Yong, 2002). This established the labor force needed for industrialization. The prosperity and the local culture, valued education and knowledge, making Suzhou people more skilled and better prepared for modern industries (Carroll, 2006).

14.2.2 *Geographical Location Was another Reason for Suzhou's Successful and Early Industrialization*

Early on, water-based transportation was essential. Because of the proximity to Tai Lake and the Yangtze Delta, sufficient rainfall and water supply, flat topography, and a long history of irrigation and canal construction, the Suzhou region had established a very advanced canal network. Additionally, Suzhou was a main port and destination along the Grand Canal. All these water transportation advantages and the prosperity of the city itself made Suzhou the main market place and commercial, cultural, and political center in its region until the rise of Shanghai approximately 150 years ago (Li Bozhong, 2002).

The rise of Shanghai brought new opportunities for Suzhou's industrialization. How Shanghai developed from a little fishing village into China's biggest and most prosperous metropolitan area is a story paralleling China's shift from an agricultural society into an industrial one. Water transportation remained the key to geographic advantage. Shanghai is located at the crossing of the two most densely populated and economically important waterways: the coast and the Yangtze River. In addition, Shanghai had conditions needed to build world-class harbors.

Further, Shanghai is located mid-way between densely populated North China and South China, and the Yangtze River helped Shanghai establish an efficient means for reaching inland areas, making it an ideal place to build the most important transportation and communication node between domestic and international markets. What's more, the Yangtze Delta region was the most prosperous area in China and could support the emergence and growth of Shanghai as a metropolitan area of national significance. All these factors combined to make Shanghai the main trade, financial and industrial center for China, and the main point of entry for international technologies, ideas, information and culture. It is also the ideal exhibition place to introduce Chinese products to overseas markets.

Suzhou is the largest city close to Shanghai. Chinese scholars describe the relationship between Shanghai and Suzhou as "A store in the front, with factories in the back," which means that Shanghai acts as the market place, and also the gateway to resources like finance, technology, information, etc. Suzhou and other adjacent areas offer lands and labor to the stores and factories of Shanghai. Though Suzhou has lost its central status in the Yangtze Delta, it benefits from the rise of Shanghai as a gateway to huge industrial resources of all kinds, and massive domestic and international markets. The proximity to Shanghai made Suzhou a competitive locale for modern manufacturing enterprises early in the industrialization of China and continuing today. Suzhou ranked as the fifth largest city economy in China in 2008.

14.2.3 Suzhou's Industrialization and Urbanization before 1984

Suzhou's industrialization and urbanization before 1984 can be divided into three phases: First, the period before the foundation of the People's Republic of China in 1949, during which the country was shifting from agriculture to an industrial society dominated by a free market. Second, from 1949 until the Economic Reform in 1978, the era of the planned economy during which there was no free market and the production and distribution of goods was tightly controlled by the central government. Third, from 1978 to 1984, the early phase of the Economic Reform that started in the countryside.

One important character of the early industrialization was that most of the enterprises were small in scale and scattered widely in the countryside, rather than concentrated in major cities, and tightly connected with agricultural production. The investment for these early industries came from the wealth associated with agriculture, and they shared resources and labor with agriculture (Fei Xiaotong, 2000). Consequently, this led to a strong tradition in the Suzhou region of small-scale industry at the town and village level. This very characteristic of early industrialization led to slow urbanization in Suzhou's central city, but many townships became stronger economically and bigger in size, and more densely populated (Huang Zongzhi, 1994).

After the foundation of the People's Republic of China, the government gradually took over private enterprises and put them under collective ownership. Private enterprises almost disappeared in the planned economy era. Many collectively owned small-scale corporations were established at town and village levels before the Cultural Revolution, during which industrialization became very slow and even stopped in most areas in China. These practices that further encouraged industrialization in the countryside, and provided the foundation for rural industrialization after the Economic Reform (see, e.g., Gu Chaolin, 2002; Li Jianbo, 2003). Meanwhile, governments of various levels created state-owned corporations, which were relatively larger in scale, and founded many technology schools and research institutes to support these industries. All these efforts helped to accumulate necessary industrial technologies and skilled engineers and industrial workers for subsequent stages of industrialization.

After 1978, the Chinese Economic Reform focused on the countryside. In the first phase of the reform, the main economic policy changes were aimed at improving agricultural production, with some slight changes in industrial sectors. For example, small private enterprises were not forbidden as capitalist elements. With the easing of tight economic restrictions in the countryside, and as a result of the tradition of rural industrial development, some small enterprises at the town and village levels began to emerge, later to become giant corporations. However, throughout this period, urban form in the Suzhou region hadn't changed very much largely because little had changed in urban areas with respect to economic policy and planning.

14.3 Phase I: "South Jiangsu Pattern" Era (1984–1991)

After the government's focus for the Economic Reform shifted from countryside to the urban area and industrial sector in 1984, Suzhou's industrialization entered a new era. This era also marked the beginning of breakneck industrialization and urbanization. The driving forces and resulting growth patterns separate this era into two main phases in Suzhou: the "South Jiangsu Pattern" Era and the "Development District" Era. The division point is 1992, when Deng Xiaoping launched his famous "South Tour Speeches," which addressed the central role of economic development in Chinese social life, and put an end to the debate, in the society and especially within the Communist Party, on the issue of whether to continue the Economic Reform and to seek a market economy.

Before the Speeches, governments at all levels removed some restrictions on economic activities but hesitated to shift completely to a market economy, the concern being the appearance of embracing a shift to capitalism. The Speeches made the direction of the reform clear: to establish a market economy. After this, the governments, especially local governments enthusiastically utilize resources of all kinds they can reach to push the industrialization and economic growth, and even get in economic activities deeply, which lead to the tight coalition of the state power and the market power, forming the "Growth Machine."

14.3.1 The Nature and Function of "South Jiangsu Pattern"

The term "South Jiangsu Pattern" describes not only the main economic growth pattern, but also the rural urbanization pattern in South Jiangsu Province (Gu Chaolin, 2002). The basic character of this pattern was that the booming township and village enterprises played a central role in socio-economic community development. Numerous small-scale, collectively-owned or private enterprises emerged throughout the countryside rather than in major urban areas. They attracted underutilized labor from the agricultural sector and efficiently incorporated them into industrial production solving the problem of unemployment and underemployment, and creating new economic growth (Hu Xuwei, 1998; Wu Tinghai, 1997; Cui Gonghao, 1999; Zhao Xinping, 2002; Gu Chaolin, 2002, etc.).

From 1978 to 1990, 2.27 million people shifted from the agricultural sector to non-agricultural production in the South Jiangsu area, with an annual shift of 181,900 persons. This "South Jiangsu Pattern" evolved quickly after the Economic Reform shifted to industrial sectors and became the primary form of economic development in this region in the late 1980s.

This raises another question: why did rural industrialization play such a major role? The location of numerous township and village enterprises was a result of various social, economic, political, and cultural factors. But among them, the special agriculture-residence-industry relationship played a major role. In this pattern, farmers were also workers, their roles shifted between agricultural and industrial work seasonally. Additionally, different members within a farm family established a kind of division of labor. For instance, older people or women would take care of the farming, and young people or men would go to factories in the villages or towns to get more income, start small businesses like grocery stores, or provide transportation services. Income from the non-agricultural businesses gradually became the largest portion of annual family income.

However, initially, farmers still preferred to live in the countryside because of the low living costs, big houses, unwillingness to give up agricultural income, and most importantly, the agricultural lands and their role as property. This organization of work and labor within farm families in Suzhou was also given a term by Chinese scholars as "partial working, partial farming (Yigong, Yinong)." The owners of these enterprises were also mostly local people or collectives, who were unwilling or restricted by social and political bonds to locate their enterprises far away from their original locations.

The "South Jiangsu Pattern" worked successfully, because this pattern built upon its rich agricultural tradition, best utilized the accumulated social wealth in the countryside and the large workforce from the agricultural sector, and gradually shifted an agricultural society to an industrial one without directly conflicting with political and policy barriers such as the urban-rural segregation policy known as the "Hukou system." The operation of the "South Jiangsu Pattern" reflected synergy of interests and aims between the government and newly born market economy, especially when the government needed to create a large number of jobs while lacking the capacity to improve urban infrastructure and social benefit systems to host a potentially massive influx of population into urban areas. In short, the "South Jiangsu Pattern" resolved major socio-economic problems with a minimal political risk.

14.3.2 Rapid Economic Growth and Contributions of the "South Jiangsu Pattern"

The rapid growth of township and village enterprises were the major symbols of South Jiangsu Pattern development. Its direct contribution to socio-economic development mainly consisted of economic and employment growth. These benefited the accumulation of wealth, including countryside household income, entrepreneur and corporate capital and reinvestment capacity, and local government revenue needed for investment in infrastructure and social services. As a consequence, social change and restructuring occurred in a manner needed to push industrial production and urbanization.

From 1978 to 1988, the gross industrial output value of township and village enterprises increased 21.6 times, and 36.6% annually, in Suzhou (Wu Tinghai, 1997). Gross Industrial Output Value increased from 4.85 billion RMB in 1978 to 56.37 billion RMB in 1990, also remarkable growth but at a relatively lower rate to that of the township and village enterprises. In 1992, industries at the town and village level contributed 58.7% of the entire industrial output in Suzhou (Qing Tingdong, 1994). Meanwhile, in the Grand Suzhou area, in the early 1980s, the Central City District's GDP accounted for 1/3 of the whole, but by the middle of 1990s, this ratio declined to less than 1/6. In 1992, for example, there were 5811 industrial enterprises in Wu County, only 151 based at the city level, 599 at the town level, 2747 at the village level, and 2314 at lower than the village level. Enterprises at the village level or lower accounted for 87.1% of the total (Wu Tinghai, 1997).

The employment data repeated the same story. In Suzhou, employees in these township and village enterprises increased steadily and rapidly from 1980 to 1987. In 1980, the total employment of these small countryside industrial corporations counted approximately 545,300. In 1985, this number increased to 1,150,300, and in 1987, this number became 1,239,400. Employment in these small countryside industrial corporations increased 12.44% annually from 1980 to 1987.

The booming township and village level employment not only contributed to economic growth, but also contributed to shifting overall economic structure. From 1978 to 1990, secondary industry contribution to GPD rose from 55.7% to 61.0%, while the same portion of primary industry dropped from 28.1% to 17.3% (Table 14.1).

Table 14.1 Changes of proportion of employees by type of industry (Source: Data drawn from Suzhou Statistical Year Book 2009 by Suzhou Statistics Bureau)

Item	1978	1990	2000	2007	2008
Primary industry	62.3	29.7	21.0	7.0	6.5
Secondary industry	27.1	51.3	49.8	63.5	61.3
Tertiary industry	10.6	19.0	29.2	29.5	32.2

14.3.3 The Main Urbanization Characteristics in the "South Jiangsu Pattern" Era

Because of the significant influence of the "partial working, partial farming (Yigong, Yinong)" pattern, a delayed urbanization process was one of the main characteristics in the "South Jiangsu Pattern" era. In Suzhou, at the end of 1992 the population living in cities and towns counted 37.34% of the total in the region, but non-agricultural work force counted for 61.68% of the total workforce. A huge portion of the industrial workforce hadn't been urbanized.

On the other hand, the fast growth of rural industrialization substantially supported the growth of small cities and townships, which became the main centers for urbanization during the same period. From 1987 to 1994, urbanized or built-up lands in Suzhou's townships increased 209.14%, on average, especially for factories, marketplaces and villager's residences (Wang Hongyang, 1999).

At the beginning of this era, farmers went to work in adjacent townships and came back to their village dwellings at night. When their lives gradually became more related to townships than villages, they moved to these townships to settle down (Gu Chaoling, 2002). The booming industrialization in the countryside contributed to the accumulation of social wealth at the township and village level, which triggered new local infrastructure construction and land development. When the farmers got wealthier, they preferred to buy new residences in adjacent townships. When the enterprises got more capital and wanted to enlarge their factories, they also went to adjacent townships rather than relatively further away and bigger cities, because they didn't want to lose their existing local employees. The push of the relocation of enterprises and worker's residences led the fast growth of small cities and townships in the late 1980s even into the 1990s (Cui Gonghao, 1999; Zhou Yixing, 1999). Meanwhile, the central government's "Small City Development Strategy" offered political support to the fast growth of small cities and townships.

14.3.4 The Downsides and Drop of the "South Jiangsu Pattern"

There were obvious limitations to these township and village level enterprises, which rooted on their scattered location and informal operation. These limitations led to the drop of "South Jiangsu Pattern" as an alternative economic mode in the

1990s. Most of the production took the form of light industrial, e.g., small agricultural implements, household appliances, clothes, building materials, and furniture, etc. Because entry thresholds were very low, the markets were soon congested with suppliers, and intense competition made scattered locations and limited technology significant disadvantages.

The countryside producers at this time relied on the technology and production methods inherited from former engineers in state-owned or collective-owned enterprises of the older command economy era. At this time, even "weekend engineers" coming from adjacent major cities like Shanghai, became an informal but important technology resource for these small enterprises. Because few of them could achieve the organizational scale needed to sustain technological systems and development, it was extremely difficult to establish any lasting competitive power in the growing market economy.

What's more, the scattered location led to dispersed pollution sources, inefficiency of industrial waste treatment and poor utilization of infrastructure (Hu Xuwei, 1998). The dropping efficiency and the rising economic and environmental costs undermined these small businesses' competitive power compared with that of the rising gigantic corporations located in newly developed districts in urban settings. Most of the downsides related directly or indirectly with their scattered location in the broad countryside.

The decline of the township and village enterprises became obvious. Employment in these enterprises dropped from 1,107,300 in 1995 to 1,006,600 in 1996, and 927,700 in 1997 (Luo Xiaolong, 2008). As Zhou Yixing (2002) said: The rural industrialization as an event and process, which supported the economic growth and small cities and townships' development in the 1980s and even 1990s, has ended, and the foundation of the nation's "Small City Development Strategy" has been significantly weakened.

14.4 Phase II: Booming of New District Development

After the early 1990s, the main character of industrialization and urbanization turned from self-organized, bottom-up, small and middle-sized enterprises in the geographically scattered "South Jiangsu Pattern," into local government driven, top-down, giant corporations located in new (urban or industrial) districts developments. The driving force for this was the combination of the liberalized political environment and the globalizing market.

14.4.1 Pro-Growth Political and Economic Coalition Structure Was Strengthened

After Deng's keynote speeches in 1992, ideological debate within the communist party on the direction of the economic reform came to an end, and it was made clear that a market economy and economic growth was the central mission at every administrative level. GDP growth became the key indicator to assess a local administrator's performance and served as the basis for their promotion, a crucial incentive for local chief executive leaders to push economic growth by all means. This combination of forces is recognized as being critically important, even to the present day.

To achieve the goals for economic growth, the state needs to choose matching partners on the market side. When the downside of the "South Jiangsu Pattern" became more obvious, municipal governments shifted their focus from local township and village enterprises to Foreign Direct Investments (FDI), large investments involving higher technology and production standards, and promising more aggressive economic growth (Wu Jiaping, 2005).

Monetary and policy incentives are widely used by municipal governments to attract important investments. For instance, municipalities offer free land, and 3 years of tax abatements to attract FDI to their jurisdictions in Suzhou. They even built new international schools to offer high quality education for the children of international employees. These kinds of economic and social incentives did attract investments to cities and created a lot of jobs. The economy grew rapidly with the coming of FDI, local firms matured quickly through cooperation with the leading firm coming from outside, more job opportunities attracted more skilled people to move to the cities, local commerce and services boomed, and the local governments' tax revenues increased quickly. As more investors wanted to come, outpacing the ability of local governments to provide services and address the negative impacts of growth, local governments instituted stricter environmental standards and new efficiency thresholds, with the result that less competitive firms began to disappear. Gains in productivity and efficiency then led to additional public resources and investment capacity for schools, public recreation facilities, public transit systems, and so on.

These incentives and preferential policies did work to attract FDIs. Suzhou became the home of FDI industrial clusters in machinery, electronic devices, and medicine, among others. In the 2000s, FDI became the dominant driving force in Suzhou. In 2002, foreign-invested enterprises counted for 73% of the gross industrial output in Suzhou, with an annual growth rate of 31.20%. Private enterprises counted for 15%, with an annual growth rate of 22.20%, and the state-owned and collective-owned enterprises counted for 12% of the gross industrial output with an annual growth rate of about 8.30%. In 2003, Suzhou attracted $6.8 billion FDI over the whole year, and exports reached $32.7 billion, with foreign-invested enterprises contributing 86% (Suzhou Institute of Development Research, 2004). From 2001 to

2005, Suzhou attracted a total of $19.8 billion in FDIs, and 107 of world top 500 enterprises had sited their branches in Suzhou (Suzhou City Government, 2006).

He Weidong (2002) identified three main elements essential to the attraction of FDI in Suzhou: First, an investment environment that included proximity to Shanghai, sound infrastructure, and a skilled workforce. Second, gradually accumulated industrial cluster eco-systems in certain fields and sectors. Third, local governments' pro-growth policies and efficiency offering strong support to the operation of FDI enterprises.

Our interviews with executive officers and managing staff in 10 enterprises of various scales and investment background in Suzhou Industrial Park (SIP) in 2006 showed similar results. Generally speaking, for these diverse enterprises, the SIP's high efficiency and high-quality infrastructure, along with strong pro-enterprise policies on the part of local government, the site's proximity to Shanghai, and the city's livability, making it able to attract skilled workers, were the four most frequently mentioned elements for their location decisions. For the semiconductor industry, the negotiation between the local government and the leading enterprise was crucial. The leading enterprise told us, when comparing various similar Development Districts in the Yangtze Delta region, the efficiency and preferential policy of the government and possibility to attract adequate skilled employees were their priorities in locational choice. After the leading enterprise's entry, more than 50 related peripheral enterprises followed, creating more than 12, 000 jobs and helping to establish and grow a strong semiconductor industry cluster.

14.4.2 New Development Districts Became the Dominant Trend

Local governments found that the creation of a "Development District" is the most effective spatial vehicle to further their economic ambitions (see, e.g., Wu Weijia, 2000; Wu Ye, 2000; Gu Chaolin, 2002; Zhou Yiyi, 2002). The so-called "Development Districts" can have various contents, themes and functions. The reasons why these new districts are so popular include at least the following:

1. They are adjacent to an existing urban area, enabling them to get utility and facility support from the city, which makes it more feasible and practicable at the very beginning. In this way, it saves a lot of cost in first infrastructure investment and minimizes the development risk.
2. They are "new," so have fewer historic burdens compared to existing urban areas. This means less displacement cost and more flexibility to implement new ideas and lay out ideal land use plans (Wu Weijia, 2000).
3. They usually have a main theme or a dominant function, which helps to cultivate certain kinds of industrial cluster development in a given location. The gathering of enterprises in similar or relevant industrial sectors established certain levels of positive competition and collaboration, which brings the benefits of economies

of scale and is essential to make the new district as a whole in the given industrial sector to compete with other regions (Lu Yuqi, 1998).

4. They get more favorable policies from upper levels of government like tax cuts, more autonomy, subsidies, and easier land use permits, etc.

In the 1990s, new development districts of various levels in Suzhou became the main vehicles to host FDIs and important areas for supporting continued economic growth. In 1998, the 5 national and 10 provincial development districts contributed 19% to the local GDP (Liu Rongzeng, 2001).

By 2003, all national and provincial development districts totaled 366.16 km^2, had attracted $22.46 billion FDI, contributed 126.65 billion RMB to local GDP, and created 790,535 jobs.

For instance, Suzhou New District (SND) began its construction in 1991 with 0.93 km^2 expropriated land and 20 million RMB, and was approved as a national high tech industrial development district in 1992. Until 1999, the urbanized area had expanded to 17.16km^2, and industrial uses occupied 58.04% of the total urbanized area. SND created over 60,000 jobs, had 75,918 residences, and its GDP reached 12 billion RMB, and the built area reached five million square meters (Tsinghua University Urban and Architecture Research Institute, 2001). Suzhou Industrial Park established in February 1994, with an existing population of 164,700, planned to build 70 km^2 at the very beginning, but by the end of 2006, the urbanized area in Suzhou Industrial Park had reached 106.9 km^2, with an urban population of 502,600.

New district development became an overwhelming trend expanding even to the town level. Toward the end of 1993, town-level new industrial districts numbered 130, occupying 32.19 km^2, and attracted $666 million of FDI, with $4.46 billion in total investment.

14.4.3 The Mutually Related Land Expense and Economic Growth Process Driven by the Growth Machine

Chinese scholars use a special term, "land finance," (Yang, 2007; Song, 2009) to describe how local governments see the developable lands as a major monetary resource. Our research shows that "land finance" is actually the driving force of the related economic growth process.

From the late 1980s, after the establishment of the land leasing system, the income from leasing land became one of the major revenue sources for local governments (Wu Weijia, 2000). Our interviews with various municipal administration officials show that regular revenue coming from taxes and fees can cover routine costs like the public employees' salary and facilities maintenance but almost all government investments in new public facilities and infrastructure construction had to be drawn from leasing land or the loans from mortgaging the developable land to national political banks or commercial banks. Suzhou's economic growth was an investment-driven one. Its GDP grew 20.5% annually from 1985 through 2008.

However, as productivity could hardly keep increasing at such a high speed for decades, GDP increases depended on new investment to create new producers.

Our study on the changes of GDP per km^2 of developed urban area in Grand Suzhou 1985–2004 reflects the changes in productivity throughout society. From 1986 to 1991, which was the "South Jiangsu pattern" era, the GDP density grew gradually from 40 million RMB/km^2 to 50 million RMB/km^2. Because the shift of the economic model increased productivity dramatically, the GDP density more than doubled from 1991 to 1996. But within the same economic model (investment-driven, FDI-leading and export-oriented), the GDP density increased 3% annually from 1996 to 2004, which fell far behind the city's GDP growth rate (Fig. 14.2).

When growth in productivity could not support economic growth at the same rapid rate, developing more lands to host more firms to contribute to GDP growth became the top priority for local governments. Municipalities lease or mortgage parcels of land, getting funds to invest in infrastructure construction on new lands, making them ready for new and more aggressive lease or mortgage, and then urban growth accelerates. As our studies show, even in 2005, the built area of Suzhou City had already exceeded the 1996 Comprehensive Plan projection for 2010 by 70.31%, mainly because industrial uses exceeded the 2010 goal by 173.34% (Fig. 14.3). With the expanding industrial lands and continual arrival of new FDI and corporations at the newly developed industrial land, the economy grew with land development (Table 14.2).

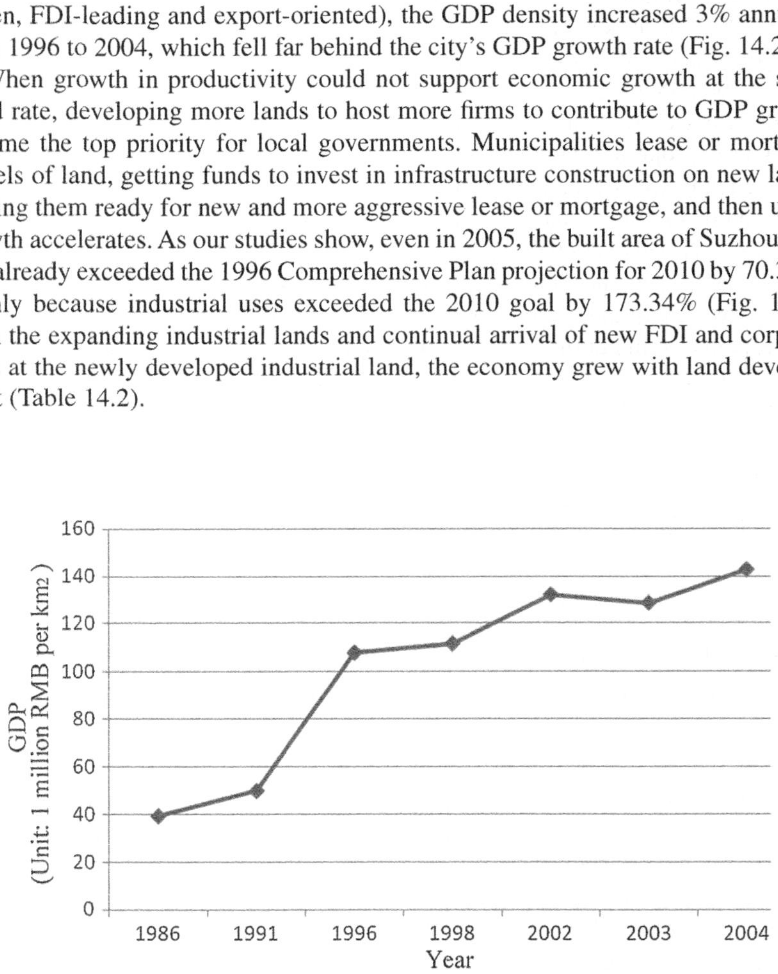

Fig. 14.2 Changes of GDP per km^2 Developed Urban Area 1986–2004 (Unit: one million RMB per km^2, Explanation: Primary industrial contribution to GDP was lower than 5%, so here was an assumption that most of the GDP were produced on urbanized lands. Source: Data drawn from Suzhou Statistical Year Book 2006 by Suzhou Statistics Bureau and Suzhou Land Use Report 2007 conducted by the Institute of Remote Sensing Applications, Chinese Academy of Sciences)

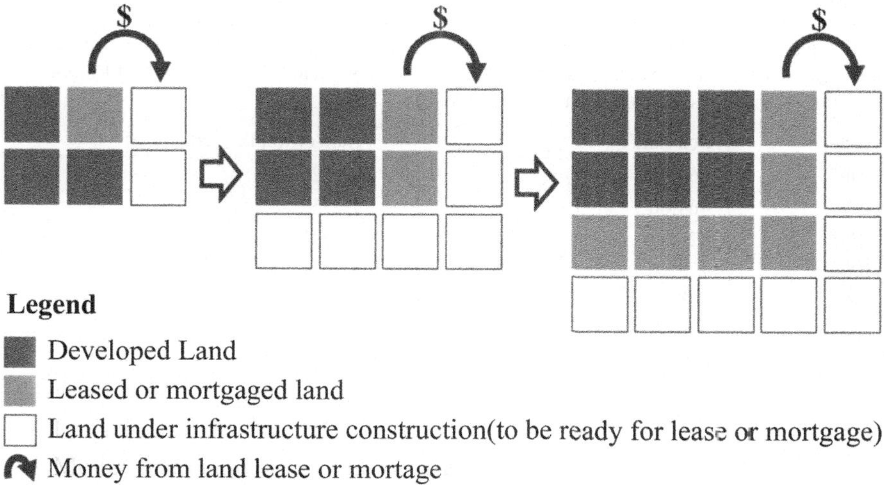

Fig. 14.3 The "Land Finance" and Urban Growth Model in Suzhou

Table 14.2 Contrast of built lands in 2005 to projections for 2010 given by Suzhou City Comprehensive Plan 1996–2010 (Explanation: "-" means lower than projections for 2010. Source: Data are drawn from Suzhou City Comprehensive Plan 1996–2010 by Suzhou Planning Bureau, and The Report for Suzhou City Comprehensive Plan 2007–2020 by China Academy of Urban Planning and Design)

	Projections for 2010 given by 1996 plan	Built lands in 2005	Exceeded (%)
Urban area (km^2)	186.6	317.8	70.31
Urban area per capita (m^2)	100	133.3	33.30
Residential (km^2)	56.47	67.08	18.79
Residential per capita (m^2)	30.5	28.14	−7.74
Public facility (km^2)	17.22	12.8	−25.67
Public facility per capita (m^2)	9.2	14.2	54.35
Industrial (km^2)	48.13	131.56	173.34
Industrial per capita (m^2)	25.8	55.19	113.91
Green land (km^2)	22	15.89	−27.77
Green land per capita (m^2)	11.8	6.7	−43.22

Table 14.3 Changes of developed and to-be-developed lands in Grand Suzhou (Explanation: Total area of Grand Suzhou is 8488 km^2 and Tai Lake was included in calculating %. The developed and to-be-developed lands were built and demolished lands picked up by analysis of the aerial maps. Source: Suzhou Land Use Report 2007 conducted by the Institute of Remote Sensing Applications, Chinese Academy of Sciences)

Year	Area (km^2)	Rate of total coverage (%)
1986	264.45	3.12
1991	469.66	5.53
1996	928.74	10.94
1998	1121.26	13.21
2002	1576.11	18.57
2004	2420.23	28.51
2006	2691.38	31.71

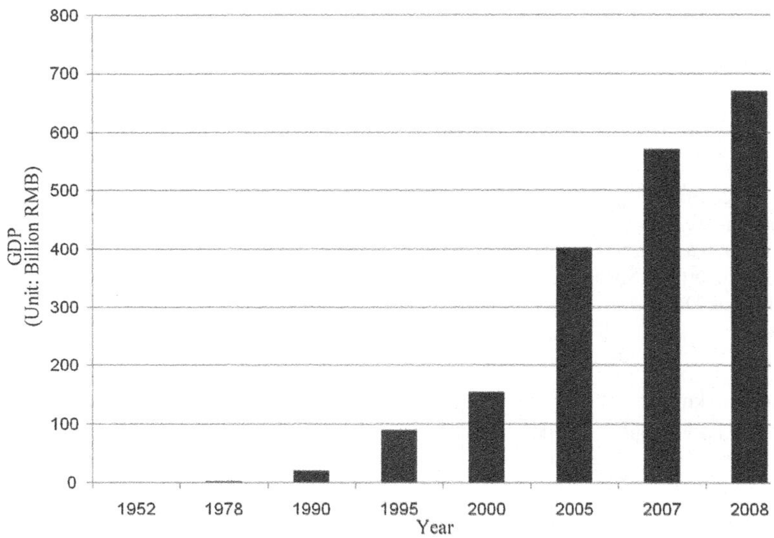

Fig. 14.4 Changes of Suzhou's GDP (Unit: Billion RMB. Source: Data drawn from Suzhou Statistical Year Book 2009 by Suzhou Statistics Bureau)

Based on the analysis of the aerial maps, we found that the developed and to-be-developed area increased by 186 km^2 annually from 1996 to 2004 (Table 14.3). If the urbanization continues at this rate, Suzhou's lands (including all vacant and farming land) will be exhausted before 2020. Lack of developable land then becomes the bottleneck for continually increasing GDP growth (Figs. 14.4, 14.5, and 14.6).

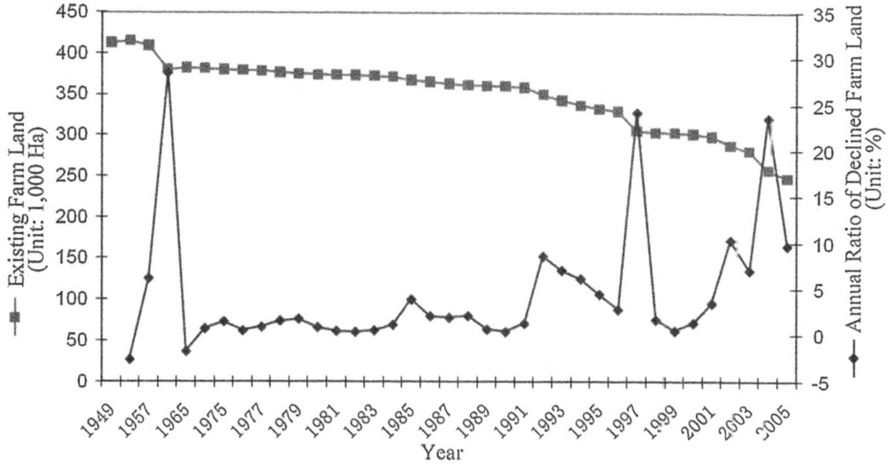

Fig. 14.5 Annual changes of farm land in Suzhou (Explanation: The left axis reflects the existing farm land, and the right axis reflects the annual decline of farm land, the minus number means increase like the year in 1949 and 1965. Source: Data drawn from Suzhou Statistical Year Book 2006 by Suzhou Statistics Bureau)

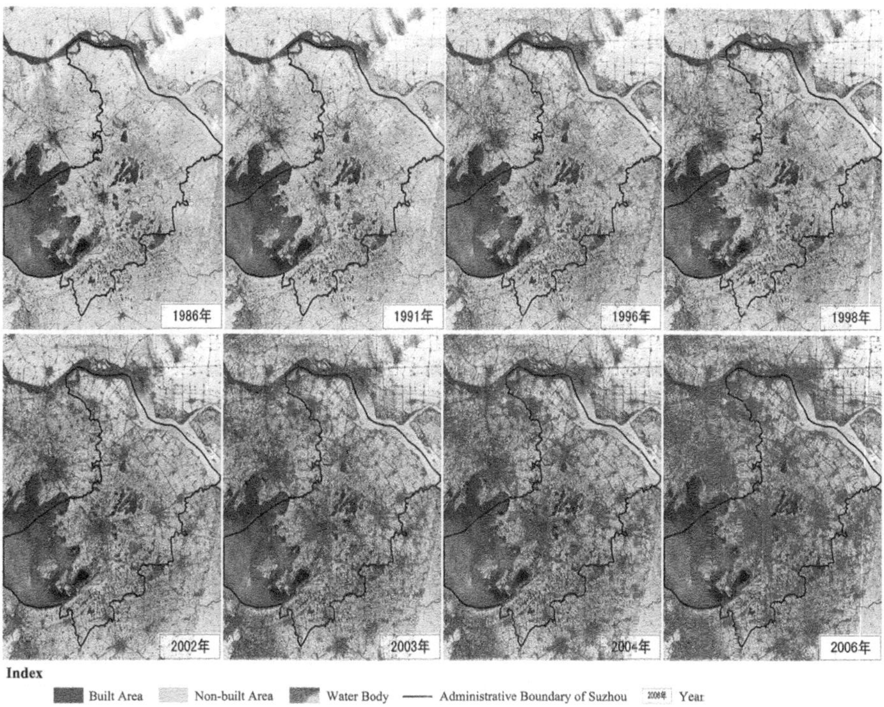

Fig. 14.6 Changes of urbanized and to-be-urbanized lands in Suzhou 1986–2006 (Explanation: These aerial maps were taken in year 1986, 1991, 1996, 1998, 2002, 2003, 2004, and 2006. The red areas are urbanized and to-be-urbanized lands. The black line is the boundary of Grand Suzhou. Data is shown in Table 14.3. Source: The Report for Suzhou City Comprehensive Plan 2007–2020 by China Academy of Urban Planning and Design, 2007a)

14.5 Emerging Trends: Expanding to a Larger Region and Shifting to a Sustainable Model?

We believe that the state and market powers which kept pushing for economic growth and urbanization will rise to the regional level in the core area of the Yangtze Delta region, and the accelerated intensive and extensive regional infrastructure will link different locations in this region more and more tightly as urbanized areas of adjacent cities begin to grow together. Hence, social and environmental problems become regional problems, requiring regional solutions.

Along the Nanjing-Shanghai transportation corridor, Wuxi, Suzhou and Shanghai's urban area have almost grown together, with less than a 3 km buffer between Wuxi and Suzhou and less than a 5 km buffer between Suzhou and Shanghai. With the opening of the 300 km/h Nanjing-Shanghai inter-city railway in 2010 and the opening of the 300 km/h Beijing–Shanghai railway in 2011, the travel time between Nanjing and Shanghai shrinks from 3 h in the freeway era to 1 h in the high-speed rail era. And the frequency of the rail services is every 15 min during rush hours. The commuting time between Wuxi or Suzhou with Shanghai shrinks to no more than half hour, less than the travel time from the west side of Shanghai to the east side. Dramatically less time distance makes the whole region work like a single city.

If we take the other two major transportation corridors, Shanghai–Hangzhou and Hangzhou–Ningbo, into consideration, the mega-regional-city hosts more than 80 million people and the two busiest ports in China, four international airports, a world class financial center, one of the most important locations for universities and scientific, technological, and social research institutions (another is Beijing), and the home of hundreds of thousands of manufacturers in various fields from electrical equipment and software to auto, ship, plane, and solar power equipment manufacturing. All of these locate in an area of more than 70,000 km^2, but all within a 1 h trip to Shanghai. A company can have their headquarters in Shanghai, use the world-class port service in Shanghai or Ningbo, get technical support from research institutes in Nanjing, Shanghai and Hangzhou, and get manufactured parts supplied from all over the region.

Social and everyday lives are more and more influenced with the rising of mega-regional-city, as experts at different locations meet more frequently, competitive pressure becomes nearer than ever but pushes people to learn quicker, which benefits the culture of innovation. People in Suzhou and Wuxi frequently go to Shanghai and Hangzhou for shopping, and people in Shanghai go to other cities in weekends more often, commercial model changes to be more splendid to attract remote regional customers. And maybe one of the most important signals of the life change induced by the mega-regional-city is that more and more people choose to commute between cities for different locations of job and residence.

With the rising of the mega-regional-city, local governments act to serve and support the trend. Suzhou city government planned to link its own subway system with that of Shanghai from the beginning. In the Suzhou City Comprehensive Plan 2007–2020, three subway and light rail lines were planned to link the Shanghai city subway and light rail network. The government of Jiangsu Province planned a provincial business center at Huaqiao town, which is near the Nanjing–Shanghai transportation corridor and just beside the Jiangsu-Shanghai boundary to attract enterprises' back offices, which need frequent communication with Shanghai but cannot afford the high rent in Shanghai. Wuxi, Suzhou and many other cities along the Beijing-Shanghai high speed rail launched aggressive new land use plans around the railway stations. The mega-regional-city is likely to come into being in the future.

Compared to the notion that the growth and urbanization in Suzhou will rise to a larger regional level, we are less certain whether the growth can shift to a more sustainable form. Even though many efforts have been or are going to be done at both the national and local levels, the dependence on the old model, which relies on "land finance" and monetary investment to gain economic growth, is still dominating at present (Li Yan, 2007).

The central government has been trying for decades to shift from economic growth as the main concern for local governments and to shift towards a more balanced set of development goals concerning social, economic and environmental outcomes; but no substantial changes have occurred. The "Scientific Outlook on Development" was launched in 2003, which proposed balanced development in the nation and was identified as the main principle guiding the development work of government (Zhongguo Gongchandang, 2003).

However, economic growth is still the major concern at the local level. In 2004, the National Bureau of Statistics and Ministry of Environmental Protection established a joint Green GDP research team, and the central government considered using Green GDP as the criteria for local administrator accomplishment. However, this effort failed because of local government opposition and technical weaknesses in the Green GDP measurement itself. Another method to assess local government, the "Development Priority Zones," was launched in 2010 as a new basis for measuring local administrator performance (The State Council P.R. China, 2010). Whether this new effort can lead to any positive change in the development approach remains to be seen.

Many municipalities have begun to emphasize redevelopment of existing urban areas to achieve higher density and host new industries with higher productivity, and gradually increase economic productivity and environmental thresholds for new enterprises. But the old approach is still dominant, and developments are spreading to some environmentally sensitive areas like those around Tai Lake.

In sum, at present the growth of the mega-regional-city is a strong trend and likely to remain so in the future. On the other hand, another desirable trend which calls for more balanced and sustainable development becomes more and more urgent as Suzhou is depleting its buildable land.

14.6 Conclusion

When we look at the situation in China today and compare it with the history in American in the 1960s, 1970s, and to the present, the driving forces behind economic growth and urbanization repeat the history in "any given locality." As Molotch put it in 1976:

I speculate that the political and economic essence of virtually any given locality, in the present American context, is growth. I further argue that the desire for growth provides the key operative motivation toward consensus for members of politically mobilized local elites, however split they might be on other issues, and that a common interest in growth is the overriding commonality among important people in a given locale-at least insofar as they have any important local goals at all. Further, this growth imperative is the most important constraint upon available options for local initiative in social and economic reform. It is thus that I argue that the very essence of a locality is its operation as a growth machine.

The Growth Machine exists in every market economy with different forms and various consequences for communities, the environment, cultural settings and political processes.

In China at present, the government takes the dominant role in the state-market coalition, which is the core of the Growth Machine, rather than the economic elites explained by Molotch in the American context. Because of the lack of balance from civil society, the state-market coalition in China is more powerful than those in western societies. And also because of the much larger population and density and the enlarged capability for the state and market elites by modern technology and globalization, the consequences brought by the Growth Machine happens at a much larger scale and at much faster speed in China than its parallels in the developed world in the past.

That is why, in Suzhou, the land expense and economic growth process rose from township and village levels to a regional level, or in other words, from rural urbanization to the emerging mega-regional-city, in no more than three decades. As the process is approaching the limits of natural resources, particularly the exhaustion of developable lands, Suzhou must shift its growth path to one that is more balanced and sustainable. Even though potentially disastrous outcomes are obvious to many political elites, and many efforts have been made to identify and address them, no substantial change can be seen at present. In this regard, Suzhou is typical of what can be observed in hundreds of cities and similar situations in China.

Hopefully, if the growth and urbanization process in Suzhou can smoothly shift to a new and better direction, we believe two closely related concepts, namely "Regionally-oriented" and "Sustainable," should be the most important notions in the new era. That new direction should include strengthening at least the following:

1. Physical connections throughout the region, leading with intensive and extensive infrastructure including regional and local transportation systems, water networks, energy frameworks and communication systems.

2. Regional socio-economic relations and collaboration. The strengthened physical connections offer necessary supports to make the whole region work more like an organic body, where one municipality can draw benefits from the strengths of others to make up for its weaknesses, especially in terms of socio-economic concerns.
3. Regional collaboration in ecological protection and restoration for a better environment as a commons for all. Common goals and principles should be settled on, and shared strategies, regulations, and standards should be established in areas where different municipalities are nearly linked together. One municipality's environmental issue can hardly be seen as an independent problem isolated from its neighboring municipalities.
4. Regional governance and management capacity building. Some institutional or semi-institutional mechanisms should be built within the region to share information, discuss issues, and negotiate regional actions, and to regulate each partner's actions. Now different municipalities rely on the central government to coordinate their activities on inter-municipal issues, which is usually inefficient and often delayed. With the demands for technical and political discussions becoming more essential and frequent, a kind of regional governance mechanism joined by different jurisdictions will become essential.

The emerging new era could be named as the "Sustainable Megaregional-city." The mega-regional-city is not an administrative arrangement; its birth is driven by the fast growth and urbanization processes in the region and the demands for a comprehensive response.

References

Bish, R. and Nourse, H. 1975. *Urban Economics and Policy* Analysis. New York: McGraw-Hill Book Company.

Broadbent, T. 1977. *Planning and Profit in the Urban Economy*. London: Methuen & Co Ltd.

Carroll, P. 2006. *Between Heaven and Modernity, Reconstructing Suzhou*, 1895-1937. Stanford: Stanford University Press.

Chen, Yong. 2002. *Study on Urban Morphological Evolution of Ancient Suzhou*. Urban Planning Forum. 5:55-60.

China Academy of Urban Planning and Design (CAUPD). 2003. *2020 Suzhoushi Chengshi Zongti Guihua Qianqi Yanjiu [Prophase Research Report for Suzhou City Comprehensive Plan 2020].*

China Academy of Urban Planning and Design (CAUPD). 2007a. *Suzhoushi Chengshi Zongti Guihua 2007-2020 [Suzhou City Comprehensive Plan 2007-2020].*

China Academy of Urban Planning and Design (CAUPD). 2007b. *Suzhou Gongyeyuanqu Fenqu Guihua 2007-2020 [Suzhou Industrial Park District Plan 2007-2020].*

Cui Gonghao. 1999. New Trends in Urban System Planning in the New Period. Planners. 3:32.

Wu, Ye. 2000. *Chengshi Tudi Liyong De Dianfan – Guanyu Suzhou Xinqu Yongdi Qingkuang De Baogao [A Model of Urban Land Use – A Report about the Land Use in Suzhou New District].* Zhongguo Tudi [China Land]. 10:21-22.

Zhongguo Gongchandang Zhongyang Weiyuanhui [Chinese Communist Party Central Committee]. 2003. *Zhonggong Zhongyang Guanyu Wanshan Shehuizhuyi Shichang Jingji*

Tizhi Ruogan Wenti De Jueding [Chinese Communist Party Central Committee's Decision on Several Problems about Improving the Socialistic Market Economy System].

Clingermayer, J. and Feiock, F. 1990. *The Adoption of Economic Development Polices by Large Cities: A Test of Economic, Interest Group, and Institutional Explanations*. Policy Studies Journal. 18(3): 539-552.

Cummings, S. 1988. *Business Elites and Urban Development: Case Studies and Critical Perspectives*. Albany: State University of New York Press.

DiGaetano, A., and Klemanski, J. 1999. *Power and City Governance: Comparative Perspectives on Urban Development*. Minneapolis: University of Minnesota Press.

Elkins, D. 1995. *The Structure and Context of the Urban Growth Coalition: The View from the Chamber of Commerce*. Policy Studies Journal 23(4):583-600.

Fei, Xiaotong. 2000. *Fei Xiaotong Lun Xiaochengzhen Jianshe [Fei Xiaotong Talking about Small City Development]*. Beijing: Qunyan Press.

He, Weidong. 2002. *Suzhou: Waizi "Tiantang" De Mijue [Suzhou: The Recipe of Foreign Investment "Heaven"]*. Huadong Jingji Guanli [East China Economic Management] 16(1):17-18.

Hu, Xuwei. 1998. *on the Spatial Agglomeration and Dispersion in Coastal Regions*. City Planning Review 22(6):22-28.

Huang, Zongzhi. 1994. *The Peasant Family and Rural Development in the Yangzi Delta, 1350-1988*. Hong Kong : Niujin University Press.

Li, Bozhong. 2002. *Industrial Growth and Urban Development in the City of Suzhou (1550 - 1850)*. Studies in Qing History. 1: 9-16.

Li, Jianbo. Zhang, Jingxing. and Cui, Gonghao. 2003. *Study of Impact of the Regionally Humanity-environment in the Developing Process of Township of Southern Jiangsu Province*. Human Geography. 18(6):5-10.

Li, Yan. 2007. *Suzhoushi 2007 Nian Zhengfu Gongzuo Baogao [Suzhou City Government Annual Work Report 2007]*.

Liu, Rongzeng and Wang, Shuhua. 2001. *Thinking on Enlargement of Opening to Domestic Regions of Development Zones in Developed Areas—A Case Study on Development Zones of Suzhou. Economic Geography. 21(1):52-55.*

Logan, J. and Crowder, K. 1997. *The Character and Consequences of Growth Regimes: An Assessment of 20 Years of Research*. Urban Affairs Review. 32(5): 603-630.

Lu, Yuqi and Se, Zhixiang. 1998. *Suzhou Gongye Yuanqu Jianshe De Quyu Xiaoying Yanjiu [A Study on the Regional Influence of the Development of Suzhou Industrial Park]*. Economic Geography. 18(3): 57-62.

Luo, Xiaolong and Shen, Jianfa. 2008. *Why City-region Planning Does not Work Well in China: The Case of Suzhou–Wuxi–Changzhou*. Cities. 25 (2008): 207–217.

Molotch, H. 1976. *The City as a Growth Machine: Toward a Political Economy of Place*. American Journal of Sociology. 82(2):309-332.

Molotch, H. and Logan, J. 1984. *Tensions in the Growth Machine: Overcoming Resistance to Value-free Development*. Social Problems. 31(5):483-499.

Qing, Tingdong. 1994. *Considerations on Urbanization Trend and Countermeasures of Su-Xi-Chang Region*. Urban Planning Forum. 2:57-61.

Song, Yang. and Ding, Chengri. 2009. *Smart Urban Growth for China*. Cambridge: Lincoln Institute of Land Policy.

Suzhoushi Wujiaju Diaochadui. 2009. *Guanyu Suzhou Nongmin Goumai Nongju Qingkuang De Diaocha Yanjiu [Analysis and Survey on the Situation of Suzhou Peasants' Purchasing Farming Stuffs]*. Source: http://www.wjj.suzhou.gov.cn/PSWeb/Web/aspx/InfoDetail.aspx?InfoID=1508. Last review: 5-16-2011. (Last review: 5-16-2011)

Suzhoushi Renmin Zhengfu. 2006. *Suzhoushi Shiyiwu Shehui Jingji Fazhan Guihua [The 11th Suzhou City 5-year Socio-economic Development Plan]*.

Suzhou Fazhan Yanjiu Yuan. 2004. *Suzhoushi 2004-2020 Nian Jingji Shehui Fazhan Zhanlue Yanjiu Baogao (Jieduan Baogao) [Suzhou City 2004-2020 Socio-economic Development Strategy Report (Draft)]*.

The State Council P.R.China.2010. *Guowuyuan Guanyu Yinfa Quanguo Zhuti Gongnengqu Guihua De Tongzhi [The State Council's Announcement on Distribution of the National Development Priority Zones Plan]*.

Qinghua Daxue Jianzhu Yu Chengshi Yanjiusuo [Tsinghua University Urban and Architecture Research Institute]. 2001. *Suzhou Xinqu Jingji Shehui, Chengshi Jianshe, Huanjing Baohu Zonghe Pingjia Suzhou New District Eco-society, Urban Development, and Environment Conservation Comprehensive Assessment]*.

Wang, Hongyang. 1999. *The Characteristics and Mechanism of the Change of Urban Land in Suzhou after 1980s*. Geographic Science. 19(2): 128-134.

Wu, Jiaping and Radbone, I. 2005. *Global Integration and the Intra-urban Determinants of Foreign Direct Investment in Shanghai*. Cities. 22(4):275–286.

Wu, Tinghai and Zhen, Hongyi. 1997. *Research on Rural Urbanization along the Side of the Yangtze River in Jiangsu Province*. Urban Planning Forum. 4:40-44.

Wu, Weijia and Guo, Zhan. 2000. *from Small Towns to Small Cities – The Development and Strategy of the Physical Environment for Southern Jiangsu Province*. City Planning Review. 24(2): 25-27.

Yang, Yifan. Deng, Dong and Dong, Ke. 2007. *"Jingcou Chengshi" Guihua Tansuo – Yi Suzhou Weili [Exploring1 Planning Strategy toward "Compact City" — a Case Study of Suzhou]*. 2007 Nian Zhongguo Chengshi Guihua Nianhui Lunwenji [China Annual National Planning Conference 2007 Symposium].

Zhang, Min. and Gu, Chaolin. 2002. *Rural-Urbanization: "South Jiangsu Model" Compared with "Pearl River Model"*. Economic Geography. 22(4):482-486.

Zhao, Xinping. Zhou, Yixing. and Chao, Guangzhong. 2002. *on the Difficulty and Practical Mistakes in Small Town Priority Strategy*. Small Town Planning. 26(10):36-40.

Zhou, Yixing. and Chao, Guangzhong. 1999. *The Urbanization Process in China in the Past 20 Years*. City Planning Review. 23(12):8-13.

Zhou, Yiyi. Zhang, Jingxiang and Chao, Ronglin. 2002. *Sunan Chengzhenhua Moshi De Huigu Yu Qianzhan [Review and Foresight on South Jiangsu Urbanization Model – A Case Study of Suzhou]*. Chengshi Wenti [Urban Problems]. 110(6):25-28.

Zhu, Kebao. [1844] 1962. *Chouren Zhuan Sabian [Biographies of Mathematicians and Astronomers, vol. 3]*. Shanghai: Shijie Shuju [Shijie Press].

Index

L. Bian et al. (eds.), *Chinese Urban Planning and Construction*,
Strategies for Sustainability, https://doi.org/10.1007/978-3-030-65562-4

T

The manufacturer's authorised representative in the EU is Springer Nature Customer Service Centre GmbH, Europaplatz 3, 69115 Heidelberg, Germany. If you have any concerns regarding our products, please contact ProductSafety@springernature.com

Printed and bound by CPI Group (UK) Ltd, Croydon, CR0 4YY
30/09/2026
02225290-0016